T0241878

The Anthropocene: Politik—Economics—Society—Science

Volume 23

Series editor

Hans Günter Brauch, Mosbach, Germany

More information about this series at http://www.springer.com/series/15232
http://www.afes-press-books.de/html/APESS.htm
http://afes-press-books.de/html/APESS_23.htm

Ma. Luisa Marván · Esperanza López-Vázquez
Editors

Preventing Health
and Environmental Risks
in Latin America

Editors
Ma. Luisa Marván
Institute of Psychological Research
Universidad Veracruzana
Xalapa, Veracruz
Mexico

Esperanza López-Vázquez
Centre of Transdisciplinary Research
  in Psychology
Universidad Autónoma del Estado
  de Morelos
Cuernavaca, Morelos
Mexico

More on this book is at: http://afes-press-books.de/html/APESS_23.htm.

ISSN 2367-4024   ISSN 2367-4032   (electronic)
The Anthropocene: Politik—Economics—Society—Science
ISBN 978-3-319-73798-0   ISBN 978-3-319-73799-7   (eBook)
https://doi.org/10.1007/978-3-319-73799-7

Library of Congress Control Number: 2017963849

Copyediting: PD Dr. Hans Günter Brauch, AFES-PRESS e.V., Mosbach, Germany
English Language Editor: Dr. Vanessa Greatorex, England

Cover photo: The photo on the book cover, river contamination, was taken by © Esperanza López
Vázquez, who has granted permission to use it here.
The photos on page iii show the Minatitlán, Veracruz flooding on October 2010, and a field worker.
© Permission to use this photo was granted by José Alfredo Rodríguez Atanacio (the flooding) and Ninfa
Ramírez Durán (the field worker).
The Esperanza's López-Vázquez picture in page 252 was granted by the professional photographer Lidia
Fitch.

Printed on acid-free paper

This Springer imprint is published by the registered company Springer International Publishing AG part
of Springer Nature
The registered company address is: Gewerbestrasse 11, 6330 Cham, Switzerland

To Abril, Elena, Yves and Marifer

Foreword

If human behaviour is a social activity derived from the dialectic of bio-psychical needs and sociocultural and eco-systemic stimuli, then behaviour cannot be understood in a vacuum, and it is essential to consider the culture and ecosystem where the behaviour takes place. It is also necessary to consider that cultural variations do not necessarily correspond to individual variations. It is important to have a separate conceptualization and study of norms and values at the macro-level and beliefs and attributes at the individual level. Cross-cultural studies show general trends derived from the insertion of individuals in their respective cultures, but there are also intra-cultural differences according to the sex, age, education and/or economic status of individuals.

As a result, the careful study and depiction of the patterns of behaviour related to risks in different eco-cultural settings are fundamental to the development of evidence-based interventions that will insure the well-being and quality of life of the people who inhabit those spaces. This book offers a fresh multidisciplinary description of research on a wide variety of health and environmental risks throughout Latin America. For aspects related to heath, the authors make a scholarly description of risks ranging from problems in pregnancy, breast and cervical cancer, neglected tropical diseases, complications derived from obesity and the lack of polyunsaturated fatty acids, mental health, consumption of alcohol and illicit drugs and antisocial or criminal behaviour. For the case of environmental and social risks, the topics covered include natural disasters, volcanic and seismic events, hydrometeorological events, bullying in schools, accidents at work in the textile industry and lack of pro-environmental behaviour.

In all chapters, the antecedents, correlates and consequences of the risk behaviours are evaluated in order to propose interventions to reduce or eradicate the negative effects of the deleterious situations. The careful analysis and inclusion of eco-systemic, social, cultural and individual variables involved in the search for appropriate and effective interventions make this book a must-read for professionals, public policy-makers, academics and the general population.

Mexico City, Mexico   Ronaldo Diaz-Loving
June 2017   Faculty of Psychology
                                          Universidad Nacional Autónoma de México

Acknowledgements

The idea for this book was raised in several Latin American forums due to the need to gather efforts to raise awareness about risk perception and risk behaviours. In these forums, researchers discussed the risks to which we are exposed and how to prevent them. This is why we assembled a group of research experts in different kinds of risks for the assemblage of this book. We would like to thank them for their contribution and expertise throughout the different chapters.

We would like to acknowledge Hans Günter Brauch from Peace Research and European Security Studies (AFES-PRESS) and Úrsula Oswald Spring from the Regional Center for Multidisciplinary Research at UNAM for their support and collaboration, which were essential to the publishing of this book. We also want to thank Vanessa Greatorex for expunging aberrations, ensuring stylistic consistency and transforming awkward syntax into lucid English.

This publication would not have been possible without the professional translation support of David Muldrew, Ricardo Lima Guimarães, Itzel Toledo García, Jacinto Terrazas, Margaret Thorpe, César Daniel González Méndez and Ernesto Zavala Álvarez. We also want to thank Fabiola Orihuela Cortés and Ma. Fernanda Marván Ramírez for providing us with technical support.

We also want to acknowledge Alejandra Núñez, Yamilet Ehrenzweig, Pedro Palacios, Ángeles Vacio, Verónica Alcalá, Sandra Cortés, Roberto Lagunes-Córdoba, Adriana Rodríguez-Barraza, Lorena Pérez, Imke Hindrichs, Doris Castellanos Simons, Jorge Damián Morán, Alexis Lorenzo, Alejandra Terán and Juan Carlos Gavilanes, who were the reviewers of the chapters.

Finally, we would like to especially acknowledge the people who voluntarily agreed to participate in the research mentioned throughout the chapters.

Veracruz, Mexico   Ma. Luisa Marván
Cuernavaca, Mexico   Esperanza López-Vázquez
November 2017

Contents

Part II Environment and Disasters

Abbreviations

AAP	American Academy of Pediatrics
AIDS	Acquired immune deficiency syndrome
Alabama HB-56	Beason-Hammon Alabama Taxpayer and Citizen Protection Act (Immigration Law)
Arizona SB 1070	Arizona Senate Bill 1070 immigration law
BCS	Brazilian Cardiology Society
BMI	Body mass index
BRCA	Breast cancer
BRCA1	Tumour suppressor gene (type 1) associated with DNA repair
BRCA2	Tumour suppressor gene (type 2) associated with DNA repair
CC	Cancer Cervicouterino (Cervical Cancer)
CDPD	Convención sobre los Derechos de las Personas con Discapacidad (Convention on the Rights of Persons with Disabilities)
CHS	Commission on Human Security
CICESE	Scientific Research and High Education Centre of Ensenada, Mexico
CIF	Clasificación Internacional del Funcionamiento, de la Discapacidad y de la Salud (International Classification of Functioning, Disability and Health)
CONACYT	Consejo Nacional de Ciencia y Tecnología de México (National Council of Science and Technology of Mexico)
CONALEP	Colegio Nacional de Educación Profesional y Técnica (National School of Professional and Technical Education)
CONAPO	Consejo Nacional de Población (National Population Council)
COP	Conference of the Parties
CVDs	Cardiovascular diseases
DHA	Docosahexaenoic acid
DRR	Disaster risk reduction
ECLAC	Economic Commission for Latin America and the Caribbean

EIRD	Estrategia Internacional para la Reducción de Desastres (International Strategy for Disaster Reduction)
EPA	Eicosapentaenoic acid
ER−	Estrogen-receptor negative
ER+	Estrogen-receptor positive
EWS	Early warning system
FA	Fatty acids
GECHS	Global Environmental Change and Human Security
GHG	Greenhouse gases
GI	Glycemic index values
GLOBOCAN	Global Cancer Observatory
HBM	Health belief model
HDL	High-density lipoprotein
HDL-c	High-density lipoprotein cholesterol
HFA	Hyogo Framework for Action
HIV/AIDS	Human immunodeficiency virus/acquired immune deficiency syndrome
HIV	Human immunodeficiency virus
HPA axis	Hypothalamic–pituitary–adrenal axis
HPV	Human papilloma virus
HTN	Blood pressure or hypertension
HUGE	Human, Gender and Environmental Security
IEBEM	Instituto de la Educación Básica del Estado de Morelos (Institute of Basic Education of the State of Morelos)
IGF-1	Insulin-like growth factor 1
IgM	Immunoglobulin M
IHDP	International Human Dimension Programme
IMSS	National Social Security Institute (Mexico)
INEGI	National Institute of Statistics and Geography
IPCC	Intergovernmental Panel on Climate Change
ISDR	International Strategy for Disaster Reduction
IUD	Intrauterine device
LDL-c	Low-density lipoprotein cholesterol
NEP	New ecological paradigm
NHC	National Hurricane Centre of the United States of America
NRC	National Research Council of the United States of America
OCHA	Office for the Coordination of Humanitarian Affairs
OECD	Organisation for Economic Co-operation and Development
OPPI	Oficina del Procurador para las Personas con Impedimento de Puerto Rico (Office of the Public Prosecutor for Persons with Disabilities in Puerto Rico)
PAP	Papanicolaou (Pap smear)
PR+	Progesterone-receptor positive
PUFAs	Polyunsaturated fatty acids
SAS	Earthquake Warning System, Mexico

SCT	Secretariat of Communications and Transportation, Mexico
SEGOB	Ministry of Government, Mexico
SIAT-CT	Early Warning System for Tropical Cyclones
SINAT	National Tsunami Warning System
SM	Secretariat of the Navy, Mexico
SS	Mexican Department of Health
T2D	Type 2 diabetes
UN	United Nations
UNAM	National Autonomous University of Mexico
UNICEF	Fondo Internacional de Emergencia de las Naciones Unidas para la Infancia (United Nations International Children's Emergency Fund)
UNISDR	UN Office for Disaster Risk Reduction
UNU-EHS	United Nationals University on Environmental and Human Security
US	United States
USAID	United States Agency for International Development
VBN	Value-belief-norm model
VLDL	Very low-density lipoproteins
VO_2 max	Maximal oxygen uptake
WHO	World Health Organization

Introduction

The book is divided into two main categories, based on the type of risk addressed by the chapters: health issues and environmental issues in Mexico and some Latin American countries.

The chapter written by Tania Romo et al. discusses high-risk pregnancy. Unlike most texts on this subject, which describe how the health problems of the mother increase the possibility of having a pregnancy of this type, the authors describe certain risk behaviours that compromise not only the perinatal health of the mother and/or the newborn, but condition the state of health of the newborn throughout his life. The authors describe how exposure to stress during gestation predisposes the individual adult to develop some types of cancer, neurological and cardiovascular diseases. The authors also describe how the nutritional conditions of the mother before and during pregnancy create a favourable or adverse *in utero* environment that will predispose the foetus to the development of disorders such as obesity, diabetes or metabolic syndrome. But the effects of stress and a poor quality maternal diet during pregnancy are not limited to children; these can be transmitted to their offspring, even for several generations.

Two chapters on cancer are presented—one is about risky behaviours that can trigger breast cancer and the other about cervical cancer. In the first chapter, written by Rosalba León et al., it is evidenced that the origin and progression of breast cancer are, to a high degree, attributable to factors related to lifestyle. In this sense, we discuss some risk behaviours such as carbohydrate-rich and low-fibre diet rich in both animal fats and trans-fatty acids obesity (mainly post-menopause) sedentary lifestyle alcohol consumption and smoking. In addition, other psychological risk factors such as personality type, chronic stress exposure and barriers to the timely detection of this type of cancer are discussed. The authors reflect on the prevention campaigns and conclude that in addition to performing diagnostic tests, the constant promotion of healthy lifestyles is necessary.

In the other chapter, written by Yamilet Ehrenzweig and Ma. Luisa Marván, some psychological variables that result in women not performing certain behaviours to prevent cervical cancer are discussed. The authors present epidemiological data which indicate that although cervical cancer is a preventable disease, it is a serious public health problem in underdeveloped countries and there is still a high death rate from this cause. The chapter also contains empirical data on some of the psychosocial factors that influence many women not to perform the Papanicolaou (Pap smear) test, which detects in a timely manner abnormal cells that give rise to cervical cancer if appropriate treatment is not received.

The chapter of Socorro Herrera and Grecia Herrera addresses the importance of the consumption of certain fatty acids to avoid a series of health risks. The authors discuss the benefits of omega-3s, which are found in fish oil, tuna, salmon, cod and some seeds such as flax and olive, among others. Results of empirical research demonstrating the protective effects of omega-3s against cardiovascular disease, diabetes and some cancers are described. Also mentioned is how it helps to relieve some skin lesions and treat some diseases, such as rheumatoid arthritis and kidney diseases. But the chapter's greatest emphasis is on describing the beneficial effects of omega-3s on the brain, improving motor, attention, learning and memory skills, and decreasing the risk of Alzheimer's and Parkinson's, among other diseases.

The chapter by Ninfa Ramírez et al., focuses on the risk behaviours related to what the World Health Organization has called Neglected Tropical Diseases, which are mainly developed in poor countries located in the tropical and subtropical strip of the world, affecting more than 100 million people. In particular, the authors discuss the conditions which foster the most common infectious disease in Latin America caused by actinomycetes (microorganisms found in natural ecosystems). This disease, called actinomycetoma, affects skin, subcutaneous tissue, fascia and bone and may spread through the thoracic cavity, abdominal or other regions of the body. It is characterized by progressive inflammation with deformation of the area where it is located, abscesses, ulcers and fistulas. It affects a specific core of the population, who are farm workers, because they have inadequate work practices such as walking barefoot, carrying agricultural products on their backs and performing labour activities lying on the soil.

The chapter by Roseane de Fátima Guimarães Czelusniak discusses the health risks caused by sedentary lifestyle in adolescents, an important issue to deal with since the lifestyle of modern societies makes it increasingly difficult to perform physical activity. The author describes the association between sedentarism and some risk indicators, such as elevated cholesterol levels. But she also talks about the association between sedentary behaviour and other behaviours considered to be risky, such as alcohol and tobacco consumption, as well as inappropriate eating habits. It is emphasized how sedentarism causes overweight and obesity, which in turn are responsible for various cardiovascular, joint, cutaneous, endocrine, gastrointestinal, respiratory and neoplastic complications, besides causing serious psychosocial problems. The chapter also describes how these complications can be prevented or even reversed, by changing lifestyle and increasing physical activity.

The chapter of Jorge Luis Arellanez describes how the process of international migration can be a risk factor for the mental health of individuals, with emphasis on the consumption of alcohol and illicit drugs. In the case of men migrating from Mexico to the USA, the post-traumatic stress associated with the experience of arrival at the destination has been studied, as well as the stress associated with the lack of social support networks that facilitate migratory stay. Increased alcohol consumption among migrants may act as a risk factor for antisocial or criminal behaviour, as well as illicit drug use. The author also talks about the emotional implications for those who remain in their place of origin. The absence of the father usually leads to a family crisis that demands a high adjustment effort and leads to the worsening of conflicts or the emergence of new problems. These situations have a widespread affective impact of stress and depression on women and contribute to the involvement of children in the use or abuse of alcohol or other illicit drugs.

The work of Iris Cátala from Puerto Rico focuses on the need to consider, in a particular way, the most vulnerable populations exposed to natural disasters. The author says it is important to consider the care of people with disabilities, children and the elderly in such a way as to show "our evolution as a society". She also emphasizes the importance of implementing risk prevention programmes, particularly with persons with disabilities, stressing that decent treatment of them is an essential part of human rights. Additionally, the author proposes some intervention strategies to attend to the physical and emotional consequences experienced by this population with disability in the face of natural catastrophes. Some of the ways she suggests to address this problem are: coordination of services between organizations, consideration of cultural beliefs related to disabilities, adequate safety, rescue and evacuation plans.

In his chapter, Jesús Manuel Macías explains the importance of early warning systems (EWSs) for disastrous events, which serve to warn the population before an event occurs (sirens, radios, volcanic activity status, cell phones, etc.), and how they tend to be designed in such a way that it is assumed that the target population is homogeneous and will be ready to respond to alerts promptly once the signal is made. However, the author shows that the population is not prepared to respond adequately, since it responds to much more complex processes, in which cultural, social and contextual factors are at stake. This situation hinders the effectiveness of EWSs and shows how those who design these systems know little of the social subjects at whom the information is directed. The author explains how these systems tend to be unused or underutilized by the false ideas that exist around them, since, to be effective, they require long and focused processes in the target population. Finally, he suggests a review of all EWSs in Mexico so that they comply with the recommendations made by the UN International Strategy for Disaster Reduction. He also suggests that these EWSs incorporate research recommendations from scientists in the social sciences, showing the complexity of the associated sociocultural factors that favour or disfavour the good use of these systems.

In the chapter by Eric Jones at al., the dynamics of social networks and the perception of risk in the face of disasters in two Latin American countries (Mexico and Ecuador) exposed to dangers of volcanic eruption are analysed. The authors analyse the population's perceptions in the past, the present and those projected expectations for the future. They are interested in knowing the influence of the social and geographical context in the perception of risk and in the community social networks. According to the authors, individuals can recover better after a disaster if the social networks to which they belong are dense. However, this can also be counterproductive if the bonds that unite them are also dense, as the same individuals are always conjugated within the network, which does not allow the flow of information from other groups, since they are less trusted than the members of the core network. The authors also note that receiving support seems more important when speaking about past experiences and perceptions than when speaking about the present, where it is more important to give support. The differences and similarities found between the two countries mark some interesting guidelines to continue analysing to discover how social networks influence this type of situations.

The chapter by Úrsula Oswald Spring analyses the problems of the risks caused by hydrometeorological events based on environmental and social vulnerability. In addition, the author succeeds in synthesizing a theoretical and practical tool that seeks to foster a process of resilience focused on: (a) the safety of populations exposed to social violence and criminality, (b) risks and disasters caused by climate change and (c) gender violence. From this comprehensive approach, the author takes into account the social problems of human security in a world of increasing violence and governments with little capacity to respond to social problems. In this way, she says, "all these factors generate high-risk scenarios that can easily become disasters with high human and material costs". The model proposed offers elements of analysis for the management of more comprehensive public policies that seek human well-being.

The chapter by Elizabeth Ojeda, Melissa Ricaurte and Esperanza López Vázquez analyses the perception of volcanic risk and the behaviour of students participating as members of the first emergency response brigades at a secondary school located at the foot of the Popocatépetl volcano (Mexico). These brigades are prepared to facilitate the evacuation, search and rescue of people first aid communication and disaster prevention and firefighting. The authors show how students and teachers of the student community in general are accustomed to observing the constant volcano activity without disrupting their daily life. In addition, the authors reflect on the general beliefs of the rest of the Tetela community about the relationship and ancestral vision of the Popocatépetl volcano, as it is regarded as a living being that even has a spirit of its own. These beliefs endure with different forms of expression. Despite being accustomed to volcanic hazards, which manifest themselves frequently, they do not deny the real dangers in which they live. Through a qualitative methodology, the work gives voice to the different visions and perceptions that the participants have, showing their postures in front of the media, their teachers, their parents, the preventive measures learned and proposals for the improvements that they consider still need to be made. This is a youthful vision, full of restlessness and expectations for the future. The authors propose some guidelines for the prevention of volcanic risk based on the knowledge found in the studied community.

The chapter by Hannia Cabezas deals with the typical behaviours of some students that lead to *bullying*, that is, a situation of harassment and intimidation, in which the student is repeatedly attacked by his peers. The different forms of aggression used by the students are favoured by some traits found in the victims, who are shy and fragile and often react with tears and isolation when they feel attacked. These conditions are interpreted by the victimizers as a weakness, fuelling the false belief that the victims have a cognitive disability that prevents them from reacting with mental agility, limiting their social skills and independence, which feeds the aggression. The victims of school bullying among peers are affected in the short and long term, as they present traits of anxiety and fear, which subsequently trigger low self-esteem in addition to anxiety and depression disorders. That is why timely intervention is required to prevent this happening.

The chapter by Nicolas Bronfman, Pamela Cisternas and Esperanza López shows the results of a study on environmental behaviour using a sample from Chile. The theoretical approach of this study is based on the Stern Value-Belief-Standard model, which considers that the different attitudinal elements of these variables influence pro-environmental behaviour when one has a moral obligation to protect society and the environment in general. It is a causal model that analyses the posture and attitudes of the participants in the face of environmental problems. This model is used by authors to understand behaviours and attitudes in terms of energy and water conservation, consumption, biodiversity, transportation and household waste handling. According to the authors, the population of Santiago has a high pro-environment ecological vision. They are aware of the environmental problems that affect them, and they also participate in caring for the environment with regard to consumption, biodiversity, transportation and the conservation of energy and water. However, citizen participation tends to be significantly lower in the case of waste handling and separation than in other environmental situations. According to the authors, this can happen, on the one hand, because the other environmental dimensions enable people to save money, and on the other hand, because of the lack of knowledge and material difficulties that exist about how to deposit the waste properly. This type of study also provides important guidelines for the development of public policies.

The chapter by Lorena Pérez Floriano and Julieta Leyva Pacheco deals with the problem of accidents at work in the textile industry. The authors argue that in order to understand the relationship between hazards and workers' health, the risk of exposure to potentially harmful labour factors should be assessed. One of the contributions of the study is the creation of a questionnaire on Risk Perception in the Textile Industry, which measures the perception of both the risk of having an accident and the ergonomic risk. Using this questionnaire, the authors observed that the workers of a textile factory perceive greater risk than their supervisors or the administrative staff. The authors emphasize the importance of knowing the perceptions and attitudes related to workers' safety in order to promote a culture of safety and responsibility. To eliminate and control the risks in the labour area, it is crucial to ensure that all employees adopt safe working practices by providing training and making improvements to the company's safety and hygiene.

Chapter 1
Introduction to Risk Psychology

Esperanza López-Vázquez and Ma. Luisa Marván

Abstract Risk perception has been studied from different approaches. From a cognitive angle, risk perception implies the cognitive evaluation that is done in an individual way but, at the same time, is impregnated and pierced by the social aspects that surround the person. Perceived risk is built throughout life and is modified according to various elements, such as the general information that is acquired from danger, as well as experience, personal and social beliefs, and emotions. Risk perception is influenced by different factors such as heuristic biases that are judgements based on experiences, values and outgoing perceptions. The culturalist approach analyses society from the base of social systems where risk appears as a result of technological and economic progress that society itself has generated. Society accepts a large number of dangers, given the great benefits many of them provide. The influence of factors such as the media and social trust are also discussed.

Keywords Risks · Risk perception · Risk behaviours · Heuristics

There are numerous examples of risk situations that we face every day, such as being subjected to conditions of chronic stress, the possibility of experiencing an earthquake, or of suffering an accident in the workplace. In order to understand, in a practical way, some of the processes that we as individuals experience when we are exposed to a risk, it is necessary to explain some theoretical concepts that have been developed from different approaches. The first part of this chapter will address some theoretical models of risk perception analysis and behaviour in the face of different hazards. A definition of the concept of risk is given to subsequently discuss some psychological factors involved in the perception of risk. We finally present a reflection on the importance of studying risks from a biopsychosocial perspective, which considers the human being as an individual and at the same time as a member of a social group.

Dr. Esperanza López-Vázquez, Research Professor, Centre of Transdisciplinary Research in Psychology, Universidad Autónoma del Estado de Morelos. Email: esperanzal@uaem.mx.

Dr. Ma. Luisa Marván, Researcher, Institute of Psychological Research, Universidad Veracruzana. Email: mlmarvan@gmail.com.

© Springer International Publishing AG, part of Springer Nature 2018
Ma. L. Marván and E. López-Vázquez (eds.), *Preventing Health and Environmental Risks in Latin America*, The Anthropocene: Politik—Economics—Society—Science 23, https://doi.org/10.1007/978-3-319-73799-7_1

1.1 Towards a Psychology of Risks

Studies on risk perception are relatively recent and have been increasing as natural, technological, and anthropogenic hazards and disasters have been increasing.

The concept of risk perception has been managed for several decades to understand how people perceive environmental risks and threats. It has been observed in different studies that this perception gives rise to a response that is reflected in a person's behaviour and attitudes.

Regardless of the sensorial act of perceiving, risk perception allows the body to prepare itself to face the eventuality of a danger. The ability to defend oneself against a threat is a necessary condition of every organism to ensure its survival, and the memory of experience is a skill that enables learning. In addition to this instinct, the human body has the possibility of altering its environment, as well as responding to it, in such a way that it is able both to create and to reduce risk (Slovic 2000).

The perception of a threat brings along a cognitive process in which environmental conditions are evaluated to decide whether a situation or circumstance is threatening or not, and what the probability is of being affected by it. According to Slovic, the results of a subject's evaluation of perceived hazards are termed *risk perceptions*, which are changing and context-related (Slovic 2000). In this way, there are no wrong perceptions, but versions according the perceived subjects' interpretation.

To better understand the concept of risk, it is important to refer to the roots of this word. 'Risk' is the anglicised version of the French word *risque*. According to the historical dictionary of the French language, the definition of "risk/risque" has changed over time. It evolved from the medieval Latin *risicus* or *riscos*, which was itself derived from the classical Latin *resex*, 'cut back', associated with the verb *resecare* "to cut back, cut short, trim". Subsequently, it acquired the connotation of "risk run by a commodity at sea". Another theory is that it could be derived from Latin *-rixor* ("to quarrel violently, brawl"), which gave rise to *rixe*, 'brawl', and ultimately 'risk'. This term in today's language conveys the idea of an inconvenience or situation that can lead to a more or less predictable eventual danger in the near or distant future. It brings along the idea of eventuality and probability, or of exposure to some possible danger or inconvenience, leading to a possible loss (Dictionary Le Robert 1999). There is no certainty that the event will really happen, or that it will not happen. In general, risk can be defined as the possibility of suffering damage that encompasses both the nature of an option and the probability of its consequences (Ayres et al. 2002).

Psychology has been widely concerned with the study of the subject's interaction with risks, and there has been a development of what some authors have called Risk Psychology (Kouabenan et al. 2006). This field of psychology relies on the principles of decision theory, which is based on theories of probability based on mathematical calculations (Munier 1992); taking a risk is based on the decision to choose between different options in a level of uncertainty of the probability of success or failure (Assally 1992). Several studies have analysed how subjects respond according to their motivations. Subjects risk gaining or losing according to how much they rely on their own abilities, or on the pleasure of risking something unknown. This guides their behaviour (Lopes 1987).

Risk Psychology reveals the effort to understand psychological mechanisms used by individuals when faced with a risk in which the ambiguous and diffuse idea of uncertainty may also be present. It can be a voluntary situation in which the subject is exposed to a certain level of risk that generates emotions that can be pleasurable (if the latter can control the situation), or unpleasant (when it exceeds his/her capacity to control the situation); it may also be situations that may be imposed by circumstances or by other people.

1.2 Cognitive Approach

From a cognitive approach, risk perception is defined as the interpretation of a stimulus of a possible or actual threat to the perceived subject. This interpretation implies the cognitive evaluation that is done in an individual way, but which in turn is impregnated and pierced by social aspects that surround the person. Perceived risk is stored in the individual memory, is built throughout life and is modified according to various elements such as the general information that is acquired from danger, the individual's experience of it, personal and social beliefs, as well as the emotions that the subject projects on to the perceived element. Common aspects that are repeated and identified as predominant can be defined as "social perceptions" of risk because they play a part in social—not just individual—thought. The social characteristic is derives not only from repetition, but also from all the cognitive as well as the subjective elements, cultural and ideological, that are shared by the members of a determined group. The interaction of knowledge, beliefs, fears and perceptual biases is influenced by elements of both personality and experience as well as by aspects of society: culture, political ideas, economic interests and the historical moment in which people exist and the social organization is found. All this results in a perception and a determined behaviour in the face of risk.

The first studies on the analysis of perceived risk go back to the 1960s. It was Starr (1969) who, analysing the risk and benefit of certain hazards, showed how subjects choose to accept or not accept the risk in terms of the benefits that a given hazard can provide. This author makes a quantitative analysis of the level of acceptance that a subject has of a certain danger, taking into account whether that danger is voluntary or involuntary. For example, he says that the subject is a thousand times more willing to accept a voluntary risk than an involuntary one; in addition, he says that death statistics caused by some types of danger will psychologically influence the level of acceptability; and that the acceptability of a risk will be proportional to the third part of the perceived benefits (real or imagined). Therefore one of Starr's main conclusions is that people more readily accept dangers that give more benefit. His quantitative model was highly criticized, but raises the need to analyse risk perception by using systematic measurement. Subsequently, other authors continued to advance the analysis of risk perception with regard to technological risks, taking into account Starr's basis and strengthening the importance of measuring perception with quantitative techniques (Fischhoff et al. 1984, 2000; Slovic 2000; Jenkin 2006; Siegrist et al. 2005).

The *psychometric paradigm* is one of the best known used by the scientific community. Based on taxonomy of different risks, a cognitive map is drawn that represents the attitudes and perceptions that the subjects have of a series of risks. Each hazard is assessed for risk, acceptability, level of control, and other variables that the researcher chooses, at a level from less to more (Slovic 2000; Fischhoff et al. 1978; Slovic et al. 2010a, b; Bronfman/López-Vázquez 2011; López-Vázquez et al. 2012). Through this method, the relationship of different variables associated with the perception of risk can be analysed. This makes it possible to see the hazard distribution on a Cartesian plane according to the risk characteristics they want to study. For example, using this methodology, the perception and acceptance of the risk of nanotechnology applied to food and its protective packaging was studied in Mexico. It was concluded that the perceived control of entering or not into contact with nanotechnologically modified products influences risk perception and, quite possibly, the consumption behaviour of these products (López-Vázquez et al. 2012).

Some risk attributes that are key in studies of risk perception have been defined from the different investigations that have been used in this paradigm. Some of these attributes are: voluntary risk, resentful fear, knowledge about a risk, how controllable the individual perceives the risk to be, perceived benefit, average number of victims a year that the risk situation has claimed, the number of victims within a year in which an event could be presented, how acceptable a risk is considered to be, the novelty, the severity of the consequences, and the catastrophic potential and chronicity of a situation.

In general, these characteristics are part of a wide range of perceptions or judgements about the preferences that an individual may have when evaluating different risks. It has been shown that risk perception can be quantified and predicted through this psychometric paradigm, which is highly valued by many researchers in various parts of the world. One of its advantages is that through this, one can analyse the perceptions and attitudes of many people at once, get their group tendency and make an analysis that allows the influence of some variables on others to be predicted.

However, this paradigm has also been criticized because of the method used, since it does not talk about the subjectivity of people and has a relative predictable power (Sjöberg 2000). Another limitation is that it does not tell us anything about the people's behaviour (Slovic 2000).

1.2.1 Judgements and Beliefs

The ability to perceive a risk or danger is complex and multifactorial. Thus, perceptions acquire, from the individual features, aspects that can be seen as descriptors of group perceptions. The prevailing social aspects will characterize the perceptions and attitudes of subjects towards a particular risk or danger. The so-called *predictive cognitive* or *heuristic biases* with regard to a risk or hazard are psychosocial factors involved in risk perception and are based on experiences, impressions, personal or group values and outgoing perceptions. Heuristics are judgements produced by a cognitive estimate of the probability calculus. Since they are individual judgements that often lack real knowledge of the situation, they can be based on misconceptions that people have when assessing the likelihood of an event. The central idea of cognitive biases is that judgement is subject to the uncertainty of the situation that is based on certain judgements that simplify the process of understanding, rather than using more formal and extensive processes (Gilovich et al. 2002; Kahneman/Tversky 1979). People tend to fix their attention on certain information which they consider important and, at the same time, ignore most relevant probabilities (Kahneman et al. 1982). Several studies describe some of the major cognitive biases used when the probability of occurrence and possible damage to people is evaluated. We note here those we consider most important.

(1) *Heuristic of availability*: the bias is the estimate of the frequency or probability of an event occurring, based on the related examples that come spontaneously to the person's mind. That is, the trial activates certain information, given the ease with which it can be evoked, making the memorable qualities perceived as very common and ignoring the actual information on the subject (Kahneman/ Tversky 1979; Morales et al. 2002; Gilovich et al. 2002).

(2) *Heuristic of representativeness*: this bias is to ignore prior probabilities of an event, leading to miscalculation of these and to have difficulty managing diagnostic information (Kahneman/Tversky 1979; Morales et al. 2002; Gilovich et al. 2002). Generally, people make use of this heuristic to the extent that the information on an event looks similar in its characteristics to those of another. When there is no precise information on the subject, this cognitive shortcut to understanding the perceived reality is used.

(3) *Anchoring bias*: the judgement is made with an idea rooted in the mind, adjusting and accommodating the information according to this basic idea. Generally, the adjustment is crude, imprecise and does no justice to the additional important information that can be obtained or that can reach us (Slovic 2000; Tversky/Kanneman 1974). An example of this would be to believe that nature's resources are inexhaustible and are to serve man. From these beliefs man can justify behaviour of misuse and pollution of non-renewable natural resources.

(4) *Overconfidence*: a typical heuristic judgement is the fact that people feel too sure of their own judgements. In a related study people were asked about the estimated frequency of two deadly events; after responding to the requested estimate, participants had to indicate the probability which they believed to be right in the previous judgement they made. Results showed that people tend to make mistakes in 8/10 of the answers given, demonstrating clear overconfidence (Fischhoff et al. 1977). Overconfidence can be dangerous in the sense that people trust to the point of not realizing how little is known or how much information is needed about the situations and dangers we face daily. Another aspect that has been observed in studies on overconfidence is that experts tend to have this bias, as well as the public (Slovic 2000).

(5) *Desire for certainty*: this is a trend that appears when someone fails to resolve a conflict of uncertainty related to a risk. The desire for certainty often leads people to deny the perceived uncertainty, so they can continue their lives (Slovic 2000). This attitude has been observed in different situations of natural hazards where the person reports low levels of perceived risk despite being in a high-risk area (López-Vázquez et al. 2008), or does not consider it a priority to be certain about the risk (López-Vázquez 2009), or explicitly denies the existence of the risk.

(6) *Heuristic of affection*: overshadowed by studies focused on the cognitive part of perception, affection was analysed in the 1980s to understand how to influence both risk perception and behaviour. It was thanks to Zajonc (1980) that affections began to be an important study element, since he claimed that everything we perceive is imbued with some sort of affection, and that emotions come before the same cognitive processes in certain circumstances, making it rationality more difficult. Fear and courage influence the perception of risk; as fear amplifies the risk, courage dims it (Lerner et al. 2003). Other studies show how judgements about risk and benefit change when elements that trigger specific emotions are introduced. In addition, people use more heuristic affection when it makes them feel safer from the risk and perceived benefit of danger. For example, in a study where information on various technologies occurred, the influence of the affections was analysed. The results that were obtained showed how heuristic affection influences the risk perception and perceived benefit, so that if there is a high risk and at the same time a positive emotion, the benefit increases; if the risk is low and the emotion is positive, the benefit is high; if the perception of risk is high and the emotion is negative, the benefit is low; if the perception is low and the emotion is negative, the benefit is low (Finucame et al. 2000). Some theorists have even given emotions a role of great importance in motivating behaviour toward risk (Slovic et al. 2010a, b).

(7) *Illusory optimism and invulnerability illusion*: the first occurs when people that they will have more positive experiences in life compared to their peers, while the second is the tendency for people to believe that the probability of having bad experiences is lower than that of their peers. This type of behaviour is analysed in the field of risk behaviours to health (Sánchez-Vallejo et al. 1998; Armor/Taylor 2002; Sharot et al. 2011), as well as natural threats. An example

of illusion of invulnerability related to natural hazards is the Popocatépetl volcano. It has been explicitly observed that people living on the slopes of the volcano think ash and volcanic materials tend to go with the winds to other cities, while inhabitants of these cities say that these materials do not affect them much because they fall mainly in the territory of the villages that are closer to the crater.

Cognitive biases, such as shortcuts thought to reduce the complexity of reality, aim to reduce anxiety and psychological stress, but in turn, perceptual errors can have serious consequences when acting against a health problem or any dangerous event.

1.3 Culturalist Approach

The culturalist approach to risk perception is headed by the studies of German sociologist Ulrich Beck (1998, 2012), who is one of the first to speak about risk society. The author discusses how, after the industrial revolution, we are faced with the emergence of technological risks that mark an era in which technology is not only synonymous with development, but new dangers. Society, seeking to outdo itself, has to some extent reduced limitations, but this in turn has faced it with new challenges. One of those big challenges is social inequality: not just unequal distribution of wealth, but also of risk. In the case of wealth, legitimization procedures have been used to justify certain groups having more wealth than others, while in the case of risks there have been attempts to minimize and relativize the consequences, justifying the risks as a consequence of modernity. This allows them to be more bearable so they do not become obstacles in the path of modernization (Montenegro 2005).

Niklas Luhmann, meanwhile, analyses society from the base of social systems where risk appears as a consequence of technological and economic progress that society itself has generated. The risk perception is quantifiable from the point of view of technology, but the risk perception of the general public would be influenced by social factors that guide the selection of an acceptable risk. This should not be understood from an individualistic point of view, but from understanding the group as a social system (Luhmann 1998).

Then we have the studies of Mary Douglas, who, by analysing the technological risks that modern society has been facing, reveals how society accepts a large number of dangers, given the high profit they provide. This acceptance masks the objective reality of the danger, leading people to become more risk tolerant. It also emphasizes how the dangers that society fears are based on various socially accepted aspects. Based on a sample of society, Douglas and Wildavsky point out how certain risks are accepted and others of equal or greater importance are disregarded. From the point of view of these authors, only the cultural approach can integrate the "'moral judgements'" of how to live with the everyday dangers of this world, and how human beings perceive them based on their culture and social structures (Douglas/Wildavsky 1983). All this is the result of a social historical process that outlines the actual and perceived risks that each human group is willing to endure. One of the ideas that these authors defend is that current society has failed to overcome primitive thinking, which consists of placing the responsibility for any wrong decision or outcome on religious or supernatural forces. Nowadays this responsibility has been shifted to technology, authorities and institutions, so people still do not take individual and collective responsibility. Considering the social organization, power relations, economic and political interests at stake, and the intentions of different groups of society, risk perception and acceptance is shaping up as a social construct (Douglas/Wildavsky 1983). An important point on which both Beck and Douglas concur is that neither the perception nor the tolerance of a risk can be reduced to knowledge about its implications and consequences. This would diminish the complexity of the problem to a simple matter of educating society, whereas other factors, such as the notion of justice, ethics, morale and the credibility of institutions, can make all the difference (Montenegro 2005). These elements in turn are crucial for deciding certain behaviour and attitudes towards the dangers we face.

Meanwhile, Wildavsky/Dake (1990) confirm that among the factors that explain the perception and acceptance of risk, cognitive biases are taking shape according to personal patterns, which can be: hierarchical, egalitarian or individualist. People defend their position against risk depending on whether they fear authority, seek social equality or ensure their personal interests. These authors explain that to understand the different positions of individuals against risk, one must take into account the theories about it. They describe some of these theories: knowledge theory (do people know about the real danger?), personality theory (perception and acceptance varies depending on the type of person), economic theory (rich people risk more because they hope to have more benefits of technology, for example), political theory (controversies about risk based on interest according to the social position people have), culturalist theories (people choose what to fear and how much to fear to sustain their way of life). These theories reveal some causal factors of the varied way people position themselves against risk.

Studies on social representations of risk are attached to this approach, and primarily defend the position to analyse the representation from a more qualitative perspective, interweaving knowledge and emotions on a topic such as risk, and considering the different areas that influence construction of representation in society, such as media and institutions, which establish a permanent and changing dialogue (Joffe 2003).

Culturalist studies have also been criticized by some authors, as it has been shown that cultural factors have very little explanatory power (Sjöberg 2000). However, the validity of the culturalist models is not comparable with that of psychometrics, since the methodologies used are very different, because we are talking about parameters and theoretical and methodological paradigms that, instead of being compared, should be complementary.

1.4 Media and Its Influence

It is important to note that the information transmitted by mass media has a strong influence on public opinion. Based on the information source, the credibility of this will also have an effect. However, many people form their opinions based on what they hear on the radio, see on television or read in some news media. People usually confront contradictory information that will be evaluated according to the credibility of the source. For a risk communication to be carried out successfully, maintaining credibility is the most important factor. The hardest thing in risk communication is creating a means of communication environment that guarantees the desired persuasive effect (Renn/Levine 1991). Access is needed to the language, aspirations and needs of the people before outlining any communication strategy.

The way information is transmitted, in this case about danger, can magnify the perception of risk or reduce it through specific communication techniques. These techniques may intensify or weaken signals that are part of the information that individuals and groups receive about a risk, filtering a multitude of signals disregarding its characteristics. Thus, it will modulate what literature has called the *social amplification of risk* (Kasperson et al. 1988). This theory has been widely studied, and it is not the purpose of this chapter to develop it in more detail, but to note that these communication strategies are an important factor influencing risk perception in public opinion and, therefore, in many behaviours of individuals.

1.5 Social Trust

It has been found in some studies (Poortinga/Pidgeon 2005; Viklund 2003; Eiser et al. 2002; Siegrist et al. 2000) that social trust in experts and authorities is a factor that impacts the perception of risk and acceptability of risks. For example, in a study conducted in developed and developing countries, it was found that trust directly and indirectly impacts the acceptability of perceived environmental hazards, but the risks are often only indirectly experienced in developed countries (Bronfman/López-Vázquez 2011). Confidence is used to manage risk through an externalized and deposited personal trust in those who are considered responsible for their management and for community protection. This occurs as a result of an increasingly complex world: people are not able to learn about all the threats they face, so they are forced to trust the authorities and experts. Confidence is used as a shortcut to reduce the need to make rational judgements based on knowledge by selecting trusted experts whose opinion can be considered accurate (Wachinger et al. 2013). This can result in a reduction of uncertainty, but because of the fundamental affective dimension of trust (involving elements such as honesty, integrity, goodwill, or lack of particular interests), people may be more at risk if their trust in experts and authorities is low or damaged by the context in which they live (Espluga et al. 2009).

On the other hand, it has also been observed that even though individuals are aware of the problems related to a particular risk, too much confidence in authorities and experts—influenced by protective measures of the government and the media—can cause the population to take no responsibility or actions for their own preparedness (Bichard/Kazmierczak 2012), or hinder the intentions of prevention (Terpstra 2011). Thus, it is observed that confidence does not always encourage people to assume their own responsibility to prepare against a risk.

1.6 Final Considerations

As discussed throughout the chapter, the approach to the study of risks, for its complexity, involves a comprehensive analysis to better understand the behaviour of people living with risk. The problem that arises from the scientific work is that most of the health and environmental resources are devoted to intervening when disease or disaster has already struck, and not to preventing these situations.

The WHO itself has stated that focusing on the study of the risks is the key to prevention. The fact that most of the deaths from chronic diseases or natural disasters occur in developing countries is alarming (WHO 2015; IPCC 2014). Overall, there is little culture of prevention in these countries, so our efforts should be directed at the modification of certain lifestyles and at reducing risk behaviours.

References

Assally J.P. (1992). *Les jeunes et le risque. Une approche psychologique de l'accident*, Paris: Vigot.

Ayres, T.J., Wood, C.T., Schmidt, R.A., McCarthy, R.L. (2002). Risk Perception and Behavioral Choice. *International Journal of cognitive ergonomics*, 2(1–2), 35–52.

Beck, U. (2012). *Risk Society. Towards a new modernity*. London: Sage.

Beck, U. (1998). *La sociedad del riesgo. Hacia una nueva modernidad* (Risk Society. Towards a new modernity). Barcelona: Paidós.

Bichard, E., & Kazmierczak, A. (2012). Are homeowners willing to adapt to and mitigate the effects of climate change? *Climatic Change*, 112(3–4), 633–654.

Bronfman, N., & López-Vázquez, E. (2011). A Cross-Cultural Study of Perceived Benefit Versus Risk as Mediators in the Trust-Acceptance Relationship. *Risk Analysis*, 31(12), 1919–1934. Douglas, M., & Wildavsky, A. (1983). *Risk and Culture. An essay on the selection of technological and environmental dangers*. California: University of California Press.

Le Robert Dictionnaire Historique de la Langue Française (Le Robert Historic Dictionary of the French Language) (1999). Paris: Diccionaires Le Robert.

Douglas, M., & Wildavsky, A. (1983). *Risk and Culture*. Berkeley: University of California Press.

Eiser, J.R., Miles, S., Frewer, L.J. (2002). Trust, perceived risk and attitudes towards food technologies. *Journal of Applied Social Psychology*, 32(11), 2423–2433.

Espluga, J., Prades, A., Gamero, N., Solá, R. (2009). El papel de la 'confianza' en los conflictos socioambientales. (The role of 'trust' in socio-environmental conflicts.) *Política y sociedad*, 46(1), 255–273.

Fischhoff, B., Watson, S., Hope, C. (1984). Defining Risk. *Policy Sciences*, 17, 123–139.
Fischhoff, B., Lichntenstein, S., Read, S., Combs, B. (2000). How safe is safe enough? A Psychometric Study of Attitudes Toward Technological Risks and Benefits. In P. Slovic (Ed.), *The perception of risk*. London: Earthscan Publications Ltd.
Fischhoff, B., Slovic, P., Lichtenstein, S. (1977). Knowing with certainty: The appropriateness of extreme confidence. *Journal of Experimental Psychology: Human Perception and Performance*, 4, 330–344.
Finucame, M. L., Alkhakami, A., Slovic, P., Jhonson, S. M. (2000). The affect heurística in judgments of risks and benefits. *Journal of Behavioral Decision Making*, 13, 1–17.
Fischhoff, B., Slovic P., Lichtenstein, S. (1977). Knowing with certainty: The appropriateness of extreme confidence. *Journal of Experimental Psychology: Human perception and perfor-mance*, 4, 330–344.
Gilovich, T., Griffin, D., Kahneman, D. (2002). *Heuristics and Biases. The Psychology of Intuitive Judgment*. New York: Cambridge University Press.
Harries, T., & Penning-Rowsell, E. (2011). Victim pressure, institutional inertia and climate change adaptation: The case of flood risk. *Global Environmental Change*, 21, 188–197.
Clinton, J. (2006). Risk Perception and Terrorism: Applying the Psychometric Paradigm. *Homeland Security Affairs*, 2, Revisado el 18 de Junio de 2015 https://www.hsaj.org/articles/169.
Joffe, H. (2003). Risk: From perception to social representation. *British Journal of Social Psychoogy*, 42, 55–73.
Kahneman, D., Slovic, P., Tversky, A. (Eds.), (1982). *Judgment under uncertainty: Heuristics and biases*. New York: Cambridge University Press.
Kahneman, D., & Tversky, A. (1979). Prospect theory: An analysis of decision under risk. *Econometrica*, 47(2), 263–291.
Kasperson, R.E., Renn, O., Slovic, P., Brown, H.S., Emel, J., Goble, R., et al. (1988). The social amplification of risk: A conceptual framework. *Risk Analysis*, 8, 177–187.
Kennedy, B.H. (1962). *The Revised Latin Primer* (revised edn.). Harlow: Longman.
Kouabenan, D.R., Cadet, B., Hermand, D., Muñoz Sastre, M.T. (2006). *Psychologie du risque. Identifier, évaluer, prévenir*. Brussels: De Boeck.
Intergovernmental Panel of Climate Change (IPCC) (2014). Climate change 2014. Impacts, Adaptation and vulnerability. *Fifth assessment Report of Intergovermental Pannel of Climate Change*. Cambridge: Cambridge UP.
Lerner, J.S., Gonzalez, R.M., Small, D.A., Fischhoff, B. (2003). Effects of fear and anger on perceived risks of terrorism: A national field experiment. *Psychological Science*, 14, 144–150.
Lopes, L.L. (1987). Between hope and fear: the psycholoy of risk, *Advances in experimental social psychology*, 20, 255–295.
López-Vázquez, E., Brunner, T., Siegrist, M. (2012). Perceived risks and benefits of nanotech-nology applied to the food and packaging sector in Mexico. *British Food Journal*, 114(2), 197–205.
López-Vázquez, E. (2009). Risk perception and coping strategies for risk from Popocatépetl Volcano, México. *Geofísica Internacional*, 48(1), 301–315.
López-Vázquez, E., Marván, M.L., Flores, F., Peyrefitte, A. (2008). Volcanic risk exposure, feelings of insecurity, stress and coping strategies in Mexico. *Journal of Applied Social Psychology*, 38(12), 2885–2902.
Luhmann, N. (1998). *Sociología del Riesgo* (Risk Sociology). Mexico: University Iberoamericana, Triena Editorials.
Montenegro, S.M. (2005). La sociología de la sociedad del riesgo de Ulrich Beck y sus críticos (The Sociology of Risk Society of Ulrich Beck and his Critics). *PAMPA Revista interuniversitaria de estudios territoriales*, 1, 41–58.
Morales, J.F., Paéz, D., Kornblit, A.L., Asún, D. (2002). *Psicología Social* (Social Psycology). Buenos Aires: Prentice Hall.

Munier, B. (1992). Psychologie du risque et cognition, Séminaire sur les aspects socio-économiques de la gestion des risques naturels 1–3 octobre 1991, Paris, France in Etudes du CEMAGREF. *Montagne*, 2, 43–52.

Poortinga, W., & Pidgeon, N.F. (2005). Trust in risk regulation: Cause or consequence of the acceptability of GM food? *Risk Analysis*, 25(1), 199–209.

Renn, O. (2008). *Risk Governance. Coping with Uncertainty in a Complex World*. London: Earthscan.

Renn, O., & Levine, D. (1991). Credibility and trust in risk communication. In R. E. Kasperson (Ed.) *Communicating Risks to the Public: Technology, Risk, and Society* (pp. 175–218). Netherlands: Kluwer Academic Publisher.

Sánchez-Vallejo, F., Rubio, J., Páez D., Blanco, A. (1998). Optimismo Ilusorio y Percepción de Riesgo (Illusory Optimism and Risk Perception). *Boletín de Psicología* (Psychology Bulletin), 58, 7–17.

Selye, H. (1950). *Stress*. Montreal: Acta Inc.

Sharot, T., Korn, C.W., Dolan, R.J. (2011). How unrealistic optimism is maintained in the face of reality. *Nature Neuroscience*, 14, 1475–1479.

Siegrist, M., Keller, C., Kiers, H.A. (2005) A new look at the psychometric paradigm of perception of hazards. *Risk Analysis*, 25(1), 211–222.

Siegrist, M., Cvetkovich, G.T, Roth, C. (2000). Salient value similarity, social trust, and risk/benefit perception. *Risk Analysis*, 20(3), 353–362.

Sjöberg, L. (2000). Factors in risk perception. *Risk analysis*, 20(1), 1–11.

Slovic, P. (2000). Perception of risk. In P. Slovic (Ed.), The perception of risk (pp. 220–231). London: Earthscan Publications Ltd.

Slovic, P., Fischhoff, B., Lichtenstein, S. (1984). Behavioral decision theory perspectives on risk and safety. *Acta Psychologica*, 56, 183–203.

Slovic, P., Finucame, M.L., Peters, E., MacGregor, D. (2010a). Risk analysis and risk as feelings: some thoughts about affect, reason, risk and rationality. In P. Slovic (Ed.), *The Feeling of risk: New perspectives on risk perception*. London: Earthscan.

Slovic, P., Peters, E., Grana, J., Berger, S., Dieck, G. S. (2010b). Risk perception of prescription drugs: results of a national Survey. In P. Slovic (Ed.), *The Feeling of risk: New perspectives on risk perception*. London: Earthscan.

Smith, W. and Lockwood, J. (1976). *Chambers Murray Latin-English Dictionary*. Edinburgh and London: Chambers and John Murray.

Starr, C. (1969). Social Benefit versus Technological Risk. *Science*, 165, 1232–1238.

Terpstra, T. (2011). Emotions, trust, and perceived risk: Affective and cognitive routes to flood preparedness behaviour. *Risk Analysis*, 31(10), 1658–1675.

Tversky, E., & Kanneman, D. (1974). Judgment under uncertainty: Heuristics and biases. *Sciences*, 185, 1124–1131.

Viklund M. (2003). Trust and risk perception in western Europe: A cross-national study. *Risk Analysis*, 23(4), 727–738.

Wachinger, G.O., Renn, C., Begg, C., Kuhlicke, C. (2013). The risk perception paradox—Implications for governance and communication of natural hazards. *Risk Analysis*, 33(6), 1–17.

Wildavsky, A., & Dake, K. (1990). Theories of risk perception: Who fears what and why?. *Daedalus*, 119(4), 41–60.

World Health Organization (2015). *Enfermedades crónicas y promoción de la salud: Prevención de las enfermedades crónicas* (Chronic Diseases and Health Promotion: Prevention of Chronic Diseases). http://www.who.int/chp/chronic_disease_report/part1/es/index4.html. Accessed 10 Dec 2016.

Zajonc, R.B. (1980). Feeling and thinking: Preferences need no inferences. *American Psychologist*, 35(2), 151–175.

Part I
Human Behaviour Affecting Health

Chapter 2
Stress and Nutrition During Pregnancy: Factors Defining Transgenerational Future Health Within the Family

Tania Romo-González, Raquel González-Ochoa, Rosalba León-Díaz and Gabriel Gutiérrez-Ospina

Abstract A normal pregnancy is defined as the physiological state of a woman from fertilization to the birth of one or more living, healthy and full-term infants. Meanwhile, the term high-risk pregnancy refers to anomalous gestational conditions (e.g. eclampsia), which increase the possibility of pathophysiological states in the mother and/or the infant. There are several biological, gynaeco-obstetric and social risk factors for a high risk pregnancy. Conditions such as malnutrition, negative emotions, perceived stress and maternal anxiety are predictors of pathological conditions in both mother and child (e.g. obesity, diabetes, different types of cancer and some neurological and cardiovascular pathology). This scenario requires a modification of the concept of 'high-risk pregnancy', and advocates taking into account the monitoring of stress, anxiety, depression and food quality during pregnancy as potential risk factors not only for the baby, but also for subsequent generations.

Keywords Pregnancy · Distress · Malnutrition · Negative emotions
Transgenerational health

Dr. Tania Romo-González, Researcher, Department of Integrated Health and Biology, Institute of Biological Research, Universidad Veracruzana. Email: romisnaider@hotmail.com.

Raquel González-Ochoa, M.Sc., Ph.D. student, Department of Integrated Health and Biology, Institute of Biological Research, Universidad Veracruzana. Email: raquelgonzlezochoa@yahoo.com.mx.

Dr. Rosalba León-Díaz, Postdoctoral Researcher, Department of Integrated Health and Biology, Institute of Biological Research, Universidad Veracruzana. Email: rossalba_leon@yahoo.com.mx.

Dr. Gabriel Gutiérrez-Ospina, Researcher, Institute of Biomedical Research, UNAM. Email: gabo@biomedicas.unam.mx.

© Springer International Publishing AG, part of Springer Nature 2018
Ma. L. Marván and E. López-Vázquez (eds.), *Preventing Health and Environmental Risks in Latin America*, The Anthropocene: Politik—Economics—Society—Science 23, https://doi.org/10.1007/978-3-319-73799-7_2

2.1 Introduction

A normal pregnancy is defined as the physiological state of a woman from fertilization to the birth of one or more living, healthy and full-term infants (Guía de Práctica clínica para el Control Prenatal con enfoque de Riesgo 2008). Unfortunately, not all pregnancies are without complications; this is when the term high-risk pregnancy is used. It is important to highlight the fact that it is common during pregnancy for changes to be observed in women's bodies that could be considered to be complications; for example, nausea, fatigue, drowsiness, liquid retention, swelling of the legs, mastodynia, frequent urination, gastroesophageal reflux, constipation and increased breathing rate are all common symptoms found in pregnant women (Cortez-Chavez 2006; Guyton/Hall 2005). These signs and symptoms, however, do not reflect real complications but rather how the body is adjusting to the pregnancy. As such, when we talk about truly high-risk pregnancies, we refer to anomalous gestational conditions (e.g., eclampsia), which increase the possibility of pathophysiological states in the mother and/or the infant, leading to the death of both or of one of the binominal parts of the mother-infant, or to permanent disabilities (e.g., psychomotor problems) in the full-term infant. The socio-medical gravity of high-risk pregnancies can be clearly seen in the fact that close to 800 women around the world die every day as a result of pregnancy-related complications, in addition to approximately three million new-born babies dying every year (World Health Organization [WHO] 2014). In Mexico, it is estimated that the leading cause of death among pregnant women is hypertensive disease (National Institute of Statistics and Geography [INEGI] 2013a), with a mortality rate for children under the age of one year old of 12.8 for every 1,000 live births (INEGI 2013).

In light of this situation, public health systems have proposed the implementation of pregnancy health programmes. For example, the campaign entitled '*Reducing Pregnancy Risks*', launched by the WHO in the year 2000, recommends opportune clinical check-ups to detect risk factors in pregnant women. A risk factor is any circumstance that increases the probability of damage occurring as a result of personal, social and environmental situations that affect the individual (Mexican Department of Health 2001). As such, obstetric risk is understood to be '*any external and/or intrinsic factor (or combination of factors) relating to women that could lead to some form of complication during pregnancy, labour, postpartum or that could, in any way, alter the normal development or survival of the product*' (Cortez-Chavez 2006: 3–4). In order to identify obstetric risk factors, the Mexican Department of Health implemented a multi-phase screening programme, entitled Prenatal Care, the goals of which are to identify, prevent and control factors that could put the life or health of the mother and the baby at risk during pregnancy (Ordoñez 2005). This programme classifies risk factors into three categories: biological, gyneco-obstetric and social; and it considers the presence of at least one of these factors to be sufficient cause for the doctor in question to categorise a pregnancy as high-risk:

Biological factors

✓ Arterial hypertension
✓ Nephropathy
✓ Diabetes Mellitus
✓ Chronic or systemic diseases
✓ Cardiopathy
✓ Aged 35 or over
✓ Psychiatric illnesses
✓ Hereditary diseases
✓ Poor nutrition
✓ Neuropathy.

Gyneco-Obstetric Background:

✓ Abortions/miscarriages
✓ Haemorrhaging during the second half of pregnancy
✓ Perinatal death
✓ Postpartum infection in previous pregnancy
✓ Prior Caesarean section
✓ Prior surgery in reproductive tract
✓ Chronic urinary tract infection
✓ Prematurity
✓ Aged 20 or under
✓ Birth defects
✓ Less than two years after termination of last pregnancy
✓ Having two or more caesareans
✓ Having two or more abortions/miscarriages
✓ Having five or more pregnancies
✓ Rhesus isoimmunisation in current or previous pregnancies
✓ Low birthweight
✓ Weight of last baby above 4,500 g
✓ Obesity and malnutrition
✓ Preeclampsia/eclampsia
✓ Gynaecological age of less than two years.

Social factors:

✓ Poverty
✓ Unwanted pregnancy
✓ Mother who is illiterate or has not finished elementary education
✓ Cultural factors
✓ Occupational factor (exposure to toxic substances, chemicals or infectious diseases; working more than 36 h per week; lifting heavy objects; standing for more than four hours per shift; mental stress; working in a cold environment or one with intense noise levels)

✓ History of sexual, physical and/or emotional abuse
✓ Risk behaviours (self-medication, smoking, alcoholism and use of other hard drugs).

(Cortez-Chavez 2006; Guía de Práctica clínica para el Control Prenatal con enfoque de Riesgo 2008; Mexican Department of Health 2001).

As such, prenatal care is a tool that helps the person charged with monitoring the health of the mother-child during pregnancy to opportunely identify risk factors during pregnancy, and thus implement the corresponding preventive and corrective measures. The following are important clinical signs that need to be monitored: vaginal bleeding during pregnancy, convulsions or blackouts accompanied by blood pressure readings higher than 90 mmHg, high temperature (>38°C), abdominal pain, ruptured membranes, severe pallor or cyanosis, marked decline or weakness, sensation of fainting, severe headaches, blurred vision, vomiting and/or respiratory difficulties. Furthermore, it is important to mention that pregnancy risk diagnoses should preferably be carried out prior to conception, continuing throughout the pregnancy, labour and postpartum (Aguilar-Moreno et al. 2001; Pérez-Rúa 2005).

During prenatal care, the doctor must inform the patient of all the physiological changes associated with pregnancy, in addition to explaining the warning signs that may present during pregnancy, labour and postpartum, emphasizing the importance of attending prenatal care sessions periodically and undertaking any laboratory and medical tests deemed necessary. He or she should also explain the importance of neonatal screening and breastfeeding, promote healthy lifestyles and encourage the mother to avoid smoking, drinking alcohol, self-medicating and consuming other substances that put the risk of her health and that of the child at risk (Aguilar-Moreno et al. 2001).

Even though prenatal care can help prevent high-risk pregnancies, this same care is notorious for not taking into account the monitoring of stress, anxiety, depression and food quality during pregnancy (Qiu et al. 2009; Winkel et al. 2015); however, it has been shown that these are potential risk factors not only for the foetus, but also for subsequent generations (Drake et al. 2011). As such, it is necessary to implement strategies that minimize the negative impact that poor nutrition and stress during pregnancy can have on the health of the mother and child.

2.2 Stress as a Risk Factor in Pregnancy and Its Impact on the Health of Different Generations

The magnitude of the response of a person to stress depends on the interaction with information from a range of different sources, both biological (e.g., genes) and psychosocial (e.g., past experiences, social support). These same interactions influence the response to stress developed by pregnant women; however, pregnant women face specific situations that the general public do not, and which are a source of information that can be stressful. These circumstances are associated with physical and hormonal changes (which are associated with mood swings) and the anxiety of the pregnancy itself, associated with concern regarding the health of the foetus and with the pain of labour (Huizink 2000, quoted by Mulder et al. 2002; Van den Bergh 1992). Additionally, it is known that age, deficient education, low socio-economic level, sexual abuse, unwanted pregnancy, an absent partner, lack of preparation for pregnancy and labour, and a history of depressive and psychiatric symptoms influence the perception of well-being and, as such, the stress levels of pregnant women (Paarlberg et al. 1996).

Furthermore, negative emotions, such as perceived stress and maternal anxiety, can predict depressive symptoms in both the mother and the baby (Milgrom et al. 2008), which is why stress has been recognized as an important etiological factor that can trigger mood disorders. This notion has been demonstrated in animal models, based on the supposition that the effects of stress on animals lead to a series of behavioural changes, some of which are reminiscent of symptoms that characterize depression and anxiety among human beings (van Dijken et al. 1992). Depression is particularly prevalent among pregnant women, affecting between 10 and 25% of them (Marcus et al. 2003; Nonacs/Cohen 2002). The greatest concern regarding these statistics lies in the fact that pregnant women with depression have a higher probability of premature labour (Orr et al. 2002); furthermore, other negative emotions, such as perceived stress and anxiety, have been associated with a reduction in the variability of foetal heartbeat (DiPietro et al. 1996), an increase in locomotor activity (DiPietro et al. 2002), and disruption to foetal habituation (Sandman et al. 1999), which, in conjunction, lead to miscarriages (Friebe/Arck 2008), low birthweight (Mutale et al. 1991) premature labour (Copper et al. 1996) and a wide variety of complications during pregnancy (Table 2.1).

Even though the pathologies present during pregnancy and around the time of labour are multifactorial, it is clear that a common element throughout these pathologies is the maternal-embryonic-foetal stress imposed on this dichotomy, both as a result of the conditions attributable to the age of the mother and the organic pathophysiological state of the mother prior to and/or during pregnancy (Muñoz/Oliva 2009; Paarlberg et al. 1995), whether or not they are associated with the socio-economic-cultural conditions of the pregnant woman. In fact, based on the previous point, a number of studies show that the expectations, fears and social conceptions that pregnant women have regarding their pregnancy have a profound impact, no matter their age, on the generation of pathophysiological states during pregnancy which are associated with psychosocial stress (Pimentel 2007; Senties/Ortiz 1993; Teixeira et al. 2009). Furthermore, it is known that women suffering from depression or anxiety associated with psychosocial stress during pregnancy show high levels of serum cortisol, which appears to lead to a state of immunological suppression (Walker et al. 1999), a decrease in serum levels of progestational hormones (Arck et al. 2007; Frye et al. 2011) and lesser neurological development (Clarke et al. 1994; Chrousos et al. 1998; Huizink 2000, quoted by Mulder et al. 2002; Schneider et al. 1999); these conditions are more pronounced if they occur during the early stages of pregnancy (Schneider et al. 1999). Moreover, exposure to prenatal stress is commonly considered to be an important factor in the

Table 2.1 Factors that trigger a high-risk pregnancy (own elaboration)

Risk factor	Complication	Reference
Gynaecological age of less than 2 years	Pre-eclampsia/eclampsia, prematurity and low birthweight	Mexican Department of Health (2001), Cortez-Chavez (2006)
Aged 35 or over	Pre-eclampsia/eclampsia, low-lying placenta and uterine atony post labour	
Less than 2 years between pregnancies	Prematurity and stunted intrauterine growth	
Having had 5 or more pregnancies	Low-lying placenta and uterine atony post labour	
Having had 2 or more abortions/ miscarriages	Greater predisposition to abortion/ miscarriage	
History of pre-eclampsia/ eclampsia	Predisposition to pre-eclampsia/ eclampsia	
Haemorrhaging during 2nd half of pregnancy	Predisposition to haemorrhaging	
Having had 2 or more caesareans	Low-lying placenta, placenta accreta and rupture of the uterine scar during labour	
Foetal death	Probability of repetition of foetal death	
Prematurity	Probability of repetition	
Arterial hypertension	Pre-eclampsia/eclampsia, stunted uterine growth, foetal and maternal death	
Diabetes Mellitus	Pre-eclampsia/eclampsia, abortions/miscarriage, birth defects, stunted intrauterine growth or foetal microsomia	
Anxiety	Pre-eclampsia, pre-term labour	Diego et al. (2006), Harville et al. (2009), Orr et al. (2007), Pimentel (2007), Romo-Gonzalez et al. (2012), Vianna et al. (2011), Wadhwa et al. (1993)
Depression	Pre-term labour	Diego et al. (2006), Field et al. (2004), Hompes et al. (2012), Paarlberg et al. (1999), Pimentel (2007), Steer et al. (1992), Vianna et al. (2011)

(continued)

Table 2.1 (continued)

Risk factor	Complication	Reference
Stress	Low birthweight, pre-term labour, miscarriage, pre-eclampsia, emotional problems in the life of the baby influence foetal programming, descendants of the mother are more susceptible to anxiety and depression	Buss et al. (2012), de Weerth et al. (2003), Diego et al. (2006), Grandia et al. (2008), Harville et al. (2009), Nepomnaschy et al. (2006), Paarlberg et al. (1999), Pawluski et al. (2011), Salacz et al. (2012), Sandman et al. (2006), van den Hove et al. (2005), Vianna et al. (2011), Wadhwa et al. (1993)
Intrauterine infections Imbalance of endocrine functions Deregulation of placental clock Deregulation of immunological tolerance	Pre-term labour	Challis (2000), Raghupathy/ Kalinka (2008), Romero et al. (1994), Ruiz et al. (2003), Smith (1998), Veenstra van Nieuwenhoven et al. (2003)

Source Cortez-Chavez (2006), Guía de Práctica clínica para el Control Prenatal con enfoque de Riesgo (2008), Mexican Department of Health (2001)

development of a number of forms of psychopathology, such as Attention Deficit Hyperactivity Disorder, schizophrenia and adult depression (Hultman et al. 1997).

These situations lead to morbid states, such as pre-eclampsia, miscarriage, compromised foetal growth, pre-term induction of labour, low birthweight, delayed post-natal psychomotor development and a greater incidence of infections during pregnancy and postpartum, among other complications (Diego et al. 2006; Field/ Diego 2008). The pathogenesis of pre-eclampsia and other hypertensive states during pregnancy is associated with exacerbated inflammatory processes and poor renal function (Sargent et al. 2007), which control not only blood pressure but also the secretion of hormones, such as cortisol and noradrenaline. Moreover, elevated concentrations of noradrenaline lead to decreased dopamine production, which can be seen in the manifestation of symptoms of depression (Field et al. 2004). Additionally, the secretion of angiogenic molecules by leukocyte cells can promote the development of hypertensive disease associated with gestational stress (Geara et al. 2009). Furthermore, neuro-immuno-endocrine alterations relating to gestational stress, be they psychosocial in nature or not, can lead to complications and ailments during pregnancy or labour, affecting the health of the child when it reaches adulthood, in the event it survives (Bastani et al. 2005; DiPietro et al. 2004; Huizink et al. 2004; Monk 2001; Wadhwa et al. 2001; Table 2.1).

It has recently been shown that exposure to stressful situations during embryonic and/or foetal life predisposes the adult to obesity, diabetes, different types of cancer and some neurological and cardiovascular pathologies, associated with epigenetic re-editing (Bastani et al. 2005; Bilbo/Schwarz 2012; Bresnahan et al. 2005; Brown et al. 2004; DiPietro et al. 2004; Ellman/Susser 2009; Huizink et al. 2004; Monk 2001; Mulder et al. 2002; Wadhwa et al. 2001). Furthermore, a number of lines of research have not only shown that stress affects the mother and the baby during pregnancy, but that the stress history of their ancestors and parents also conditions the evolution and result of the pregnancy, given that maternal glucocorticoids can regulate the transcription of a number of the product's genes. These are not limited to individuals exposed to the stress of the mother, as they can also be transmitted to the descendants several generations into the future (Drake et al. 2011). As such, in 2013, Crudo et al., suggested that epigenetic re-editing, associated with the methylation of DNA, the acetylation of histones and the transcription stemming from different environments in the early lives of humans and rodents, is not limited to a small number of genes, but rather involves a range of genetic networks and functional genomic patterns.

2.3   Nutrient Consumption During High-Risk Pregnancy: The Transgenerational Impact

NOM-007-SSA2-1993, a standard focusing on the care offered to women during pregnancy, labour and postpartum and to newborns, states that the majority of obstetric damage and health risks facing the mother and the offspring can be prevented, detected and treated successfully. As such, it is important to consider that the nutritional conditions of the mother before and during pregnancy will create a favourable or adverse in utero environment that will predispose the foetus to disorders such as obesity, diabetes or metabolic syndrome (Barker 1995; Nathanielsz et al. 2007; Remacle et al. 2007; Reusens et al. 2007).

While foods greatly define health, growth and development from foetus to old age, it is important to consider that both undernutrition and overnutrition have a harmful influence on the development of diseases. This is why diet and nutrition during pregnancy and early postnatal life are important in ensuring health (Darnton-Hill 2013). There is a wide range of experimental evidence that highlights the impact that eating habits have on health. For example, studies in rodents have shown that consuming large quantities of fat during pregnancy can lead to symptoms of metabolic syndrome among offspring when they reach adulthood (Brown et al. 1990; Chechi/Cheema 2006; Guo/Jen 1995). It would seem that this is due to the fact that a maternal diet that is high in fats leads to the accumulation of lipids, inflammation and oxidative stress in the foetal liver (McCurdy et al. 2009), which, in turn, leads to the development of metabolic disorders, such as diabetes, dyslipidaemia and hypertension. Moreover, experiments on rats fed on a 'cafeteria' type diet during pregnancy showed that offspring acquire a preference for 'junk' food that is high in fats, sugars and energy density, combined with a lack of control over food intake as adults (Bayol et al. 2007).

It has also been shown that offspring whose mothers were submitted to a severe decrease in food intake during pregnancy presented with hyperphagia and decreased physical activity (Vickers et al. 2003). In both cases, it is clear that the subjects developed immediate adaptive metabolic responses that mean they are predisposed to developing metabolic diseases in adulthood (Langley-Evans 2006). Further studies, also on rodents, corroborated the fact that high-fat diets consumed by pregnant females 'programme' the hypothalamic pathways that regulate food intake (Bouret 2009). Although we do not know as much about the fine mechanisms that underlie the establishment of patterns of risk behaviour and metabolic states in human beings as we do in animal models, it is interesting that a large number of epidemiological studies on humans suggest that maternal obesity contributes to the incidence of obesity, type 2 diabetes and metabolic syndrome among their descendants (Levin 2006; Martin-Gronert/Ozanne 2005; Oken/Gillman 2003), not to mention that maternal malnutrition during pregnancy also increases the incidence of obesity among descendants (Levin 2009). In humans, as in animal models, the damaging effects associated with poor maternal nutrition during pregnancy are not only observed in the trends for metabolic disorders, they also have a neurological impact. Poor nutrition during pregnancy is associated with an elevated risk of depression (Rofey et al. 2009), attention deficit with or without hyperactivity (Waring/Lapane 2008), and anxiety (Kelley et al. 2005; Kiyohara/Yoshimasu 2009).

In summary, the type of nutrients consumed during pregnancy is an important factor to be taken into consideration in preventing high-risk pregnancy, as these nutrients lead to persistent changes in the body, changes that can also be transmitted from generation to generation. It has been postulated that these changes occur by means of epigenetic reprogramming (Sandovici et al. 2011), which makes neuroendocrine and metabolic conditioning possible so as to control food consumption, use and breakdown. The epigenetic mechanisms that condition hormonal and metabolic systems include the modification of DNA methylation levels and the modification of histone structure through methylation, acetylation and ubiquitination (Blewitt et al. 2006; Waterland et al. 2007).

Based on this, it is clear that nutritional problems lead to a series of epigenetic modifications that can have a negative impact on the mother-child dyad and on subsequent generations. However, it is important to remember that epigenetic marks and the phenotypical changes associated with them can be modified and even reversed by employing 'positive' factors, be they hormonal, psychological, nutritional, environmental, social and even cultural (Szyf et al. 2008), that function as translators for a 'healthy' environment (Gabory et al. 2009; Lelievre 2009).

2.4 Conclusions

The Secretary General of the United Nations, Ban Ki-moon (2009), referred to maternal health as 'the mother of all health challenges … perhaps there is no other issue that better links safety, prosperity and progress as women's health…' (Castro-Santoro 2011, Par. 1)

In keeping with these views, we believe that, in order to tackle the problem of maternal and child morbi-mortality, it is important to start by redefining high-risk pregnancy. This new concept should not merely be considered an anomalous condition that increases the possibilities of developing pathological states or sequels that, in the short-, mid- and long-term, lead to the death of the mother and/or the offspring, but rather it should incorporate the risk that this implies for future generations, and, moreover, promote the message that pregnant women should avoid situations that generate stress, anxiety or depression, in addition to tackling poor nutrition, given that, as we have seen, these are factors that have a strong influence on the development of high-risk pregnancy.

Pregnancy is a crucial time, during which both a living being and the immediate and future health of subsequent generations is nurtured. As numerous studies have shown, the inadequate progress of the pregnancy conditions and/or promotes the development of psychiatric illnesses, such as schizophrenia, behavioural problems, such as attention deficit disorder or hyperactivity, emotional problems, such as depression, and even obesity, diabetes, different types of cancer, neurological pathologies and cardiovascular problems. This is why it is important to emphasize that proper stress management and a good diet can help reverse and/or prevent the incidence of these health problems among future generations.

Finally, it is important to recognize that stress and diet must be added to the list of factors that should be monitored in order to prevent high-risk pregnancy, and it is also important to modify the concept of prenatal care, focusing on monitoring pregnant women not only in terms of physical health, but also in mental and nutritional terms, through measures aimed at developing strategies that promote proper nutrition.

References

Aguilar-Moreno, V., Muñoz-Soto, R., Velasco-Vite, J., Cabezas-García, E., Ibargüengoitia-Ochoa, F., Nuñez-Urquiza, R.M., et al. (2001). Control prenatal con enfoque de riesgo.Una herramienta indispensable para el médico general. *Práctica Médica Efectiva,* 3(9), 1–4.

Arck, P., Hansen, P.J., Mulac Jericevic, B., Piccinni, M.P., Szekeres-Bartho, J. (2007). Progesterone during pregnancy: endocrine-immune cross talk in mammalian species and the role of stress. *American Journal of Reproductive Immunology,* 58(3), 268–279 (2007). https://doi.org/10.1111/j.1600-0897.2007.00512.x.

Barker, D.J.P. (1995). The fetal and infant origins of disease. *European Journal of Clinical Investigation,* 25(7), 457–463.

Bastani, F., Hidarnia, A., Kazemnejad, A., Vafaei, M., Kashanian, M. (2005). A randomized controlled trial of the effects of applied relaxation training on reducing anxiety and perceived stress in pregnant women. *Journal of Midwifery & Women's Health,* 50(4), e36–40 (2005). https://doi.org/10.1016/j.jmwh.2004.11.008.

Bayol, S.A., Farrington, S.J., Stickland, N.C. (2007). A maternal 'junk food' diet in pregnancy and lactation promotes an exacerbated taste for 'junk food' and a greater propensity for obesity in rat offspring. *British Journal of Nutrition,* 98(4), 843–851(2007). https://doi.org/10.1017/s0007114507812037.

Bilbo, S.D. & Schwarz, J.M. (2012). The immune system and developmental programming of brain and behaviour. *Frontiers in Neuroendocrinology*, 33(3), 267–286 (2012). https://doi.org/10.1016/j.yfrne.2012.08.006.

Blewitt, M.E., Vickaryous, N.K., Paldi, A., Koseki, H., Whitelaw, E. (2006). Dynamic reprogramming of DNA methylation at an epigenetically sensitive allele in mice. *Plos Genetics*, 2(4), e49 (2006). https://doi.org/10.1371/journal.pgen.0020049.

Bouret, S.G. (2009). Early life origins of obesity: role of hypothalamic programming. *Journal of Pediatric Gastroenterology & Nutrition*, 48(Suppl 1), S31–38 (2009). https://journals.lww.com/jpgn/Abstract/2009/03001/Early_Life_Origins_of_Obesity__Role_of.6.aspx.

Bresnahan, M., Schaefer, C.A., Brown, A.S., Susser, E.S. (2005). Prenatal determinants of schizophrenia: what we have learned thus far? *Epidemiology e Psichiatria Sociale*, 14(4), 194–197.

Brown, A.S., Begg, M.D., Gravenstein, S., Schaefer, C.A., Wyatt, R.J., Bresnahan, M., et al. (2004). Serologic evidence of prenatal influenza in the etiology of schizophrenia. *Archives of General Psychiatry*, 61(8), 774–780 (2004). https://doi.org/10.1001/archpsyc.61.8.774.

Brown, S. A., Rogers, L. K., Dunn, J. K., Gotto, A. M. Jr., Patsch, W. (1990). Development of cholesterol homeostatic memory in the rat is influenced by maternal diets. *Metabolism - Clinical and Experimental*, 39(5), 468–473.

Buss, C., Davis, E.P., Shahbaba, B., Pruessner, J.C., Head, K., Sandman, C.A. (2012). Maternal cortisol over the course of pregnancy and subsequent child amygdala and hippocampus volumes and affective problems. *Proceedings of the National Academy of Sciences of United States of America*, 109(20), E1312–E1319 (2012). https://doi.org/10.1073/pnas.1201295109.

Castro-Santoro, R. (2011). Discurso del Representante de la Organización Panamericana de la Salud. http://www.omm.org.mx/index.php/lanzamiento-del-observatorio?id=155. Accessed 10 Jun 2014.

Clarke, A.S., Wittwer, D.J., Abbott, D.H., Schneider, M.L. (1994). Long-term effects of prenatal stress on HPA axis activity in juvenile rhesus monkeys. *Developmental Psychobiology*, 27(5), 257–269 (1994). https://doi.org/10.1002/dev.420270502.

Copper, R.L., Goldenberg, R.L., Das, A., Elder, N., Swain, M., Norman, G., et al. (1996). The preterm prediction study: maternal stress is associated with spontaneous preterm birth at less than thirty-five weeks' gestation. National Institute of Child Health and Human Development Maternal-Fetal Medicine Units Network. *American Journal of Obstetrics & Gynecology*, 175 (5), 1286–1292.

Cortez-Chavez, J.A. (2006). Diez principales causas de alto riesgo en el Hospital General de zona no. 1 de la ciudad de Colima en el año 2004. (Tesis no publicada, Especialidad en Medicina Familiar). Colima: University of Colima.

Crudo, A., Suderman, M., Moisiadis, V.G., Petropoulos, S., Kostaki, A., Hallett, M., et al. (2013). Glucocorticoid programming of the fetal male hippocampal epigenome. *Endocrinology*, 154 (3), 1168–1180 (2013). https://doi.org/10.1210/en.2012-1980.

Challis, J.R.G. (2000). Mechanism of parturition and preterm labor. *Obstetrica & Gynecological Survey*, 55(10), 650–660.

Chechi, K., & Cheema, S.K. (2006). Maternal diet rich in saturated fats has deleterious effects on plasma lipids of mice. *Experimental & Clinical Cardiology*, 11(2), 129–135.

Chrousos, G.P., Torpy, D.J., Gold, P.W. (1998). Interactions between the hypothalamic-pituitary-adrenal axis and the female reproductive system: clinical implications. *Annals of Internal Medicine*, 129(3), 229–240.

Darnton-Hill, I. (2013). Asesoramiento sobre nutrición durante el embarazo. Fundamento biológico, conductual y contextual. http://www.who.int/elena/titles/bbc/nutrition_counselling_pregnancy/es/. Accessed 5 Apr 2014.

de Weerth, C., van Hees, Y., Buitelaar, J.K. (2003). Prenatal maternal cortisol levels and infant behaviour during the first 5 months. *Early Human Development*, 74(2), 139–151.

Diego, M.A., Jones, N.A., Field, T., Hernandez-Reif, M., Schanberg, S., Kuhn, C., et al. (2006). Maternal psychological distress, prenatal cortisol, and fetal weight. *Psychosomatic Medicine,* 68(5), 747–753 (2006). https://insights.ovid.com/crossref?an=00006842-200609000-00015.

DiPietro, J.A., Ghera, M.M., Costigan, K., Hawkins, M. (2004). Measuring the ups and downs of pregnancy stress. *Journal of Psychosomatic Obstetrics & Gynaecology*, 25(3–4), 189–201.

DiPietro, J.A., Hilton, S.C., Hawkins, M., Costigan, K.A., Pressman, E.K. (2002). Maternal stress and affect influence fetal neurobehavioral development. *Developmental Psychology*, 38(5), 659–668.

DiPietro, J.A., Hodgson, D.M., Costigan, K.A., Hilton, S.C., Johnson, T.R.B. (1996). Fetal Neurobehavioral Development. *Child Development*, 67(5), 2553–2567.

Drake, A.J., Liu, L., Kerrigan, D., Meehan, R.R., Seckl, J.R. (2011). Multigenerational programming in the glucocorticoid programmed rat is associated with generation-specific and parent of origin effects. *Epigenetics*, 6(11), 1334–1343 (2011). https://doi.org/10.4161/epi. 6.11.17942.

Ellman, L.M., & Susser, E.S. (2009). The promise of epidemiologic studies: neuroimmune mechanisms in the etiologies of brain disorders. *Neuron*, 64(1), 25–27 (2009). https://doi.org/ 10.1016/j.neuron.2009.09.024.

Field, T., & Diego, M. (2008). Cortisol: the culprit prenatal stress variable. *International Journal of Neuroscience*, 118(8), 1181(2008). https://doi.org/10.1080/00207450701820944.

Field, T., Diego, M., Hernandez-Reif, M., Vera, Y., Gil, K., Schanberg, S., et al. (2004). Prenatal maternal biochemistry predicts neonatal biochemistry. *International Journal of Neuroscience*, 114(8), 933–945.

Friebe, A., & Arck, P. (2008). Causes for spontaneous abortion: what the bugs 'gut' to do with it? *The International Journal of Biochemistry & Cell Biology*, 40(11), 2348–2352 (2008). https:// doi.org/10.1016/j.biocel.2008.04.019.

Frye, C.A., Paris, J.J., Osborne, D.M., Campbell, J.C., Kippin, T.E. (2011). Prenatal Stress Alters Progestogens to Mediate Susceptibility to Sex-Typical, Stress-Sensitive Disorders, such as Drug Abuse: A Review. *Frontiers in Psychiatry,* 17(2), 52 (2011). https://doi.org/10.3389/ fpsyt.2011.00052.

Gabory, A., Ripoche, M.A., Le Digarcher, A., Watrin, F., Ziyyat, A., Forne, T., et al. (2009). H19 acts as a trans regulator of the imprinted gene network controlling growth in mice. *Development*, 136(20), 3413–3421(2009). https://doi.org/10.1242/dev.036061.

Geara, A.S., Azzi, J., Jurewicz, M., Abdi, R. (2009). The renin-angiotensin system: an old, newly discovered player in immunoregulation. *Transplantation Reviews*, 23(3), 151–158 (2009). https://doi.org/10.1016/j.trre.2009.04.002.

Grandia, C., González, M. A., Naddeo, S., Basualdo, M.N., Salgado, M.P. (2008). Relación entre estrés psciosocial y parto prematuro. Una investigación interdisciplinaria en el área urbana de Buenos Aires. *Revista del Hospital Materno Infantil Ramón Sarda,* 27(2), 51–69. http://www. redalyc.org/articulo.oa?id=91227202. Accessed 29 May 2014.

Guía de Práctica clínica para el Control Prenatal con enfoque de Riesgo. (2008). México: Secretaría de Salud. http://www.cenetec.salud.gob.mx/descargas/gpc/CatalogoMaestro/028_ GPC__PrenatalRiesgo/IMSS_028_08_EyR.pdf. Accessed 7 Jan 2014.

Guo, F., & Jen, K.L. (1995). High-fat feeding during pregnancy and lactation affects offspring metabolism in rats. *Physiology & Behavaior*, 57(4), 681–686.

Guyton, A.C., & Hall, J.E. (2005). Embarazo y Lactancia. In A.C. Guyton, & J.E Hall (Eds.). *Tratado de Fisiología Médica* (10 ed.): España: McGraw-Hill.

Harville, E.W., Savitz, D.A., Dole, N., Herring, A.H., Thorp, J.M. (2009). Stress questionnaires and stress biomarkers during pregnancy. *Journal of Women's Health*, 18(9), 1425–1433.

Hompes, T., Vrieze, E., Fieuws, S., Simons, A., Jaspers, L., Van Bussel, J., et al. (2012). The influence of maternal cortisol and emotional state during pregnancy on fetal intrauterine growth. *Pediatric Research*, 72(3), 305–315 (2012). https://doi.org/10.1038/pr.2012.70.

Huizink, A.C., Mulder, E.J., Buitelaar, J.K. (2004). Prenatal stress and risk for psychopathology: specific effects or induction of general susceptibility? *Psychological Bulletin*, 130, 115–142 (2004). https://doi.org/10.1037/0033-2909.130.1.115.

Hultman, C.M., Ohman, A., Cnattingius, S., Wieselgren, I. M., Lindstrom, L. H. (1997). Prenatal and neonatal risk factors for schizophrenia. *The British Journal of Psychiatry*, 170, 128–133.

Kelley, A.E., Schiltz, C.A., Landry, C.F. (2005). Neural systems recruited by drug- and food-related cues: Studies of gene activation in corticolimbic regions. *Physiology & Behavior,* 86 (1–2), 11–14 (2005). https://doi.org/10.1016/j.physbeh.2005.06.018.

Kiyohara, C., & Yoshimasu, K. (2009). Molecular epidemiology of major depressive disorder. *Environmental Health and Preventive Medicine,* 14(2), 71–87 (2009). https://doi.org/10.1007/s12199-008-0073-6.

Langley-Evans, S.C. (2006). Developmental programming of health and disease. *Proceedings of the Nutrition Society,* 65, 97–105.

Lelievre, S.A. (2009). Contributions of extracellular matrix signaling and tissue architecture to nuclear mechanisms and spatial organization of gene expression control. *Biochimica et Biophysica Acta (BBA) – General Subjects,* 1790(9), 925–935 (2009). https://doi.org/10.1016/j.bbagen.2009.03.013.

Levin, B.E. (2006). Metabolic imprinting: critical impact of the perinatal environment on the regulation of energy homeostasis. *Philosophical Transactions of the Royal Society of London. Series B: Biological Sciences,* 361(1471), 1107–1121(2006). https://doi.org/10.1098/rstb.2006.1851.

Levin, B.E. (2009). Synergy of nature and nurture in the development of childhood obesity. *International Journal of Obesity,* 33(Suppl 1), S53–56 (2009). https://doi.org/10.1038/ijo.2009.18.

Marcus, S.M., Flynn, H.A., Blow, F.C., Barry, K.L. (2003). Depressive symptoms among pregnant women screened in obstetrics settings. *Journal of Women's Health,* 12(4), 373–380 (2003). https://doi.org/10.1089/154099903765448880.

Martin-Gronert, M.S., & Ozanne, S.E. (2005). Programming of appetite and type 2 diabetes. *Early Human Development,* 81(12), 981–988 (2005). https://doi.org/10.1016/j.earlhumdev.2005.10.006.

McCurdy, C.E., Bishop, J.M., Williams, S.M., Grayson, B.E., Smith, M.S., Friedman, J.E., et al. (2009). Maternal high-fat diet triggers lipotoxicity in the fetal livers of nonhuman primates. *The Journal of Clinical Investigation,* 119(2), 323–335 (2009). https://doi.org/10.1172/jci32661.

Mexican Department of Health (2001). Manual de atención. Embarazo saludable, parto y puerperio seguros, recién nacido sano. http://www.paho.org/mex/index.php?option=com_docman&task=doc_download&gid=576&Itemid=. Accessed 11 Feb 2014.

Milgrom, J., Gemmill, A.W., Bilszta, J.L., Hayes, B., Barnett, B., Brooks, J., et al. (2008). Antenatal risk factors for postnatal depression: a large prospective study. *Journal of Affective Disorders,* 108(1–2), 147–157 (2008). https://doi.org/10.1016/j.jad.2007.10.014.

Monk, C. (2001). Stress and mood disorders during pregnancy: implications for child development. *Psychiatric Quarterly,* 72(4), 347–357.

Mulder, E.J., Robles de Medina, P.G., Huizink, A.C., Van den Bergh, B.R., Buitelaar, J.K., Visser, G. H. (2002). Prenatal maternal stress: effects on pregnancy and the (unborn) child. *Early Human Development,* 70(1–2), 3–14.

Muñoz, M.P., & Oliva, P.M. (2009). Los estresores psicosociales se asocian a síndrome hipertensivo del embarazo y/o síntomas de parto prematuro en el embarazo adolescente. *Revista Chilena de Obstetricia y Ginecología,* 74(5), 281–285 (2009). https://doi.org/10.4067/s0717-75262009000500003.

Mutale, T., Creed, F., Maresh, M., Hunt, L. (1991). Life events and low birthweight–analysis by infants preterm and small for gestational age. *British Journal of Obstetrics & Gynaecology,* 98 (2), 166–172.

Nathanielsz, P.W., Poston, L., Taylor, P.D. (2007). In utero exposure to maternal obesity and diabetes: animal models that identify and characterize implications for future health. *Obstetrics and Gynecology Clinics of North America,* 34(2), 201–212 (2007). https://doi.org/10.1016/j.ogc.2007.03.006.

National Institute of Statistics and Geography (2013a). Estadísticas a propósito del día mundial de la salud. Instituto Nacional de Estadística y Geografía, Aguascalientes, Ags. a 7 de abril del. Sala de Prensa.

National Institute of Statistics and Geography (2013b). Indicadores de demografía y población. http://www3.inegi.org.mx/sistemas/temas/default.aspx?s=est&c=17484. Accessed 7 Jan 2014.

Nepomnaschy, P.A., Welch, K.B., McConnell, D.S., Low, B.S., Strassmann, B.I., England, B.G. (2006). Cortisol levels and very early pregnancy loss in humans. *Proceedings of the National Academy of Sciences of the United States of America,* 103(10), 3938–3942 (2006). https://doi.org/10.1073/pnas.0511183103.

Nonacs, R., & Cohen, L.S. (2002). Depression during pregnancy: diagnosis and treatment options. *The Journal of Clinical Psychiatry,* 63(Suppl 7), 24–30.

Norma Oficial Mexicana NOM-007-SSA2-1993. Atención de la mujer durante el embarazo, parto y puerperio y del recién nacido. Criterios y procedimientos para la prestación del servicio 1994.

Oken, E., & Gillman, M.W. (2003). Fetal origins of obesity. *Proceedings of the Nutrition Society,* 11(4), 496–506 (2003). https://doi.org/10.1038/oby.2003.69.

Ordoñez, J. (2005). Evaluación del riesgo materno-neonatal durante el embarazo. *Investigaciones Andina,* 7(10), 38–47.

Orr, S.T., James, S.A., Blackmore Prince, C. (2002). Maternal prenatal depressive symptoms and spontaneous preterm births among African-American women in Baltimore, Maryland. *American Journal of Epidemiology,* 156(9), 797–802.

Orr, S.T., Reiter, J.P., Blazer, D.G., James, S.A. (2007). Maternal Prenatal Pregnancy-Related Anxiety and Spontaneous Preterm Birth in Baltimore, Maryland. *Psychosomatic Medicine,* 69 (6), 566–570 (2007). https://insights.ovid.com/crossref?an=00006842-200707000-00013.

Paarlberg, K.M., Vingerhoets, A.J., Passchier, J., Dekker, G.A., Heinen, A.G., van Geijn, H. P. (1999). Psychosocial predictors of low birthweight: a prospective study. *British Journal of Obstetrics and Gynaecology,* 106(8), 843–841.

Paarlberg, K.M., Vingerhoets, A.J., Passchier, J., Dekker, G.A., Van Geijn, H.P. (1995). Psychosocial factors and pregnancy outcome: a review with emphasis on methodological issues. *Journal of Psychosomatic Research,* 39(5), 563–595.

Paarlberg, K.M., Vingerhoets, A.J., Passchier, J., Dekker, G.A., van Geijn, H.P. (1996). Psychosocial factors as predictors of maternal well-being and pregnancy-related complaints. *Journal Psychosomatic Obstetrics & Gynaecolgy,* 17(2), 93–102.

Pawluski, J.L., van den Hove, D.L.A., Rayen, I., Prickaerts, J., Steinbusch, H.W.M. (2011). Stress and the pregnant female: Impact on hippocampal cell proliferation, but not affective-like behaviors. *Hormones and Behavior,* 59(4), 572–580 (2011). https://doi.org/10.1016/j.yhbeh.2011.02.012.

Pérez-Rúa, Y.A. (2005). Embarazo de alto riesgo. In F. Espinoza (Ed.). *La Neonatología en la Atención Primaria* (pp. 35–39). La Paz: M.S.D.

Pimentel, B. (2007). Ansiedad, depresión y funcionalidad familiar en embarazo de alto riesgo obstétrico en el hospital materno infantil de la C.N.S., la Paz- Bolivia. *Revista Paceña de Medicina Familiar,* 4(5), 15–19.

Qiu, C., Williams, M.A., Calderon-Margalit, R., Cripe, S.M., Sorensen, T.K. (2009). Preeclampsia risk in relation to maternal mood and anxiety disorders diagnosed before or during early pregnancy. *American Journal of Hypertension,* 22(4), 397–402 (2009). https://doi.org/10.1038/ajh.2008.366.

Raghupathy, R., & Kalinka, J. (2008). Cytokine imbalance in pregnancy complications and its modulation. *Frontiers in Bioscience: a Journal and virtual library,* 13, 985–994.

Remacle, C., Dumortier, O., Bol, V., Goosse, K., Romanus, P., Theys, N., et al. (2007). Intrauterine programming of the endocrine pancreas. *Diabetes, Obesity & Metabolism,* 9(2), 196–209 (2007). https://doi.org/10.1111/j.1463-1326.2007.00790.x.

Reusens, B., Ozanne, S.E., Remacle, C. (2007). Fetal determinants of type 2 diabetes. *Current Drug Targets,* 8, 935–941.

Rofey, D.L., Kolko, R.P., Iosif, A.M., Silk, J.S., Bost, J. E., Feng, W., et al. (2009). A Longitudinal Study of Childhood Depression and Anxiety in Relation to Weight Gain. *Child Psychiatry and Human Development,* 40(4), 517–526 (2009). https://doi.org/10.1007/s10578-009-0141-1.

Romero, R., Mazor, M., Munoz, H., Gomez, R., Galasso, M., Sherer, D.M. (1994). The preterm labor syndrome. *Annals of the New York Academy of Sciences, 734*, 414–429.

Romo-Gonzalez, T., Retureta, Y.B., Sánchez-Rodríguez, E.N., Martínez, J.A., Chavarría, A., Gutiérrez-Ospina, G. (2012). Moderate anxiety in pregnant women does not compromise gestational immune-endocrine status and outcome, but renders mothers to be susceptible for diseased states development: A preliminary report. *Advances in Bioscience and Biotechnology, 3*, 101–106 (2012). https://doi.org/10.4236/abb.2012.31015.

Ruiz, R.J., Fullerton, J., Dudley, D.J. (2003). The Interrelationship of Maternal Stress, Endocrine Factors and Inflammation On Gestational Length. *Obstetrical & Gynecological Survey, 58*(6), 415–428 (2003). https://doi.org/10.1097/01.ogx.0000071160.26072.de.

Salacz, P., Csukly, G., Haller, J., Valent, S. (2012). Association between subjective feelings of distress, plasma cortisol, anxiety, and depression in pregnant women. *European Journal of Obstetrics, Gynecology and Reproductive Biology, 165*(2), 225–230 (2012). https://doi.org/10.1016/j.ejogrb.2012.08.017.

Sandman, C.A., Glynn, L., Schetter, C.D., Wadhwa, P., Garite, T., Chicz-DeMet., A., et al. (2006). Elevated maternal cortisol early in pregnancy predicts third trimester levels of placental corticotropin releasing hormone (CRH): Priming the placental clock. *Peptides, 27*(6), 1457–1463 (2006). https://doi.org/10.1016/j.peptides.2005.10.002.

Sandman, C.A., Wadhwa, P.D., Chicz-DeMet, A., Porto, M., Garite, T.J. (1999). Maternal corticotropin-releasing hormone and habituation in the human fetus. *Developmental Psychobiology, 34*(3), 163–173.

Sandovici, I., Smith, N.H., Nitert, M. D., Ackers-Johnson, M., Uribe-Lewis, S., Ito, Y., et al. (2011). Maternal diet and aging alter the epigenetic control of a promoter–enhancer interaction at the Hnf4a gene in rat pancreatic islets. *Proceedings of the National Academy of Sciences of the United States of America, 108*(13), 5449–5454 (2011). https://doi.org/10.1073/pnas.1019007108.

Sargent, I.L., Borzychowski, A.M., Redman, C.W.G. (2007). NK cells and pre-eclampsia. *Journal of Reproductive Immunology, 76*(1–2), 40–44 (2007). https://doi.org/10.1016/j.jri.2007.03.009.

Schneider, M.L., Roughton, E.C., Koehler, A.J., Lubach, G.R. (1999). Growth and Development Following Prenatal Stress Exposure in Primates: An Examination of Ontogenetic Vulnerability. *Child Development, 70*(2), 263–274.

Senties, M., & Ortiz, G. (1993). Evaluación de los niveles de ansiedad y el estado emocional en mujeres embarazadas de bajo nivel socioeconómico. *Psicología y Salud. Nueva Época, 2*, 47–54.

Smith, R. (1998). Alterations in the hypothalamic pituitary adrenal axis during pregnancy and the placental clock that determines the length of parturition. *Journal of Reproductive Immunology, 39*(1–2), 215–220.

Steer, R.A., Scholl, T.O., Hediger, M.L., Fischer, R.L. (1992). Self-reported depression and negative pregnancy outcomes. *Journal of Clinical Epidemiology, 45*(10), 1093–1099.

Szyf, M., McGowan, P., Meaney, M.J. (2008). The social environment and the epigenome. *Environmental and Molecular Mutagenesis, 49*, 46–60 (2008). https://doi.org/10.1002/em.20357.

Teixeira, C., Figueiredo, B., Conde, A., Pacheco, A., Costa, R. (2009). Anxiety and depression during pregnancy in women and men. *Journal of Affective Disorders, 119*(1–3), 142–148 (2009). https://doi.org/10.1016/j.jad.2009.03.005.

van den Bergh, B.R.H. (1992). Maternal emotions during pregnancy and fetal and neonatal behaviour. In J.G. Nijhuios (Ed.). *Fetal behaviour. Developmental and perinatal aspects* (pp. 157–174). Oxford: Oxford Univ. Press.

van Dijken, H.H., Mos, J., van der Heyden, J.A., Tilders, F.J. (1992). Characterization of stress-induced long-term behavioural changes in rats: evidence in favor of anxiety. *Physiology & Behavior, 52*(5), 945–951.

Veenstra van Nieuwenhoven, A.L., Heineman, M.J., Faas, M.M. (2003). The immunology of successful pregnancy. *Human Reproduction Update, 9*(4), 347–357.

Vianna, P., Bauer, M.E., Dornfeld, D., Chies, J.A.B. (2011). Distress conditions during pregnancy may lead to pre-eclampsia by increasing cortisol levels and altering lymphocyte sensitivity to glucocorticoids. *Medical Hypotheses,* 77(2), 188–191(2011). https://doi.org/10.1016/j.mehy. 2011.04.007.

Vickers, M.H., Breier, B.H., McCarthy, D., Gluckman, P.D. (2003). Sedentary behaviour during postnatal life is determined by the prenatal environment and exacerbated by postnatal hypercaloric nutrition. *American Journal of Physiology - Regulatory, Integrative and Comparative Physiology,* 285, R271–R273 (2003). https://doi.org/10.1152/ajpregu.00051. 2003.

Wadhwa, P., Culhane, J., Rauh, V., Barve, S. (2001). Stress and Preterm Birth: Neuroendocrine, Immune/Inflammatory, and Vascular Mechanisms. *Maternal and Child Health Journal,* 5(2), 119–125.

Wadhwa, P.D., Sandman, C.A., Porto, M., Dunkel-Schetter, C., Garite, T.J. (1993). The association between prenatal stress and infant birthweight and gestational age at birth: A prospective investigation. *American Journal of Obstetrics and Gynecology,* 169(4), 858–865.

Walker, J.G., Littlejohn, G.O., McMurray, N.E., Cutolo, M. (1999). Stress system response and rheumatoid arthritis: a multilevel approach. *Rheumatology,* 38(11), 1050–1057.

Waring, M.E., & Lapane, K.L. (2008). Overweight in Children and Adolescents in Relation to Attention-Deficit/Hyperactivity Disorder: Results From a National Sample. *Pediatrics,* 122, e1–e6 (2008). https://doi.org/10.1542/peds.2007-1955.

Waterland, R.A., Travisano, M., Tahiliani, K.G. (2007). Diet-induced hypermethylation at agouti viable yellow is not inherited transgenerationally through the female. *The FASEB Journal: official publication of the Federation of the American Societies for Experimental Biology,* 21 (12), 3380–3385.

Winkel, S., Einsle, F., Pieper, L., Höfler, M., Wittchen, H.U., Martini, J. (2015). Associations of anxiety disorders, depressive disorders and body weight with hypertension during pregnancy. *Archives of Women's Mental Health,* 18(3), 473–483 (2015). https://doi.org/10.1007/s00737-014-0474-z.

World Health Organization (2000). Trabajando con individuos, familias y comunidades para mejorar la salud materna y neonatal. http://whqlibdoc.who.int/hq/2010/WHO_MPS_09.06_spa.pdf. Accessed 26 Mar 2015.

World Health Organization (2014). Mortalidad Materna.http://www.who.int/mediacentre/factsheets/fs348/es/. Accessed 7 Jan 2014.

Chapter 3
Risk Behaviours for Developing Breast Cancer: A Multi-disciplinary Approach

Rosalba León-Díaz, Yamilet Ehrenzweig, Tania Romo-González and Carlos Larralde

Abstract Breast cancer (BRCA) is the most common neoplasia among women around the world. The Official Mexican Standard states that health services must provide opportune information for the prevention of BRCA. However, this information focuses on the reduction of biological, iatrogenic or environmental risk factors and those related to reproductive history, neglecting those risk factors relating to lifestyle and other psychosocial risk factors (e.g. personality type, stress management). In this chapter we point out some recent findings that link not only obesity but also dietary habits, glycaemic index, alcohol intake and sedentary lifestyle as risk factors for developing BRCA. We also highlight the importance of addressing the psychological and social aspects, such as taboos of patients, to provide a multidisciplinary approach that allows an accurate prevention and treatment of this disease.

Keywords Breast cancer · Psychosocial risk factors · Personality type
Stress management · Obesity · Eating habits

3.1 Introduction

Around the world, breast cancer (BRCA) is the most common neoplasia among women. In Mexico, guidelines for its diagnosis and treatment are stipulated in Official Mexican Standard NOM-041-SSA2-2011, which outlines that prevention

Dr. Rosalba León-Díaz, Posdoctoral Researcher, Department of Integrated Health and Biology, Institute of Biological Research, Universidad Veracruzana. Email: rossalba_leon@yahoo.com.mx.

Dr. Yamilet Ehrenzweig, Researcher, Institute of Psychological Research, Universidad Veracruzana. Email: yamiletehrenzweig@hotmail.com.

Dr. Tania Romo-González, Researcher, Department of Integrated Health and Biology, Institute of Biological Research, Universidad Veracruzana. Email: romisnaider@hotmail.com.

Dr. Carlos Larralde (1938–2015), Researcher, Department of Immunology, Institute of Biomedical Research, National Autonomous University of Mexico.

© Springer International Publishing AG, part of Springer Nature 2018   31
Ma. L. Marván and E. López-Vázquez (eds.), *Preventing Health and Environmental Risks in Latin America*, The Anthropocene: Politik—Economics—Society—Science 23, https://doi.org/10.1007/978-3-319-73799-7_3

measures must include: (1) educational communication for the general public regarding symptomology of BRCA and risk factors; (2) the promotion of healthy lifestyles that help to decrease morbidity resulting from BRCA; and, (3) the advancement of demand for detection services.

Regarding point number 3, for BRCA detection, this Standard stipulates three types of intervention aimed at women based on their age and level of vulnerability: self-examination, clinical examination and mammogram. Self-examination is recommended from the age of 20 onwards. For self-examination, patients must have information regarding the symptoms and signs of the disease and information about when to seek medical advice. A clinical examination is necessary every year from the age of 25, and a mammogram after the age of 40 and until the age of 69, on a bi-annual basis. The purpose of the latter is to identify BRCA ideally during the preclinical stage of the disease. This is why women must receive the clinical results in writing no more than 21 working days after their examination. In the event of any abnormalities, the patient must also receive the date and location for a diagnostic evaluation at a public institution.

Despite these guidelines, mammograms, the primary method for detecting and diagnosing BRCA in Mexico, reduce mortality by only 7–23% (Chávarri-Guerra et al. 2012). This is due mainly to a lack of adequate infrastructure for the procedure, in addition to a shortage of trained and certified radiologists needed to interpret the tests. According to the Mexican Department of Health (SS), in 2010 there were 415 mammography technicians at public health institutions and hospitals and 366 within the private sector (Chávarri-Guerra et al. 2012; Mexican Department of Health [SS] 2011). This number of mammography technicians does not provide the coverage needed to undertake the number of tests recommended by international organizations, such as the National Cancer Institute (19.9 mammograms per million inhabitants). Moreover, the majority of mammograms carried out in Mexico are for diagnostic purposes and not for screening, which means that only 6% of women are diagnosed during the early stages of the disease, leading to higher treatment costs and reduced survival rates (SS 2011). Furthermore, studies of private and public hospitals in the metropolitan area of Mexico City show that the technical performance of operational mammography technicians is inadequate, not to mention that 53–82% of them do not comply with quality control measures and evaluations (Brandan/Villaseñor 2006).

These difficulties in opportunely diagnosing BRCA can be clearly seen in the statistical data available. For example, in 2012, 1.67 million new cases of BRCA were diagnosed around the world, representing 25% of all cancers. Furthermore, BRCA led to 522,000 deaths, making it the fifth leading cause of death by cancer in the world. It is the leading cause of death by cancer in developing countries, with 324,000 deaths, and the second cause in developed countries, with 198,000 (GLOBOCAN 2012). In Mexico, the mortality rate for BRCA is 9.7/100,000 inhabitants (GLOBOCAN 2012), a figure that has doubled over the past 20 years (Knaul et al. 2008).

This data highlights that, despite the inclusion of a clause within the Standard specifically for the opportune detection of BRCA, the morbi-mortality for this

Table 3.1 Risk factors associated with developing breast cancer

Factor	Description
Biological	Female; ageing (the older the person, the greater the risk); personal or family history of breast cancer among mother, daughters or sisters; history of atypical ductal hyperplasia, radial or star-shaped marks, lobular carcinoma *in situ* by biopsy; menstrual life of more than 40 years (menarche prior to the age of 12 and menopause after the age of 52), mammary density and known carrier of BRCA1 or BRCA2 genes
Iatrogenic or environmental	Exposure to ionizing radiation, mainly during development or growth (*in utero* or during adolescence) and thoracic radiotherapy treatment
Reproductive history	Nullipara or having had first pregnancy after the age of 30; hormone therapy during peri- or post-menopause for more than 5 years
Lifestyles	Diet rich in carbohydrates and low in fibre; diet rich in animal fats and trans fatty acids; obesity, mainly during post-menopause; sedentary lifestyle; alcohol consumption of more than 15 g/day; and smoking

Source Official Mexican Standard NOM-041-SSA2-2011. For the prevention, diagnosis, treatment, control and monitoring of breast cancer

disease has not decreased. Furthermore, points 1 and 2 of the Standard, which focus on providing education about risk factors and the promotion of healthy lifestyles, have not been effective in preventing and opportunely diagnosing the disease. This could be as a result of the fact that the measures stemming from the application of the Standard focus mainly on decreasing the prevalence of biological, iatrogenic or environmental risks factors and those related to reproductive history (Table 3.1), neglecting those risk factors relating to lifestyle (even when it contemplates these factors) and other psychosocial risk factors. With regard to the latter, a number of authors have associated personality type (Morris/Greer 1980; Temoshok 1987; Kune et al. 1991; Zetu et al. 2013), stress management (Wayner et al. 1979; Kiecolt-Glaser/Glaser 1991; Kruk/Aboul-Enein 2004; Andersen et al. 2008) and cultural aspects (Carvalho et al. 2005), among others, with the onset of BRCA. This wide range of factors associated with the development of this neoplasia makes it necessary to create and promote materials, such as this chapter, which provide accurate information regarding those risk behaviours and factors that the Standard does not overtly mention or focus on, in order to ensure their inclusion in healthcare strategies to prevent BRCA.

3.2   Eating Habits

In Mexico, the lifestyles and eating habits of the population have changed dramatically over the past 20 years, and this shift can be seen in the increased prevalence of overweight and obesity: 33.4% in 1988 and 71.9% in 2006. There is a greater prevalence of obesity among adult women (36.9%) compared to 23.5% among men (Olaiz-Fernández et al. 2006; Romieu et al. 2011).

Furthermore, the analysis of dietary patterns of the Mexican population shows that 35.1% of women are at risk of excessive carbohydrate consumption and 12.6% demonstrate excessive fat intake. The excessive consumption of these two elements (fats and carbohydrates) has been associated with overweight and obesity and a number of neoplasias, including BRCA, mainly among post-menopausal women (De Pergola/Silvestris 2013; Olaiz-Fernández et al. 2006; Romieu et al. 2011).

The exact manner in which obesity favours the development of cancer has not yet been pinpointed, but the following proposals have been made: (a) the insulin—insulin-like growth factor 1 [IGF-1] axis; (b) endogenous reproductive hormones; and, (c) chronic inflammation (De Pergola/Silvestris 2013; Patterson et al. 2013; Vucenik/Stains 2012). The first proposal is one of the most widely described and is explained below: obesity commonly leads to a state of insulin resistance and, consequently, a state of hyperinsulinemia. This hyperinsulinemic state increases the bio-availability of the IGF-1 protein, a protein that alters the cellular micro-environment, favouring the development of tumours. In cases of BRCA among post-menopausal women, it has been observed that IGF-1 favours the rapid progression and aggressiveness of cancerous cells. Furthermore, an epidemiological association between the presence of IGF-1 and BRCA in pre-menopausal women has been researched (De Pergola/Silvestris 2013; Patterson et al. 2013; Sundaram et al. 2013; VanSaun 2013; Vucenik/Stains 2012).

Another event favoured by obesity and linked to the appearance of cancer is the dysregulation of adipokines, proteins that are tasked with maintaining metabolic homeostasis, modulating inflammation, angiogenesis, proliferation and apoptosis. One of the major adipokines is leptin, a protein that controls the sensation of fullness and modulates the homeostasis between glucose and insulin, in addition to increasing the proliferation, migration and invasion of cancerous cells. A number of studies have shown an 83% increase in the leptin receptor among people with BRCA, while the receptor of 34% of patients with high leptin levels and high expression presented with metastasis (Alegre et al. 2013; Lima et al. 2009; Minatoya et al. 2013; VanSaun 2013).

With regard to the analysis of dietary patterns and their association with obesity and BRCA, it is important to take Glycaemic Index values (GI) into consideration. A prospective study in France highlights the association between elevated total GI and BRCA in overweight women, suggesting that the intake of fast-absorbing carbohydrates could be relevant in the development of this pathology in the presence of an underlying resistance to insulin. Furthermore, it found a direct association between carbohydrate intake, glycaemic load and oestrogen-receptor negative BRCA [ER-] (Lajous et al. 2008). A similar study among people in Mexico showed that there is a direct link between the development of BRCA (among both pre- and post-menopausal women) and carbohydrate intake, and, specifically, the elevated glycaemic load of the diet (Romieu/Lajous 2009).

In addition to the intake of food with a high glycaemic value, in Uruguay, for example, a strong link between the presence of BRCA and a Western diet was found (fried and grilled meat and processed meats). Moreover, the increased risk of developing BRCA is associated with the consumption of red meat, and the risk is significantly higher among post-menopausal women (Torres-Sánchez et al. 2009; Coronado et al. 2011; Romieu et al. 2011).

On the other hand, in countries such as Argentina and Brazil, the consumption of green, leafy vegetables, non-citrus fruits, apples and water melon has been associated with a decrease in the risk of BRCA. Furthermore, in Mexico, the consumption of onion and spinach, polyunsaturated fats and vitamin E, vitamin B12 and folic acid, as well as flavonoids and phytoestrogens considerably reduces the risk of BRCA (Torres-Sánchez et al. 2009).

3.3 Sedentary Lifestyle

Another risk behaviour that has more recently been associated with the development of BRCA is a lack of physical activity, which represents the fourth major risk factor in terms of global mortality. It is also estimated to be the leading cause of approximately 21–25% of colon and breast cancers (World Health Organization [WHO] 2010).

For the specific case of BRCA, there is evidence that post-menopausal women who do light exercise reduce the risk by 20%. This effect is attributed to the regulation that physical exercise brings to body energy balance (calories consumed vs. energy expended) and the consequent decrease in resistance to insulin and hyperinsulinemia, predisposing factors to the development of BRCA (Alegre et al. 2013; Neilson et al. 2014; Romieu/Lajous 2009; Sundaram et al. 2013). However, only 16% of Mexican women regularly exercise, and the average amount of time dedicated to recreational physical activity is 5 min per day, despite the fact that there is evidence to show that risk decreases by 9% per hour of moderate exercise per week, and the Official Mexican Standard proposes at least 30 min of physical activity per day (Romieu/Lajous 2009).

3.4 Alcohol Intake

The harmful use of alcohol causes 2.5 million deaths every year, being ranked as the third major risk factor of the global burden of disease (WHO 2011). Carcinogenesis is one of the most significant health consequences attributed to alcohol consumption, not to mention the fact that approximately 3.6% of all cancers (5.2% among men and 1.7% among women) around the world, in addition to 3.5% of deaths linked to cancer, are related to chronic alcohol consumption (Seitz/Becker 2007; Varela-Rey et al. 2013).

A number of epidemiological studies have associated dose-dependent alcohol intake with BRCA, and, despite the fact that a specific mechanism has not been identified, it is known that alcohol consumption: (1) increases endogenous oestrogen levels; (2) produces carcinogenic metabolites, such as acetaldehyde and reactive oxygen species; and, (3) decreases the capacity to absorb essential nutrients (Coronado et al. 2011; Seitz/Becker 2007; Lu et al. 2014; Varela-Rey et al. 2013). Moreover, a link between alcohol consumption and the presence of oestrogen receptor-positive mammary tumours (ER+) has been found, mainly among post-menopausal women. Specific studies have shown that an alcohol intake greater than 27 g/day increases the risk of presenting BRCA ER+/PR+ among post-menopausal patients (Enger et al. 1999).

3.5 Emotional Suppression and Chronic Exposure to Psychological Stress

Studies carried out by Pennebaker (1999) have shown that expressing emotions and recognizing traumatic events have a positive effect on an individual's physical and mental health. This is because emotional suppression is associated with more anguish or emotional stress (Iwamitsu et al. 2005a, b), while emotional expression is associated with improved psychological adjustment and, as such, a decrease in distress and an increase in quality of life.

Furthermore, different studies have tried to establish the recognition of certain character traits that are predisposed to specific kinds of illnesses. Specifically, people with behavioural patterns of negation, emotional suppression (mainly anger), conflict avoidance, who are socially agreeable, harmonizing and highly rational, with a C personality type, have a greater possibility of developing infectious diseases and cancer (Herranz et al. 1997; Temoshok 1987; Zozulya et al. 2008), in addition to BRCA (Herranz et al. 1997).

Emotional suppression, specifically, is considered to be the psychological variable that most affects the psychosocial adjustment of people with cancer, and it can be related to pathological clinical outcomes [death or improvement] (Cordova et al. 2003). More specifically, repression (low anxiety and high defensiveness) and suppression (deliberate retention of the expression of negative emotions) are associated with the faster progression of BRCA and a smaller likelihood of survival (Giese-Davis et al. 2002, 2006; Weinberger 1990).

Moreover, psychological alterations, such as a state of chronic stress, mean that the organism activates physiological systems to conserve homeostasis, leading to a negative impact on the nervous system through the activation of biochemical changes and a hormonal imbalance, such as the inefficient handling of catecholamine and corticosteroid hormones, which, in turn, has an impact on the endocrine and immune systems as well as on health in general (Dhabhar/McEwen 1997; Andersen et al. 2008).

The possible contribution of psychological stress to the development of BRCA has been widely studied (Antonova et al. 2011; Michael et al. 2009; Priestman et al.

1985), and it has been associated with immunosuppression generated by high levels of corticosteroids and conducts to tackle stress, such as, among others: an increase in the intake of foods rich in carbohydrates and high in fats (McGregor/Antoni 2009), alcohol consumption, smoking and poor sleep quality (Torres/Nowson 2007). These conducts favour increases in Body Mass Index (BMI), waist circumference, weight, abdominal adiposity (McGregor/Antoni 2009) and, in general, obesity, which we have already described as a risk factor for developing BRCA.

3.6   Socio-cultural Barriers to the Opportune Detection of BRCA

Other risk factors that favour morbidity and mortality stemming from BRCA focus more on socio-cultural barriers relating to beliefs, attitudes, behaviours or social situations that do not permit self-monitoring and reduce access to early detection and treatment programmes for this disease. Examples of these barriers include embarrassment, fear of losing their role as an object of desire, fear of losing their partner, reluctance to abandon the household or children during the treatment process, social context, the age of the women or their family situations, their ideas regarding the incurability of cancer and the generation of negative feelings about the disease, among others. All of these barriers can, at any given time, become more decisive than physical access or financial barriers (Stein et al. 1998; Grana 1998; Hewitt et al. 2004; Angus et al. 2006; Daly/Collins 2007).

There is also a serious lack of knowledge surrounding self-care measures, as demonstrated by Nigenda et al. (2008), who interviewed women with BRCA and found that the majority knew how useful self-examination was for detecting 'abnormalities' in their breasts; however, almost none of them knew the technique, nor did they use it on a monthly basis after menstruating. Furthermore, they mentioned that before being diagnosed, they had 'vague' information about the changes the disease causes, such as 'redness', 'splitting of the nipple' or the appearance of 'lumps'. Giraldo/Arango (2009) comment that, despite women knowing that self-examination is important in detecting BRCA in its early stages, they do not know how or when to do it. They also found that women with higher levels of education carried out self-examination on a periodic basis and were more aware of the risk and the symptoms of BRCA. These self-care measures are also undertaken by women who are caring for a relative with cancer, and their main motivation is the fear of developing the disease themselves, which is why they search for information about how to prevent it.

In terms of the barriers perceived by women within the healthcare system, Hewitt et al. (2004) found obstacles such as the lack of communication between healthcare professionals and patients, failures in clinical practice, failures in diagnosis, errors in the treatment prescriptions, the inexperience of healthcare professionals in evaluating the urgency of the situation and the anxiety stemming from the latter, inadequate coordination and fragmentation of the care patients require, lack of knowledge of healthcare professionals regarding the resources available in the community, and, finally, an underdeveloped system that guarantees quality and responsibility for healthcare services. Furthermore, Nigenda et al. (2008) found the lack of diagnostic equipment and insufficient specialized training for doctors to be issues that women viewed as being decisive during the diagnostic process for BRCA. Moreover, representatives from civil society organizations recognized, through testimonials shared by women in self-help groups, the lack of clinical training and low levels of sensitivity of some healthcare professionals, both at public and social security institutions. These factors led women to lack confidence in healthcare services and feel embarrassed or scared by the doctor, especially when accepting some diagnostic procedures [breast exploration and mammogram] (Cumpián-Loredo 2000; Wiesner 2007; Sánchez/Dos Santos 2007).

3.7   Conclusions

There is more and more academic literature to support the fact that the origin and progress of BRCA is 50% attributable to factors associated with lifestyle (Khan et al. 2010; Antonova et al. 2011; Macon/Fenton 2013). As such, recognizing and studying these factors is crucially important in developing new methods of preventing, diagnosing and treating BRCA, especially in Mexico, where diet and lifestyles have changed radically.

There are few BRCA prevention campaigns supported by the official healthcare system that constantly and perceptibly reinforce the importance of monitoring eating habits, our lifestyles and risk behaviours (such as alcoholism or a sedentary lifestyle). Almost all of the campaigns focus on risk factors relating to biology or reproductive history. This clearly reflects the clinical procedures currently in place in Mexico, given that women within the age range for a mammogram are not given an insulin-resistance test, their BMI data is not taken and they are not asked about their alcohol intake, dietary habits or physical activity. It is not difficult to see why the only mention of these risk factors in the Standard means that a greater focus on these elements is required. As such, more is needed than just mass campaigns that focus on self-examination or mammograms; the constant promotion of healthy lifestyles is important, not so as to avoid death (a natural process for any human being at any time of their lives), but rather to help people have a positive impact on their personal well-being. This task not only involves the healthcare and education sectors, but also people in general.

Some risk factors discreetly mentioned in this chapter, such as a personality that is susceptible to cancer, are not as widely recognized by the healthcare sector and are unknown within the elementary education sector, which is why these concepts must be established within our cultural backgrounds. Other factors, such as psychosocial stress, are not officially considered to be risks despite the notorious (and, in many cases, costly) consequences for our health. The most notorious of these factors is obesity, which leads to a number of diseases, including BRCA.

Moreover, it has been recognized that socio-cultural factors greatly affect prevention. This is due to the fact that they are not truly identified as risk factors, but rather as part of our cultural background. The behaviours stemming from these factors are seen as normal, preventing them from being identified as a health risk for women. As such, it is fundamentally important to, firstly, understand them and break inherent cultural and health paradigms, and drive a true culture of breast cancer prevention.

The epidemiological panorama and the multi-causality of BRCA make incorporating earlier detection testing into primary prevention indispensable. This requires multi-disciplinary studies that offer effective alternatives to save more lives than self-examination, detection by mammogram, surgery, radiotherapy and chemotherapy. Furthermore, these options must promote changes in risk behaviours and the adoption of healthy lifestyle through health education, a fundamental tool in acquiring healthy habits that prevent cancer (Bayes 1990).

References

Alegre, M.M., Knowles, M.H., Robison, R.A., O'Neill, K.L. (2013). Mechanics behind breast cancer prevention - focus on obesity, exercise and dietary fat. *Asian Pacific Journal of Cancer Prevention*, 4, 2207–2212.

Andersen, B.L., Yang H.C., Farrar, W.B., Golden-Kreutz, D.M., Emery, C.F., Thornton L.M., et al. (2008). Psychological interventions improve survival for breast cancer patients. *Cancer*, 113, 3450–3458 (2008). https://doi.org/10.1002/cncr.23969.

Angus, J., Miller, K.L., Pulfer, T., McKeever, P. (2006). Studying delays in breast cancer diagnosis and treatment: critical realism as a new foundation for inquiry. *Oncology Nursing Forum*, 33, E62–E70 (2006). https://doi.org/10.1188/06.onf.e62-e70.

Antonova, L., Aronson, K., Mueller, C.R. (2011). Stress and breast cancer: from epidemiology to molecular biology. *Breast Cancer Research: BRC*, 13, 208 (2011). https://doi.org/10.1186/bcr2836.

Bayes, R. (1990). Psicología y cáncer: prevención. Psicología Oncológica. Barcelona: Martínez Roca Edition.

Brandan, M.E., & Villaseñor, N.Y. (2006). Detección del Cáncer de Mama: Estado de la Mamografía en México. *Cancerología*, 1, 147–162. http://incan-mexico.org/revistainvestiga/elementos/documentosPortada/1172289111.pdf. Accessed 14 March–May 2018.

Carvalho, F.A.F., Mesquita, M.E., de Almeida, A.I.M., de Figueiredo, C.Z.M. (2005). Aspectos culturales en el proceso de padecer cáncer de mama. *Avances en Enfermería*, 23, 228–35. https://revistas.unal.edu.co/index.php/avenferm/article/view/37559/39902. Accessed 15 Jan 2014.

Chávarri-Guerra, Y., Villarreal-Garza, C., Liedke, P.E., Knaul, F., Mohar, A., Finkelstein, D.M., et al. (2012). Breast cancer in Mexico: a growing challenge to health and the health system. *The Lancet Oncology*, 13, e335–43 (2012). https://doi.org/10.1016/s1470-2045(12)70246-2.

Cordova, M.J., Giese-Davis, J., Golant, M., Kronnenwetter, C., Chang, V., McFarlin, S., et al. (2003). Mood disturbance in community cancer support groups. The role of emotional suppression and fighting spirit. *Journal of Psychosomatic Research*, 55, 461–7.

Coronado, G.D., Beasley, J., Livaudais, J. (2011). Alcohol consumption and the risk of breast cancer. *Salud Pública de México,* 53, 440–447 (2011).

Cumpián-Loredo, B.P. (2000). Conocimiento sobre detección oportuna de CaCu y mamario. *Revista de Enfermería del Instituto Mexicano del Seguro Social,* 8, 129–132.

Daly, H., & Collins, C. (2007). Barriers to Early Diagnosis of Cancer in Primary Care: A Needs Assessment of GPs. *Irish Medical Journal,* 100(10), 624–626.

De Pergola, G., & Silvestris, F. (2013). Obesity as a major risk factor for cancer. *Journal of Obesity,* 291546 (2013). https://doi.org/10.1155/2013/291546.

Dhabhar, F.S., & McEwen, B.S. (1997). Acute stress enhances while chronic stress suppresses cell-mediated immunity in vivo: A potential role for leukocyte trafficking. *Brain, Behavior and Immunity,* 11, 286–306 (1997). https://doi.org/10.1006/brbi.1997.0508.

Enger, S.M., Ross, R.K., Paganini-Hill, A., Longnecker, M.P., Bernstein, L. (1999). Alcohol consumption and breast cancer oestrogen and progesterone receptor status. *British Journal of Cancer,* 79, 1308–14 (1999). https://doi.org/10.1038/sj.bjc.6690210.

Giese-Davis, J., Koopman, C., Butler, L.D., Classen, C., Cordova, M., Fobair, P., et al. (2002). Change in emotion-regulation strategy for women with metastatic breast cancer following supportive-expressive group therapy. *Journal of Consulting and Clinical Psychology,* 7, 916–25 (2002). http://chc.medschool.ucsf.edu/pdf/8.%20Giese-Davis,%20et%20al.,%20jccp%202002.pdf.

Giese-Davis, J., Wilhelm, F.H., Conrad, A., Abercrombie, H.C., Sephton, S., Yutsis, M., et al. (2006). Depression and stress reactivity in metastatic breast cancer. *Psychosomatic Medicine,* 68, 675–83 (2006). https://insights.ovid.com/crossref?an=00006842-200609000-00006.

Giraldo, M.C.V., & Arango, R.M.E. (2009). Representaciones sociales frente al autocuidado en la prevención del cáncer de mama. *Investigación y Educación en Enfermería,* 27, 191–200. http://www.scielo.org.co/pdf/iee/v27n2/v27n2a04. Accessed 26 Nov 2013.

GLOBOCAN. (2012). Estimated Cancer Incidence, Mortality and Prevalence Worldwide in 2012. http://globocan.iarc.fr. Accessed 21 Oct 2013.

Grana, G. (1998). Ethnic Differences in Mammography Use among Older Women: overcoming the Barriers. *Annals of Internal Medicine,* 128, 773–775.

Herranz, J.S., Mateos de la Calle N., Bueno, C.J. (1997). Expresión emocional y personalidad tipo C: diferencias entre mujeres con patología mamaria maligna, benigna y normales. *Revista de Psicología de la Salud,* 9, 93–126. https://www.uam.es/gruposinv/esalud/Articulos/Genero/expresion-emocional-personalidad-tipo-c.pdf. Accessed May 2014.

Hewitt, M., Herdman, R., Holland, J. (2004). *Meeting psychosocial needs of women with breast cancer.* National Research Council.

Iwamitsu, Y., Shimoda, K., Abe, H., Tani, T, Okawa, M., Buck, R. (2005a). The relation between negative emotional suppression and emotional distress in breast cancer diagnosis and treatment. *Health Communications,* 18, 201–215 (2005). https://doi.org/10.1207/s15327027hc1803_1.

Iwamitsu, Y., Shimoda K., Abe, H., Tani, T., Okawa, M., Buck R. (2005b). Anxiety, emotional suppression, and psychological distress before and after breast cancer diagnosis. *Psychosomatics,* 46, 19–24 (2005). https://doi.org/10.1176/appi.psy.46.1.19.

Khan, N., Afaq, F., Mukhtar, H. (2010). Lifestyle as risk factor for cancer: evidence from human studies. *Cancer Letters,* 293, 133–143 (2010). https://doi.org/10.1016/j.canlet.2009.12.013.

Kiecolt-Glaser, J.K., & Glaser, R. (1991). Stress and immune function. In R. Ader, D.L. Felten, & N. Cohen (Eds.), Psychoneuroinmunology (pp. 849–868). New York: USA Academic Press.

Knaul, F.M., Nigenda, G., Lozano, R., Arreola-Ornelas, H., Langer, A., Frenk, J. (2008). Breast cancer in Mexico: a pressing priority. *Reproductive Health Matters,* 16, 113–123 (2008). https://doi.org/10.1016/s0968-8080(08)32414-8.

Kruk, J., & Aboul-Enein, H.Y. (2004). Psychological stress and the risk of breast cancer: a case-control study. *Cancer Detection and Prevention,* 28, 399–408 (2004). https://www.sciencedirect.com/science/article/pii/S0361090X04001266?via%3Dihub.

Kune, G.A., Kune, S., Watson, L.F., Bahnson, C.B. (1991). Personality as a risk factor in large bowel cancer: data from the Melbourne Colorectal Cancer Study. *Psychological Medicine,* 21, 29–41.

Lajous, M., Boutron-Ruault, M. C., Fabre, A., Clavel-Chapelon, F., Romieu I. (2008). Carbohydrate intake, glycemic index, glycemic load and risk of postmenopausal breast cancer in a prospective study of French women. *American Journal of Clinical Nutrition,* 87, 1384–1391. http://ajcn.nutrition.org/content/87/5/1384.full.pdf+html. Accessed 12 May 2014.

Lima, M.M., Velásquez, E., Unshelm, G., Torres, C., Rosa, F., Lanza, L. (2009). Asociación de la insulina y el factor de crecimiento semejante a la insulina tipo 1 (IGF-1) en el cáncer de mama. *Gaceta Médica de Caracas,* 117, 226–231.

Lu, Y., Ni, F., Xu, M., Yang, J., Chen, J., Chen, Z., et al. (2014). Alcohol promotes mammary tumor growth through activation of VEGF-dependent tumor angiogenesis. *Oncology Letters,* 8, 673–678 (2014). https://doi.org/10.3892/ol.2014.2146.

Macon, M.B., & Fenton, S.E. (2013). Endocrine disruptors and the breast: early life effects and later life disease. *Journal of Mammary Gland Biology and Neoplasia,* 18, 43–61 (2013). https://doi.org/10.1007/s10911-013-9275-7.

McGregor, B.A., & Antoni, M.H. (2009). Psychological intervention and health outcomes among women treated for breast cancer: a review of stress pathways and biological mediators. *Brain, Behavior and Immunity,* 23, 159–66 (2009). https://doi.org/10.1016/j.bbi.2008.08.002.

Mexican Department of Health (2011). Observatorio del Desempeño Hospitalario 2011. http://www.dged.salud.gob.mx/contenidos/dess/descargas/odh/ODH_2011.pdf. Accessed 8 Jan 2014.

Michael, Y.L., Carlson N.E., Chlebowski R.T., Aickin M., Weihs K.L., Ockene J.K., et al. (2009). Influence of stressors on breast cancer incidence in the Women's Health Initiative. *Health Psychology,* 28, 137–46 (2009). https://doi.org/10.1037/a0012982.

Minatoya, M., Kutomi, G., Asakura, S., Otokozawa, S., Sugiyama, Y., Nagata, Y., et al. (2013). Equol, adiponectin, insulin levels and risk of breast cancer. *Asian Pacific Journal of Cancer Prevention,* 14, 2191–9.

Morris, T. & Greer, S.A. (1980). A 'Type C' for cancer? Low trait anxiety in the pathogenesis of breast cancer. *Cancer Detection and Prevention,* 3, Abstract No. 102.

Neilson, H.K., Conroy, S.M., Friedenreich, C.M. (2014). The Influence of Energetic Factors on Biomarkers of Postmenopausal Breast Cancer Risk. *Current Nutrition Reports,* 3, 22–34 (2014). https://doi.org/10.1007/s13668-013-0069-8.

Nigenda, G., Caballero, M., González-Robledo, L.M. (2008). Barreras de acceso al diagnóstico temprano del cáncer de mama en el Distrito Federal y en Oaxaca. *Salud Pública de México,* 51, S254–S262.

Official Mexican Standard NOM-041-SSA2-2011. Ministry of the Interior Official. Gazette of the Federation. DOF: 09/06/2011. http://dof.gob.mx/nota_detalle.php?codigo=5194157&fecha=09/06/2011. Accessed 15 Oct 2013.

Olaiz-Fernández, G., Rivera-Dommarco, J., Shamah-Levy, T., Rojas, R., Villalpando-Hernández, S., Hernández-Avila, M., et al. (2006). Encuesta Nacional de Salud y Nutrición 2006. Cuernavaca, México: Instituto Nacional de Salud Pública. http://ensanut.insp.mx/informes/ensanut2006.pdf. Accessed 15 Oct 2013.

Patterson, R.E., Rock, C.L., Kerr, J., Natarajan, L., Marshall, S.J., Pakiz B., et al. (2013). Metabolism and breast cancer risk: frontiers in research and practice. *Journal of the Academy of Nutrition and Dietetics,* 113, 288–96 (2013). https://doi.org/10.1016/j.jand.2012.08.015.

Pennebaker, J.W. (1999). The effects of traumatic disclosure on physical and mental health: the values of writing and talking about upsetting events. *International Journal of Emergency Mental Health,* 1, 9–18.

Priestman, T.J., Priestman, S.G., Bradshaw, C. (1985). Stress and breast cancer. *British Journal of Cancer,* 51, 493–498.

Romieu, I., & Lajous, M. (2009). The role of obesity, physical activity and dietary factor son the risk for breast cancer: Mexican experience. *Salud Publica de México,* 51, S172–S180 (2009). https://doi.org/10.1590/s0036-36342009000800007.

Romieu, I., Escamilla-Núñez, M.C., Sánchez-Zamorano, L.M., Lopez-Ridaura, R., Torres-Mejía, G., Yunes, E.M., et al. (2011). The association between body shape silhouette and dietary pattern among Mexican women. *Public Health Nutrition, 15*, 116–125 (2011). https://doi.org/10.1017/s1368980011001182.

Sánchez, R.P., & Dos Santos, M. (2007). Cáncer de mama, pobreza y salud mental: respuesta emocional a la enfermedad en mujeres de camadas populares. *Revista Latino-Americana de Enfermagen, 15*, 786–91. http://www.scielo.br/pdf/rlae/v15nspe/es_11.pdf. Accessed 14 May 2014.

Seitz, H.K., & Becker, P. (2007). Alcohol metabolism and cancer risk. *Alcohol Research & Health, 30*, 38–41, 44–47. http://pubs.niaaa.nih.gov/publications/arh301/38-47.pdf. Accessed 19 Nov 2013.

Stein, J., Fox, S., Murata, P. (1998). The influence of ethnicity, socioeconomic status, and psychological barriers on use of mammography. *Journal of the Health and Social Behavior, 32*, 101–113. http://www.jstor.org/discover/10.2307/2137146?uid=3738664&uid=2&uid=4&sid=21106803976023. Accessed 9 Jan 2014.

Sundaram, S., Johnson, A.R., Makowski, L. (2013). Obesity, metabolism and the microenvironment: Links to cancer. *Journal of Carcinogenesis, 12*, 19 (2013). https://doi.org/10.4103/1477-3163.119606.

Temoshok, L. (1987). Personality, copying style, emotion and cancer: Towards and integrative model. *Cancer Survey, 6*, 545–567. http://industrydocuments.library.ucsf.edu/tobacco/docs/kqdy0192. Accessed 21 Oct 2013.

Torres, S.J., & Nowson, C.A. (2007). Relationship between stress, eating behaviour, and obesity. *Nutrition, 23*, 887–894 (2007). https://doi.org/10.1016/j.nut.2007.08.008.

Torres-Sánchez, L., Galván-Portillo, M., Lewis, S., Gómez-Dantes, H., López-Carrillo, L. (2009). Dieta y cáncer de mama en Latinoamérica. *Salud Publica de Mexico, 51*, S181–S190. http://www.scielo.org.mx/scielo.php?script=sci_arttext&pid=S0036-36342009000800008. Accessed 21 Oct 2013.

VanSaun, M.N. (2013). Molecular pathways: adiponectin and leptin signaling in cancer. *Clinical Cancer Research, 19*, 1926–32 (2013). https://doi.org/10.1158/1078-0432.ccr-12-0930.

Varela-Rey, M., Woodhoo, A., Martinez-Chantar, M.L., Mato, J.M., Lu, S.C. (2013). Alcohol, DNA methylation, and cancer. *Alcohol Research, 35*, 25–35. http://pubs.niaaa.nih.gov/publications/arcr351/25-35.htm. Accessed 19 Aug 2014.

Vucenik, I., & Stains, J.P. (2012). Obesity and cancer risk: evidence, mechanisms and recommendations. *Annals of the New York Academy of Sciences, 1271*, 37–43 (2012). https://doi.org/10.1111/j.1749-6632.2012.06750.x.

Wayner, L, Cox, T., Mackay, C. (1979). Stress, immunity and cancer. In D. J. Oborne, M. M. Gruneberg, J.R. Eiser (Eds.). *Research in Psychology and Medicine* (pp. 253–259). London: Academic Press.

Weinberger, D.A. (1990). The construct validity of the repressive coping style. In: J. L., Singer (Ed). *Repression and dissociation: Implications for personality theory, psychopathology, and health* (p. 337–386). Chicago: University of Chicago Press.

Wiesner, C. (2007). Determinantes psicológicos, clínicos y sociales del diagnóstico temprano del cáncer de mama en Bogotá, Colombia. *Revista Colombiana de Cancerología, 11*, 13–22. http://www.cancer.gov.co/documentos/revistas/2007/pub1/3.%20Resumen.pdf. Accessed 12 Jan 2013.

World Health Organization (2010). Recomendaciones mundiales sobre actividad física para la salud. http://apps.who.int/iris/bitstream/10665/44441/1/9789243599977_spa.pdf. Accessed 8 May 2014.

World Health Organization (2011). Nota descriptiva No. 349. Febrero de 2011. http://www.who.int/mediacentre/factsheets/fs349/es/. Accessed 21 Oct 2013.

Zetu, I., Iacob, M., Dumitrescu, A.L., Zetu, L. (2013). Type C coping, self reported oral health and oral health related behavior. *Procedia - Social and Behavioral Sciences*, 78, 491–495. http://www.sciencedirect.com/science/article/pii/S1877042813009063. Accessed 21 Aug 2014.

Zozulya, A.A., Gabaeva, M.V., Sokolov, O.Y., Surkina, I.D., Kost, N.V. (2008). Personality, coping style, and constitutional neuroimmunology. *Journal of Immunotoxicology,* 5, 221–225 (2008). https://doi.org/10.1080/15476910802131444.

Chapter 4
Behaviour Related to Cervical Cancer Risks

Yamilet Ehrenzweig and Ma. Luisa Marván

Abstract Cervical cancer (CC) is a serious public health problem for women in developing countries due principally to a lack of preventive behaviours. CC has a slow evolution that needs years to develop, which allows early detection of abnormal cells that can be treated to prevent CC but otherwise can become cancer cells. This early detection can be accomplished via the Papanicolaou test. In Mexico, we have found that between 20 and 30% of women with primary or middle school education had never undergone a Papanicolaou, and, among those who had done so, only about the half had undergone the test prior to reaching 30 years of age. In addition to low schooling, other variables that we have found to be associated with failure to undergo the test are lack of knowledge about CC and the usefulness of Papanicolaou, as well as the fact of thinking that health depends on luck, destiny or random chance.

Keywords Cervical cancer · Papanicolaou · Lack of knowledge
Locus of control · Barriers · Fatalistic beliefs

4.1 Introduction

When speaking about health, one often thinks only in terms of its physical aspects and rarely about its sociocultural and psychological factors, even though there is evidence that such factors profoundly influence our health. Among many issues, the psychology of health studies the form in which attitudes and emotions are related to the processes of health and illness including specific behaviours focused on protecting health, avoiding illness, or those that endanger one's health. Behaviours that contribute to maintaining or promoting health and preventing illness are known as

Dr. Yamilet Ehrenzweig, Researcher, Institute of Psychological Research, Universidad Veracruzana. Email: yamiletehrenzweig@hotmail.com.

Dr. Ma. Luisa Marván, Researcher, Institute of Psychological Research, Universidad Veracruzana. Email: mlmarvan@gmail.com.

The authors are grareful to Ma. Fernanda Marván Ramírez who assisted us in collecting information.

healthy or immunological behaviours, whereas those that contribute to the development or appearance of illness are known as unhealthy or pathological behaviours (Rodríguez-Marín/Neipp 2008; Oblitas 2010).

Among healthy behaviours are those that promote prevention of disease at any of its three levels, that is: (a) primary prevention that avoids the appearance of disease, (b) secondary prevention focused on detecting disease in its early stages when adequate measures can impede its progression, and (c) tertiary prevention including measures for the treatment and rehabilitation of diseased patients (Vignolo et al. 2011).

In this chapter, we will discuss some psychosocial factors that influence the primary and secondary prevention of cervical cancer (CC)—behaviours that protect against or avoid its appearance and those directed towards early detection of the disease when it becomes evident.

4.2   Cervical Cancer

Cervical cancer is an important public health problem for women in developing nations such as those of Latin America. As we will discuss throughout this chapter, CC can be avoided in large measure but efforts to prevent it have not succeeded in lowering its incidence in these regions. At global level, CC has the third largest risk of incidence in all women but it is more prevalent in developing nations. In these nations, the rate is 15.7/100,000 inhabitants in contrast to the rate of 9.9/100,000 inhabitants of developed nations. In Latin America, CC has the second largest risk of incidence in women (21.2/100,000 inhabitants) and is exceeded only by breast cancer. This situation is similar to that of Mexico although in this country the rate is slightly higher (23.3/100,000 inhabitants) and almost one and a half times that observed in developed nations (GLOBOCAN 2012).

At global level, CC is responsible for 7.5% of all cancer deaths among women. But the risk of dying from this disease is approximately 2.5 times greater in developing nations than in developed nations. In developed nations, the mortality rate is 3.3/100,000 inhabitants, whereas in Latin America the rate is 8.7/100,000 inhabitants and in Mexico it is 8.1/100,000 inhabitants (GLOBOCAN 2012). Epidemiological studies in Mexico have demonstrated that the average mortality rate in northern states is 8.5/100,000 inhabitants and 11.96/100,000 in southern states (Palacio-Mejía et al. 2009). Our investigations, described later in this chapter, were conducted in Veracruz where the CC mortality rate is 13.2/100,000 women age 25 and over (INEGI 2011) making it one of the states with the greatest mortality risk for CC (Sánchez-Barriga 2012).

It is important to note that the differences mentioned for both morbidity and mortality[1] due to CC are principally caused by a lack of preventive behaviours towards CC, thus making it essential to know the aetiology[2] and risk factors associated with the disease.

One of the most significant advances in recent decades in the study of CC has been the clarification of the role of the human papilloma virus (HPV) in its aetiology since practically all cases of CC present with this virus (Walboomers et al. 1999).

More than 100 types of HPV have been identified, of which about 18 are catalogued as 'high risk' given their association with the cancer. Among these, HPV-16 and HPV-18 have the greatest risk and cause nearly 75% of all cases. HPV infection is sexually transmitted and is the most frequent sexually transmitted disease in the entire world. Any person who has sexual contact with an infected person can be a virus carrier, even when many years have passed since contact occurred. Men and women can become infected and transmit the virus unknowingly due to the possible absence of symptoms. Nevertheless, in the great majority of women, the infection disappears spontaneously and does not progress; only persistent infections carry a greater risk of developing into precancerous lesions which, if not treated opportunely, can progress to cancer (Espinosa-Romero et al. 2014; Manzo-Merino et al. 2014).

CC has a slow evolution and progression needing years to develop. Abnormal cells appear initially in the form of precursor lesions called cervical dysplasias. If these cells aren't detected and treated during this initial stage, they may change into precancerous cervical cells that, after a period of time without treatment, can further change into cancer (Kumar et al. 2007). This long period of evolution provides an opportunity for the pathology to be detected in its first incipient stage and adequate treatment given to prevent the disease. Hence the supreme importance of early detection that can be accomplished via a study of vaginal cytology (Papanicolaou or PAP test), as will be extensively discussed in this chapter.

In spite of the fact that nearly all cases of CC show infection by HPV, not all cases of HPV infections develop precancerous lesions. This leads us to suppose that risk factors may exist that are associated with a high probability of developing precancerous lesions or cancer. Apart from biological factors that include having a persistent HPV infection, having a family history of the CC, being between 25 and 64 years old, having HIV or a weakened immune system, there are other risk factors that concern lifestyle. These include:

- Tobacco usage
- Malnutrition
- Deficiency of antioxidants (vitamins C and E)
- Use of oral contraceptives
- Frequent untreated vaginal infections

---

[1]Morbidity refers to the number of people suffering from the disease in a given place and time, whereas mortality is the number of people who die from the disease in a given place and time.
[2]Aetiology refers to the origin or causes of a disease.

- Elevated number of childbirths
- Use of an intrauterine device (IUD)
- Never having had a vaginal cytology (PAP test).

On the other hand, there are risky behaviours that increase the probability of contracting HPV but are not considered factors that facilitate the evolution of HPV infection into CC. These factors include:

- Initiating sexually active behaviour at an early age
- Having multiple sexual partners
- Having sexual contact with a male partner who currently has or previously has had many female sexual partners.

Because HPV is sexually transmitted, the manner of preventing the infection would be the practice of preventive behaviours such as monogamy, but both members of the couple should practise such behaviours. Fortunately, the recent development of vaccines against HPV today offers an alternative for preventing the infection. To date, two vaccines have been developed that demonstrate efficacy against the two types of virus that cause the major part of the cases of CC. Longitudinal follow-up studies have shown that the protection conferred by these vaccines lasts for at least 5 years. Different clinical studies have shown the efficacy of these vaccines in women between 9 and 26 years (Muñoz et al. 2008), although the recommended age for vaccination varies from country to country and depends greatly on the average age at which young women begin to have sexually active lives. In Mexico, the Federal Commission for Protection against Health Risks (*Comisión Federal para la Protección contra Riesgos Sanitarios* or COFEPRIS) authorized use of the vaccine in June, 2006, with the recommendation that it be included in the vaccination registry booklet for girls 11–12 years old since the vaccine should be administered prior to having sexual relations in order to obtain its maximum benefit.

However, these girls do not make the decision themselves to become vaccinated; it is their parents who decide for them. The acceptability of the vaccine varies in relation to the sociocultural context. For example, in one study in Colombia, researchers found that the majority of parents of girls enrolled in public schools were in favour of vaccinating their daughters, whereas parents of girls enrolled in private schools were more critical and less accepting. Some parents expressed concern about the short length of the follow-up studies and lack of knowledge about long-term adverse effects. Other parents viewed the vaccine as simply a way of promoting sexual activity in their daughters (Wiesner et al. 2010).

These vaccines have been recently created and will protect young people. Thus, we hope that epidemiological studies in the near future will demonstrate less alarming rates of morbidity and mortality from CC even in developing nations like our own. But in order to increase the probability of this occurring, it is important to conduct informative campaigns to sensitize the general public, and especially preadolescent girls and their parents, to the importance of vaccination.

Vaccination is a healthy behaviour for primary prevention of CC but it does not eliminate the need to employ other healthy behaviours such as the PAP test since, as mentioned above, the vaccine is only effective against the two types of virus that cause about 75% of the cases of CC and infections caused by the remaining types of viruses are not covered by the vaccine.

The Papanicolaou or PAP test is a cytological study used for early detection of cancer and consists of taking a sample of cells from the uterine cervix in order to identify dysplasia or lesions that are precursors of CC. It is a screening test designed to recognize different types of cervical lesions that, in their initial development phases, are easy to eliminate, thus avoiding their progression to cancer. The PAP is a simple, painless test and obtaining the sample takes less than five minutes. Women should request it in any health institution. The test is low-cost, which allows its application to large populations. Since the introduction of this test into clinical practice in 1943, the morbidity and mortality for CC has diminished considerably although this has been less evident in developing nations.

Since 1974, the PAP test has been offered free of charge in all public institutions of the Mexican Health Service (*Sector Salud*) and is obligatory in government public support programmes. Additionally, the Health Service maintains a series of permanent campaigns so that all women can periodically avail themselves of this test. As part of these campaigns, women are given information about CC, its risk factors and protective factors as well as the importance of having the PAP, which, if the results are positive, offers women the opportunity to receive adequate early treatment that can mean the difference for them between life and death.

The activities of the Health Service concerning the prevention, detection, diagnosis, treatment, control and epidemiological surveillance are established in the Official Mexican Norm *NOM-014-SSA2-1994* (2007). Regarding prevention, the Norm mandates that the first PAP test should be made when a woman becomes sexually active or by age 25 even if she has not had sexual relations. It further states that all women should be informed of the cytological result less than four weeks after the date of taking the sample. The Norm also establishes that in the case of women with two consecutive annual cytological studies that resulted negative for intraepithelial lesions or cancer, the PAP should be repeated every three years. Lastly, when the cytological result is infection for HPV, mild dysplasia, or cervical intraepithelial neoplasia grade 1 with a satisfactory colposcopic examination. the PAP should be repeated after one year.

Regarding other preventive actions, the Norm dictates that health personnel have the responsibility of informing the general population about the importance of the disease, its risk factors, and the possibilities for prevention, diagnosis, and early intervention. In accordance with the Norm, health personnel should animate the population to demand medical attention that is opportune, periodic, of high quality, and guarantees the patient is sufficiently informed concerning any procedures performed. To achieve the former, counselling about CC has included, as an indispensable strategy since 2000, the use of a communication process that actively involves two people and takes into consideration their attitudes, feelings, beliefs, values, cultural identity, and conflicts. It is a listening process that begins with detecting the needs of the user population and establishes the goal of facilitating accurate decision-making, the process of acceptance, and management of the problem while simultaneously stimulating its prevention (SSA 2007).

Through these measures, usage of the PAP in Mexico has increased in a gradual and sustained manner from 33.3% in 2000 to 49.1% in 2012 (ENSANUT 2012). Nevertheless, these efforts have been insufficient and participation in taking the test remains low. This low participation is due to an infinite number of various factors, which are psychosocial in nature.

4.3   Obstacles to the Prevention of Cervical Cancer

Some investigators have conducted studies in Mexico to discover which factors predispose utilization of the PAP and have found that the women who most use the PAP are those with a history of vaginal infections, those who have used family planning methods, and those with at least one child but, more than anything else, they are women with a higher level of education. The investigators have also found other predisposing factors of a cultural nature, such as spousal acceptance of gynaecological examinations and cognitive factors, such as having adequate knowledge of the PAP's utility (Aguilar-Pérez et al. 2003; Hernández-Hernández et al. 2007). Similarly, a Peruvian study also showed that women who failed to correctly utilize PAP tests were those with limited knowledge and unfavourable attitudes towards the PAP (Bazan et al. 2007).

Our working group carried out a study in the city of Xalapa (Veracruz) with women 35 years or older and with different levels of education (Ehrenzweig et al. 2013). We found that 20–30% of women with primary or middle school education had never undergone a PAP test and, of those that had done so, only about half had undergone the test prior to reaching 30 years of age. These findings differ from those of women with a higher level of education, since in that group only three women had never undergone a PAP test and the majority of all women in the group had begun testing prior to reaching 30 years of age.

We also studied knowledge concerning the prevention of CC. When asked if they knew why PAP tests were useful, we found that women with middle school or higher education showed a similar level of knowledge that was significantly greater than that of women with only primary education. When questioned about the risk factors for CC and when the first PAP test should be performed, we found the greatest number of correct responses were from women with higher education followed by women with middle school and lastly by those with only primary education, who demonstrated limited knowledge.

Finally, we calculated the total number of points for knowledge concerning the prevention of CC (utility of the PAP, timing of the first PAP, and risk factors for CC) and found a correlation between this point score and usage of the PAP. As expected, the number of women who had undergone the PAP test was greater in the group of women with the highest scores on the test of knowledge. Also, in this group there was a greater percentage of women who had undergone the PAP prior to reaching 30 years of age.

Another variable we included in the study was the 'locus of control in health versus chance.' The location or *locus* of control is a construct that refers to an individual's perception of where the causal agent for what happens in his/her life is located. People with an internal locus of control see themselves as capable of influencing their own destiny, of changing an adverse situation and increasing their probability of success—all of which increases their motivation and efforts to confront any given situation. In contrast, people with an external locus of control believe that the consequences of their actions depend on external factors beyond their control, such as luck or destiny or the actions of other persons—which leads them to not recognize their own capacity to alter the course of events or to influence them by their own actions. In the case we are dealing with, the locus of control for health refers to the degree to which a subject perceives that her health is the result of her own actions (internal locus of control), or is controlled by other powers, or even depends on luck, destiny or random chance (external locus of control) (Oros 2005).

We found that women who had the highest external locus of control had less knowledge about prevention of CC than the remainder of the participants. This result led us to conduct other studies in which we explored women's beliefs about things that might cause CC and the relation of these beliefs to the women's utilization of the PAP test. But before presenting these studies, we consider it pertinent to explain what constitutes a belief in health and to briefly review the literature on why it is that certain beliefs constitute an obstacle to the prevention of CC.

The perception of risk held by a person implies their individual cognitive evaluation of such risk, which is constructed throughout the course of their whole life, is modified by their life experiences, their accumulating fund of knowledge, and their emotions generated by what they perceive (López 2013). These cognitive evaluations are called beliefs and constitute the explanatory frame of reference conceived by the person with respect to his or her surroundings and may or may not conform to reality; but it is in accordance with their beliefs that people tend to search, accommodate and classify. Beliefs may facilitate or limit change depending on the degree of adjustment of newly acquired information to the person's extant frame of reference (Dilts et al. 1996). Health beliefs are derived from the perceptions and classifications made by social groups concerning illnesses and the persons suffering from them. In this manner, cognitive schemes are formed that guide the immediate behaviour of people, allowing them to make a rapid evaluation of an illness (Álvarez 2002). These schemes or beliefs are constant cognitive factors that delineate the behaviour of people and are acquired through the process of socialization (Conner/Norman 2005).

The Health Belief Model (HBM) was developed by a group of psychologists through their efforts to explain the lack of public participation in programmes for the early detection and prevention of illnesses. The general hypothesis is that people do not execute healthy behaviours (demonstrating prevention, participation, adherence, or rehabilitation) unless they have at least minimal levels of motivation and relevant information concerning their health, they see themselves as potentially vulnerable, they see the disease as threatening, they are convinced of the intervention's efficacy, and they see little difficulty in putting into practice the healthy behaviour (Rosenstock 1974; Rosenstock et al. 1988). One of the components of this model is the perception of threat that is interpreted based on two fundamental beliefs: the perceived risk of the disease and the anticipated gravity of its consequences. Another of the model's components is the evaluation of the conduct to be carried out. Such evaluation encompasses two groups of beliefs—those concerning the benefits of the recommended health behaviour and others concerning the costs or barriers involved in its execution (Abraham/Sheeran 2005). Based on multiple studies performed using this model, it has been concluded that the collection of beliefs involved and the perceived barriers to executing the action are the most important individual predictors of unhealthy behaviour (Valenzuela/Miranda 2001; Abraham/Sheeran 2005).

The barriers are all those factors that impede healthy behaviour. In addition to structural barriers such as economic barriers, lack of access to health care services, or lack of information, psychosocial barriers also play an important role in the failure of prevention of CC. The principal barriers of this type that have been associated with reluctance to undergo PAP testing in different parts of the world, particularly in developing nations, are:

– The belief that the PAP test is a painful procedure.
– The fear of contracting an infection by being tested.
– The fear of becoming sterile due to manipulation during the test.
– The shame of having a gynaecological exploration, particularly when a male doctor performs the test.
– The lack of approval from male partners to have the test performed.
– Fear that the test result may indicate the presence of cancer.
– In addition to the barriers described above, young and single women also report fearing that their parents will find out about the testing (López et al. 2013; Menard et al. 2010; Abdullahi et al. 2009; Lovell et al. 2007; Ybarra et al. 2012).

Another cognitive factor we consider involved in behavioural actions related to risks is fatalism, which is defined as the generalized tendency to believe that events are predetermined or caused by external forces and that little or nothing can be done to change their course. It is implicit in this vision that willpower or individual action exercise little power to change the course of destiny (Flórez et al. 2009). These beliefs are associated with negative or pessimistic attitudes regarding preventive behaviours and the consequences of an illness (Espinosa/Gallo 2011). Thus, when a person believes that the development of an illness is in the hands of God or is due to luck or destiny, then she can do nothing to prevent it and therefore fails to adopt the behaviours that would lead to its early detection (Straughan/Seow 2000).

The term 'cancer fatalism' has been used to study fatalistic beliefs in the different stages of development of this disease from primary prevention up to survival (Niederdeppe/Levy 2007). Many authors have demonstrated that women with low levels of education and/or low income tend to demonstrate more fatalistic beliefs (Niederdeppe/Levy 2007; Russell et al. 2006). Beyond that, some have come to the conclusion that fatalism is a trait of Latin American culture and is a passive response that dissuades women from adopting early detection behaviours (Abraído-Lanza et al. 2007). In fact, in the United States, the belief that destiny is a risk factor for the development of CC is more common in immigrant Latinas than in Latinas born in the US or in Caucasians (Chávez et al. 1997). In a literature review on empirical studies of Latina women concerning fatalism and the usage of PAP tests and mammograms, Espinosa/Gallo (2011) concluded that a relationship does exist between fatalism and the adoption of early detection behaviours, although further studies are needed and as yet it is not possible to know if fatalism operates in a different manner among subpopulations of Latin American women.

Given our interest in uncovering distinct cognitive factors related to the perception of risk of CC among Mexican women, and therefore healthy or unhealthy behaviours, our working group carried out two investigations—one on psychosocial barriers and the other on fatalistic beliefs. Both studies were carried out in the state of Veracruz and involved surveying adult women. In the first study, participants included women from an urban area (the city of Xalapa) and women from various rural areas of the state (Marván et al. 2013). Three groups were defined according to the educational level and place of residence of the participants: (a) women living in a rural area with primary school as their maximum educational level; (b) women living in an urban area with primary school as their maximum educational level; and (c) women living in an urban area with a higher than primary educational level. No women living in the rural areas had a higher than primary educational level. Furthermore, when questioned concerning the PAP test and in order to explore the presence of psychosocial barriers, the women were presented with the following affirmations to which they were asked to respond either yes or no: 'I would not like to know if I had cervical cancer', 'It is very embarrassing to undergo a PAP test', 'The PAP test is painful', 'Undergoing the PAP test will only lead to more worries', and 'My partner would not like me to undergo a PAP test.'

We found that in spite of the majority of women having undergone a PAP test (79%), only 29.1% had undergone their first PAP in accordance with guidelines in the Mexican Official Norms. In comparing rural and urban women, we found that more urban women had undergone the PAP test and had followed the Mexican Official Norms. On the other hand, rural women with primary schooling were those who demonstrated the most psychosocial barriers against undergoing the PAP: 37% of these women affirmed that having a PAP test would only bring about greater worries; 46% claimed it is very embarrassing to undergo a PAP test; and 17% affirmed that their partners would not like them to have a PAP test. It is interesting to note that there were no differences in the percentage of women in the three groups who affirmed that the PAP test is painful (24% on average).

In this study, we also applied the test of knowledge concerning the prevention of CC as mentioned above and we analysed the predictive value of both the psychosocial barriers and the reported knowledge in undergoing the PAP test. We found that not undergoing the PAP test was predicted by thinking that having the test performed was embarrassing, by male partners not liking a woman to have the test, and by the belief that the PAP test is painful. But the most important predictors were the levels of knowledge about the prevention of CC: women with greater knowledge about the PAP had 2.8 times greater probability of undergoing the test, while women with greater knowledge about CC risk factors had 7.4 times greater probability of being tested.

In the second study, we also questioned women in the city of Xalapa and in some rural zones in the state of Veracruz; but in contrast to the previous study, all the women participants had only a primary school level of education (Marván et al. 2014). Besides questioning them about undergoing the PAP, these women were presented with four fatalistic beliefs about CC to which they were asked to respond if they were true or false: 'Cervical cancer is due to bad luck', 'If one gets CC it's due to destiny', 'God gives women CC because they have lived a bad life', and 'There is little I can do to prevent CC'.

There were more rural than urban women who believed that CC is due to bad luck (35%) or to destiny (47%). On the other hand, there was a similar percentage of rural and urban women who believed that God gives women CC for having lived a bad life (20% on average) and who considered there was little they could do to prevent CC (31% on average).

As in the previous study, we analysed the predictive value of fatalistic beliefs about CC in undergoing the PAP test. In the case of rural women, we found that those who believed that CC is due to bad luck or destiny had almost four times lower probability of undergoing the test than women without such beliefs. Among urban women, the only predictor of not undergoing the PAP test was the belief that one couldn't do much to prevent the disease. These differences in the predictive factors of PAP usage between rural and urban women may be due to the far greater extension in rural populations of the belief that luck and destiny have a determining role in the health of individuals.

4.4 Final Considerations

It is lamentable that a significant number of deaths due to CC are continuing to occur in spite of CC being a disease susceptible to early detection and opportune treatment due to the lengthy time that elapses between the appearance of the first lesions and their evolution into cancer.

Clearly, much of our findings on women's acquired knowledge, barriers to taking action, and their beliefs and attitudes play an important part in the prevention of CC both for the avoidance of risky behaviours that foster the development of the disease and for the early detection of lesions that, if not treated opportunely, may later convert into cancer. The term cancer per se carries many interpretations that are permeated by multiple psychosocial factors as well as those presented in this chapter. For example, witnessing the development of cancer in someone close (who may or may not die from the disease) changes one's perception of the risk and consequently changes their behaviour such that they may present new healthy or unhealthy behaviours.

Speaking about the prevention of CC is complicated since it requires changes in lifestyle. In Mexico, there exists a National Programme for the Prevention and Control of Cervical Cancer through which preventive actions are taken, such as administering the HPV vaccine and giving PAP tests for early detection, and which mentions counselling as an indispensable strategy for the transmission of information. Nevertheless, in the operation of the programme difficulties have been encountered in reaching adequate coverage, in guaranteeing the quality of the test, and in achieving appropriate follow-up and opportune treatment of women with precancerous lesions and invasive cancer (OPS 2010). In this sense, it is urgent to design health education programmes for each stage of peoples' lives, programmes in which discussions are held about the disease and useful behaviours are suggested to facilitate its primary and secondary prevention, and programmes that promote public awareness that adopting such behaviours may signify the difference between life and death for individuals.

The benefits of a vaccine or the early detection of CC cannot be integrated as protective behaviours to reduce the risk of death if the person has deeply ingrained beliefs that lead to risky behaviours and to barriers for adopting healthy behaviours. This is why health professionals involved in this and other types of grave diseases should not neglect the basic perceptions and personal frames of reference of individuals in order to facilitate adequate communication with them on the three levels of prevention: primary, secondary and tertiary.

References

Abdullahi, A., Copping, J., Kessel, A., Luck, M., Bonell, C. (2009). Cervical screening: Perceptions and barriers to uptake among Somali women in Camden. *Public Health, 123,* 680–5.

Abraham, C., & Sheeran, P. (2005). The health belief model. In M. Conner and P. Norman. *Predicting health behavior* (2nd Edition). England: Open University Press.

Abraído-Lanza, A.F., Viladrich, A., Flórez, K.R., Céspedes, A., Aguirre, A.N., de la Cruz, A. A. (2007). Commentary: Fatalismo reconsidered: A cautionary note for health-related research and practice with Latino populations. *Ethnicity & Disease*, 17, 153–158.

Aguilar-Pérez, J.A., Leyva-López, A.G., Angulo-Nájera, D., Salinas, A., Lazcano-Ponce, E. C. (2003). Tamizaje en cáncer cervical: conocimiento de la utilidad y uso de citología cervical en México. (Screening for cervical cancer: knowledge of the utility and usage of cervical cytology in Mexico). *Revista Saúde Pública*, 37,100–106.

Álvarez, B.J. (2002). *Estudio de la creencias, salud y enfermedad. Análisis psicosocial.* (A study of beliefs, health and sickness. A psychosocial analysis.) Mexico: Trillas.

Bazan, F., Posso, M. y Gutiérrez, C. (2007). Conocimientos, actitudes y prácticas sobre la prueba de Papanicolaou (Knowledge, attitudes and practices about the PAP test). *En Anales de la Facultad de Medicina*. 68(001), 47–54. http://dx.doi.org/10.15381/anales.v68i1.1238.

Chávez, L.R., Hubbell, F.A., Mishra, S.I., Valdez, R.B. (1997). The influence of fatalism on self-reported use of Papanicolaou smears. *American Journal of Preventive Medicine*, 13, 418–424.

Conner, M., & Norman, P. (2005). Predicting health behavior: A social cognition approach. In M. Conner & P. Norman *Predicting health behavior*. *2nd Edition* (pp. 1–27). England: Open University Press.

Dilts, R., Hallbom, T., Smith, S. (1996). *Las Creencias. Caminos hacia la salud y el bienestar*. (Beliefs: Roads to health and wellbeing). Argentina: Urano Editions.

Ehrenzweig, Y., Marván, M.L., Acosta, E.A. (2013). Conocimientos sobre la prevención del cáncer cervicouterino, locus de control y realización del Papanicolaou. (Knowledge about the prevention of cervical cancer, locus of control and usage of the Papanicolaou). *Revista Psicología y Salud*, 23(2), 161–169.

ENSANUT (2012). *Prevención y diagnóstico temprano de cáncer en la mujer: soluciones a su alcance. Encuesta Nacional en Salud y Nutrición. Evidencia para la política pública en salud.* (Prevention and early diagnosis of cancer in women: solutions within reach. National Survey on Health and Nutrition. Evidence for public health policy.) Resource document. México: Secretaria de Salud e Instituto Nacional de Salud Pública. http://ensanut.insp.mx. Accessed 25 Nov 2016.

Espinosa, K., & Gallo, L.C. (2011). The relevance of fatalism in the study of Latinas' cancer screening behavior: A systematic review of the literature. *International Journal of Behavioral Medicine*, 18, 310–318.

Espinosa-Romero, R., Arreola-Rosales, R.L., Velázquez-Hernández, N., Rodríguez-Reyes, E. R. (2014). Métodos de detección oportuna del cáncer cervicouterino. (Early detection methods for cervical cancer.) *Gaceta Mexicana de Oncología*, 13(4), 48–52.

GLOBOCAN (2012). Estimated Cancer Incidence, Mortality and Prevalence Worldwide in 2012. Resource document. http://globocan.iarc.fr. Accessed 25 Nov 2016.

Flórez, K.R., Aguirre, A.N., Valadrich, A., Céspedes, A., De la Cruz, A.A., Abraído-Lanza, A.F. (2009). Fatalism or destiny? A qualitative study and interpretative framework on Dominican women's breast cancer beliefs. *Journal of Immigrant and Minority Health*, 11, 291–301.

Hernández-Hernández, D.M. et al. (2007). Factores asociados con incumplimiento para tamizaje en cáncer de cerviz. (Factors associated with non-compliance in cervical cancer screening.) *Revista Médica Instituto Mexicano del Seguro Social*, 45(4), 313–320.

INEGI. (2011). *Comunicado No. 267/11*. México.

Kumar, V.K., Abbas A.K., Fasuto, N., Aster, J.C. (2007). *Patología estructural y funcional* (Structural and functional pathology) (7th Ed.). Spain: Elsevier.

López, V.E. (2013). Percepción de riesgo y respuesta psicosocial ante desastres naturales y tecnológicos. (Risk perception and psychosocial response to natural and technological disasters). In G. R. Ortiz. *Tópicos selectos en psicología de la salud. Aportes Latinoamericanos* (pp. 139–164). Mexico: Ducere.

López, C.A., Calderón, M.A., González, M.M. (2013). Conocimientos, actitudes y prácticas respecto al cáncer de cuello uterino de mujeres entre 14 y 49 años de un barrio de la comuna 10 de la ciudad de Armenia, Colombia. [Knowledge, attitudes and practices concerning cervical cancer in women between 14 and 19 years in a neighborhood in the 10th commune of the city of Armenia, Colombia.] *Revista Médica Risaralda*, 19(1), 14–20.

Lovell, S., Kearns R.A., Friesen, W. (2007). Sociocultural barriers to cervical screening in South Auckland, New Zealand. *Social Science Medicine*, 65, 138–50.

Manzo-Merino, J., Jiménez-Lima, R., Cruz-Gregorio, A. (2014). Biología molecular del cáncer cervicouterino. (Molecular biology of cervical cancer). *Gaceta Mexicana de Oncología*, 13(4), 18–24.

Marván, M.L., Ehrenzweig, Y., Castillo-López, R.L. (2013). Knowledge about cervical cancer prevention and psychosocial barriers to screening among Mexican women. *Journal of Pyschosomatic Obstetrics & Gynecology*, 34(4), 163–169.

Marván, M.L., Ehrenzweig, Y., Castillo, R.L. (2014). Fatalistic beliefs and cervical cancer screening among Mexican women. *Health Care for Women International*, 25, 1–15.

Menard, J., Kobetz, E., Maldonado, J.C., Barton, B., Blanco, J., Diem, J. (2010). Barriers to cervical cancer screening among Haitian immigrant women in Little Haiti, Miami. *Journal of Cancer Education*, 25(4), 602–8.

Muñoz, N., Reina, J.L., Sánchez, G.I. (2008). La vacuna contra el virus del papiloma humano: una gran arma para la prevención primaria del cáncer de cuello uterino. (The vaccine against human papilloma virus: a great weapon for the primary prevention of cervical cancer). *Colombia Médica, 39*(2), 196–204.

Niederdeppe, J., & Levy, A.G. (2007). Fatalistic beliefs about cancer prevention and three prevention behaviors. *Cancer Epidemiology Biomarkers & Prevention,* 16, 998–1003.

Oblitas, L.A. (2010). *Psicología de la salud.* (Psychology of health). Mexico: Cengage Learning Editores.

OPS (2010). *Situación de los Programas para la Prevención y el Control del Cáncer Cervicouterino: Evaluación rápida mediante encuesta en 12 países de América Latina.* (Status of Programs for the Prevention and Control of Cervical Cancer: A rapid evaluation via survey in 12 Latin American countries). Resource document. http://www.paho.org/hq/index.php?option=com_docman&task=doc_view&gid=17788&Itemid=270&lang=en. Accessed 25 Nov 2016.

Oros, L. (2005). Locus de control: Evolución de su concepto y operacionalización. (Locus of control: Evolution of the concept and its implementation). *Revista de Psicología,* 14, 89–98.

Palacio-Mejía, L.S., Lazcano-Ponce, E., Allen-Leigh, B. Hernández-Ávila, M. (2009). Diferencias regionales en la mortalidad por cáncer de mama y cérvix en México entre 1979 y 2006. (Regional differences in the mortality from breast and cervical cancers in Mexico between 1979–2006). *Salud Pública de México,* 51(2):S208–S219.

Rodríguez-Marín, J., & Neipp L., M.C. (2008). *Manual de Psicología Social de la Salud.* Madrid: Síntesis.

Rosenstock, I.M. (1974). Historical origins of the health belief model. *Health Education Monographs,* 2, 328–335.

Rosenstock, I.M., Strecher, V.J., Becker, M. H. (1988). Social learning and the health belief model. *Health Education Quarterly,* 5(2), 175–183.

Russell, K.M., Perkins, S.M., Zollinger, T.W. Champion, V.L. (2006). Sociocultural context of mammography screening use. *Oncology Nursing Forum,* 33, 105–112.

Sánchez-Barriga, J.J. (2012). Tendencias de mortalidad por CaCu en las siete regiones socioeconómicas y en las 32 entidades federativas de México en los años 2000–2008. (Mortality trends for cervical cancer in the seven socioeconomic regions and the 32 federal states of Mexico in the years 2000–2008.) *Gaceta Médica de México,* 148, 42–51.

Secretaría de Salud (2007). Modificación a la Norma Oficial Mexicana 014 SSA2-1994 Para la Prevención, detección, diagnóstico, tratamiento, control y vigilancia epidemiológica del cáncer cérvico uterino. (Modification of the official Mexican Norm 014 SSA2-1994 for the prevention, detection, diagnosis, treatment, control and epidemiological oversight of cervical cancer). Resource document. Diario Oficial Mexicano de la Federación. www.salud.gob.mx. Accessed 25 Nov 2016.

Straughan, P.T. & Seow, A. (2000). Attitudes as barriers in breast screening: A prospective study among Singapore women. *Social Science and Medicine,* 51, 1695–1703.

Valenzuela, M.T. & Miranda, A. (2001). ¿Por qué no me hago el Papanicolaou? Barreras psicológicas de mujeres de sectores populares de Santiago de Chile. (Why won't I have a Papanicolaou test? Psychological barriers of women from popular sectors of Santiago, Chile.) *Revista Chilena de Salud Pública,* 5(2–3),75–80.

Vignolo, J., Vacarezza, M., Álvarez, C., Sosa, A. (2011). Niveles de atención, de prevención y atención primaria de la salud. (Levels of attention, of prevention and primary healthcare attention.) *Archivos de Medicina Interna.* 33, 7–11.

Walboomers J.M., Jacobs M.V., Manos M.M., Bosch F.X., Kummer J.A., Shah K.V., et al. (1999). Human papillomavirus is a necessary cause of invasive cervical cancer worldwide. *The Journal of Pathology,* 189, 12–9.

Wiesner, C., Piñeros, M., Trujillo, L.M., Cortés, C., Ardila, J. (2010). Aceptabilidad de la vacuna contra el virus papiloma humano en padres de adolescentes en Colombia. (Acceptability of the vaccine against human papilloma virus in parents of adolescents in Colombia.) *Revista de Salud Pública de Colombia,* 12 (6), 961–973.

Ybarra, J.L., Pérez, B.E., Romero, D. (2012). Conocimientos y creencias sobre el Papanicolaou en estudiantes universitarios. (Knowledge and beliefs about the Papanicolaou in university students.) *Psicología y Salud*, 22(2), 185–194.

Chapter 5
Omega-3: An Intelligent Decision for Brain Nutrition

Socorro Herrera Meza and Grecia Herrera Meza

Abstract Chronic degenerative diseases put the health of the general population at risk. However, the information given to the population to prevent or decrease the incidence of illness does not always reach its intended audience or is inaccurate, as in the case of information regarding polyunsaturated fatty acids such as ω-3, found in aquatic animals and some seeds. It has been observed that supplementation with ω-3 prevents cardiovascular diseases and strokes; reduces triglycerides, blood pressure, formation of thrombi and arrhythmia; alleviates cutaneous lesions resulting from psoriasis, rheumatoid arthritis, nephropathy and asthma; and decreases the metastasis of prostate, breast and colon cancer. Decrease in ω-3 levels, on the other hand, is associated with dementia, migraines, post-partum depression bipolarity, symptoms of depression and aggression, and neurological disorders relating to memory, attention and information processing. If the information on ω-3 were widely available, people would be able to choose what they eat and create a culture of prevention, thus decreasing the risk factors.

Keywords ω-3 · Supplementation · Cardiovascular problems · Neurological disorders · Prevention

5.1 Introduction

Despite major technological and scientific advances, chronic degenerative diseases, such as diabetes mellitus, cardiovascular diseases and cancer, are on the rise and putting the health of the general population at risk. These serious illnesses have been associated with risk factors, such as: obesity, poor eating habits, sedentary lifestyles, ignoring medical indications and the absence of a culture of prevention in

Dr. Socorro Herrera Meza, Researcher, Institute of Psychological Research at Universidad Veracruzana. Email: soherrera@uv.mx.

Dr. Grecia Herrera Meza, Researcher, Food Development and Research Unit at Instituto Tecnológico de Veracruz. Email: greehem@gmail.com.

© Springer International Publishing AG, part of Springer Nature 2018
Ma. L. Marván and E. López-Vázquez (eds.), *Preventing Health and Environmental Risks in Latin America*, The Anthropocene: Politik—Economics—Society—Science 23, https://doi.org/10.1007/978-3-319-73799-7_5

Mexico. These factors have led to an increase in health risks affecting the general public.

Furthermore, information regarding how to prevent or decrease the incidence of some of these risk factors never reaches its intended audience or is inaccurate, as in the case of information that has been created regarding the consumption of certain foods, specifically fats.

Fats, better known as fatty acids (FA), are compounds that play a fundamental role in human nutrition, given that, depending on the type of fatty acid, they can accelerate or reverse diseases that are caused by their consumption.

FAs not only play a major role in foods, such as giving them flavour or texture, they also have important functions in the body, such as, for example, acting as our main energy reserve, transporting a range of nutrients, regulating our metabolism, forming vitamins and hormones (Badui 2006) and creating cell membranes and neuron structures, i.e. they create a layer that divides the cell, the strength of which is determined mainly by the proportion and types of fatty acids from which they are composed. This function is fundamentally important in our bodies as this membrane maintains the balance between the interior and exterior of the cells, regulating the flow of substances or nutrients (Flores et al. 2007), in addition to improving contact among cells and promoting the transmission of nerve impulses, leading to better memory and learning processes (Valenzuela et al. 2011).

There are two types of fatty acids in cell membranes: saturated fatty acids (those fats that we know to be bad or harmful) and polyunsaturated fatty acids (PUFAs) (those fats that help keep us healthy). PUFAs include omega-3 and 6 (ω3 and ω6) fatty acids, which have been classified as essential, i.e. these should be obtained as part of our daily diet and cannot be substituted by any other nutrient, nor can the body produce them naturally. On the other hand, there are non-essential fatty acids, which our bodies can get from what we eat or can synthesize, i.e. produce from the carbohydrates we ingest (Muriana 2004).

Once the digestion, absorption and transportation of the fats have been completed, the ω-6 and ω-3 are introduced into the adipose cells and form fatty compounds that have a range of different functions in the body. When these cells receive specific stimuli, different enzymes are activated, i.e. proteins that regulate and favour chemical reactions in human beings are activated (Muriana 2004).

The omega-3 and 6 compete for these enzymes, each of them generating different compounds. The consumption of ω-6 fatty acids generates, through reactions within the body, compounds called eicosanoids, which can lead to platelet aggregation (the formation of thrombi), the formation of blood clots and vasoconstriction (the thinning of the blood vessels). In contrast, ω-3 fatty acids favour the production of another type of eicosanoid that has the opposite effect, i.e. they prevent the formation of clots and cause vasodilation (the expansion of the blood vessels) (Krummel 2001). As such, it is extremely important to ensure a balance between the intake of ω-6 and ω-3 fatty acids in order to avoid any imbalance leading to the creation of undesirable compounds that pose a risk to our health.

Omega-6 is generally found in seeds, such as sunflower, corn, safflower and peanuts. Omega-3 is found in aquatic animals and some seeds. In this chapter, we focus specifically on ω-3: what foods it is found in, what functions it has in the body, what benefits it provides as part of our diet, and what illnesses are associated with a lack of ω-3, at a metabolic, cerebral and cognitive level.

We reflect on our eating habits and compare a Mediterranean diet to Western eating habits. The Mediterranean diet, rich in polyunsaturated fatty acids, is based on the combination of a high intake of vegetables, fruits, nuts, whole grains, olive oil, fish, cheese and moderate wine consumption with lower quantities of milk and meat. On the other hand, the Western diet consists of high quantities of trans-fatty acids, saturated fats, sugars, flour, meat and a large amount of industrialized food, which can be easily acquired on the market (Haast/Kiliaan 2014).

The Western diet has a number of deficiencies (and also excesses) with regard to its nutritional properties. The most fundamental deficiency, and perhaps the most relevant, is that of omega-3 (Valenzuela 2005). ω-3 fatty acids are composed of docosahexaenoic acid (DHA) and eicosapentaenoic acid (EPA), which are mainly found in fish oil, tuna, salmon, cod and some seeds, such as flax and olive, among others. The bullet-pointed list at the end of this section provides a detailed overview of omega-3 content in a variety of foods (Muriana 2004; Flores et al. 2007).

Despite the fact that food rich in omega-3 is widely available to the general public, it is sad to see that, in Mexico, the consumption of seafood is extremely low, compared to that in eastern countries. For example, annual per capita consumption of fish in an eastern country, such as Japan, is 61 kg per capita per year; however, in Mexico, consumption stands at 10 kg per capita per year (FAO 2009). It is important to highlight the fact that both countries have the same access to seafood; however, sociocultural and marketing factors play an important role in promoting unhealthy eating habits among the general population (Flores et al. 2007), mainly diets rich in sugars and meat but low in omega-3.

The nutritional importance of these fatty acids has led health authorities to establish a daily recommended dose of ω-3. Although there is no exact dosage, the World Health Organization (WHO) recommends that adults consume between 1.2 and 1.5 g per day (Valenzuela 2005). Other authors recommend a dose of 2–4 g per day (Pirillo/Catapano 2013), while some experts recommend eating fish or food with a high omega-3 content at least once or twice a week.

The health benefits of DHA are numerous, and including them in our daily diets is an easy task. We can find them in what we eat on a daily basis, which is a practical, tasty and pleasurable way of incorporating them into our diets, and also as food supplements. The food sources richest in omega-3 are the following (in g/100 g):

- Flaxseed oil (55.3)
- Salmon oil (35.30)
- Cod liver oil (19.75)
- Herring oil (11.86)
- Walnut (11.50)
- Soybean (7.30)
- Soybean oil (7.60)

- Wheat germ (5.30)
- Caviar (3.74)
- Sardine (3.0)

Other foods that contain omega-3, although in less quantity, are: herring, fish from the Pacific, salmon, fish from the Atlantic, anchovy, butter, fresh tuna, crab, trout (rainbow), olive oil, corn oil, sunflower oil, parmesan cheese, hen's egg (white), prickly pear fruit, strawberry, canned tuna, cow's milk (Flores et al. 2007).

5.2 Omega-3 and Chronic Degenerative Diseases

It has been proven that genetic and environmental factors determine how susceptible an individual is to becoming ill; therefore, nutrition is a determining environmental factor and one that has a major impact on us as human beings. Studies of the evolution of diets indicate that the major changes to our eating habits have occurred mainly in the type and amount of essential fatty acids and antioxidants found in foods. The changes that have taken place over the past hundred years are major catalysts in some of the most prevalent chronic diseases in our society, such as cancer, diabetes mellitus, obesity, metabolic syndrome and cardiovascular diseases (Mata et al. 2004).

Cardiovascular diseases (CVDs) represent a serious global public health issue as they are the leading cause of death around the world. Unfortunately, CVDs are becoming more and more prevalent among the younger generations (Alwan 2011). According to the World Health Organization (WHO), some 17.5 million people died as a result of CVDs in 2012. In Mexico, according to the National Social Security Institute (IMSS), 283,732 Mexicans died of CVDs in the same year (De la Peña 2012).

CVDs are strongly associated with a number of risk factors, including: increases in levels of serum total cholesterol (TC), triglycerides (TG), hypertension, strokes, diabetes and obesity (Carrero et al. 2005). Many of these risk factors are associated with diet, and they can be modified through proper eating habits. The consumption of ω-3 polyunsaturated acids can have a positive effect on cardiovascular health; even eating a small amount of fish (once a week) can reduce any type of risk relating to this disease (Carrero et al. 2005).

There are a number of studies that recommend the consumption of DHA to help prevent cardiovascular problems and strokes (Pirillo/Catapano 2013). In fact, since 1990, these fatty acids have been used in Japan to treat arteriosclerosis and hyperlipidaemia (Arab-Tehrany et al. 2012).

Carrero et al. (2005) discuss a study entitled 'The Seven Countries', in which, over a period of 20 years, they observed how men who ate 30 g of fish per day reduced their risk of death from coronary disease by 50% compared to volunteers who did not eat any fish.

Recent research has shown that omega-3 reduces the level of triglycerides in the bloodstream among patients suffering from hypertriglyceridemia (an increase in the level of triglycerides in the blood), with a daily intake of 2 g, as well as increasing high density lipoproteins (HDL, good fats). Other studies highlight a decrease in triglycerides of between 25 and 35% with high doses of between 3 and 4 g/day, in addition to a 36% decrease in low-density lipoproteins (VLDL), which cause damage to arteries (Hartweg et al. 2007; Pirillo/Catapano 2013).

Increase in blood pressure or hypertension (HTN) is the cause of 13% of deaths around the world (Alwan 2011). A number of studies have shown that consuming ω-3 from fish on a daily basis is associated with a decrease in blood pressure (Mata et al. 2004). Some studies suggest that supplementing a dose of 2–3 g/day decreases blood pressure among individuals suffering from hypertension (Arab-Tehrany et al. 2012).

Thrombosis (which drives the formation of clots in blood vessels and can cause acute myocardial infarctions) and arrhythmia (the alteration in cardiac rhythm) are key factors in the development of clinical manifestations of coronary heart disease. It has been shown that consuming omega-3 reduces platelet aggregation, the formation of thrombi and arrhythmia (Mata et al. 2004).

It has also been shown that groups with a high fish intake have seen decreases in the incidence of Type 2 diabetes (T2D) as well as positive impacts on the effect of insulin (Pirillo/Catapano 2013). However, this is still controversial; there are a number of studies that prove a positive direct association between omega-3 intake and a decrease in T2D, while others refute this association (Kaushik et al. 2009; Wu et al. 2012; Zhou et al. 2012).

The supplementation of DHA has also been studied in a wide range of other diseases. For example, it has been shown that consuming these fatty acids helps to alleviate cutaneous lesions resulting from psoriasis (a chronic skin disease stemming from autoimmune complications) (Rahman et al. 2013), in addition to having a positive effect on the treatment of rheumatoid arthritis (Arab-Tehrany et al. 2012), nephropathy (kidney damage or disease) (Bell et al. 2012) and asthma (D'Auria et al. 2014). Some researchers have studied the positive effects of omega-3 on Crohn's disease, but the information available remains contradictory: some studies show that consumption of omega-3 can help mitigate the symptoms of the disease, while other maintain that the administration of omega-3 aggravates the problem (Swan/Allen 2013).

Mexico is ranked first in the world for childhood obesity (ENSANUT 2012). Obesity is a complex disorder with a number of different causes, the most significant of which is inadequate diet. Obesity and overweight involve the chronic inflammation of cells. i.e. adipose tissue (fat that accumulates in the body), as a result of body weight, producing substances known as cytokines, which are directly related to the occurrence of cardiovascular diseases, diabetes, hypertension and dyslipidaemias. It has been shown that supplementing DHA can reduce the production of cytokines, and, as such, the risk of suffering from the aforementioned diseases (Jalbert 2013; Lee et al. 2013; Legrand-Poels et al. 2014).

Scientific evidence highlights the association between omega-3 intake and a protective effect from a range of different cancers. More specifically, delays in tumour growth, reductions in tumour size and a decrease in the metastasis of prostate, breast and colon cancer have been observed (Valenzuela et al. 2011a; Ballesteros-Vásquez et al. 2012; Merendino et al. 2013; Vasudevan et al. 2014).

5.3   Omega-3 and the Brain

Our brain tissue is composed mainly of fatty acids, and it has been estimated that between 50 and 60% is comprised of different types of fat, 35% of which are omega-3 fatty acids (Flores et al. 2007).

Brain development occurs mainly during the last trimester of pregnancy, finalizing at the age of 3. Omega-3 plays a fundamental and specific role in brain structure and function, given that it is the major component of the brain membranes. DHA is found in neuron membranes, creating greater fluidity within the cell and driving the increased synapsis and transmission of nerve impulses among neurons, which, in turn, favours learning processes (Valenzuela et al. 2011b).

The omega-3 fatty acids present in the neuronal membranes also regulate and increase the expression (creation) of important proteins, which help to improve neuroplasticity, neuronal protection, memory, learning and the prevention of illnesses, such as Alzheimer's and Parkinson's (Pasinettia et al. 2014; Porqueta et al. 2014).

There are specific stages of human development during which DHA is needed by the body in significant quantities: pregnancy, breastfeeding and childhood. How does DHA reach us when we are in our mother's womb? ω-3 is provided by the mother from her own reserves and her diet (Innis 2008). These fatty acids pass through the placenta and accumulate in the brain of the foetus (Dutta-Roy et al. 2004; Muriana 2004) prior to birth. After birth, omega-3 is delivered via breast milk (Das 2003).

During pregnancy, the nervous and visual systems are created. DHA accumulates mainly in the retina cells, called rods and cones, helping ensure visual acuity (Muriana 2004).

During the formation of the nervous system, which begins between the third and fourth week of pregnancy (Lagercrantz/Ringstedt 2001), the amount of omega-3 increases; however, these requirements increase even further during the tenth week of pregnancy, when the brain hemispheres are formed. It is during this period when brain cells called neurons and glial cells are actively created, with around 200,000 neurons being formed every minute (Sanhueza et al. 2004).

In order to maintain a balance of the amount of omega-3 between the foetus and the mother, the latter must ingest around 100 mg/day of these fatty acids during the last trimester of pregnancy (Das 2003), as it is during this stage when the mother's ω-3 levels are particularly depleted as she transfers these fatty acids to the foetus and the breastfeeding child (Montgomery et al. 2003).

As such, studies into humans and animals show that a lack of DHA during the perinatal period, which comprises the period encompassing the 22nd week of gestation until 7 days after birth, has an impact on visual acuity (Muriana 2004), capacity for learning, concentration and the child's IQ, alterations that do not become evident until they reach adulthood (Innis 2008; Levan et al. 2004).

It has been shown that DHA is not only needed during the first stage of life. An unbalanced diet, specifically a lack of omega-3, has negative effects on neural functions (the formation of the neural tube and of the nervous system), both during infancy and adulthood (Flores et al. 2007). Studies conducted with infant rats show that a deficit of DHA leads to an increase in locomotive activity and induced cataleptic states (muscular immobility) (Ikemoto et al. 2001; Haubner et al. 2002).

Furthermore, over the past three decades, psychiatric and neurodegenerative illnesses, such as bipolar disorders, depression (Valenzuela et al. 2009), Alzheimer's (Nussbaum/Ellis 2003; Sontrop/Campbell 2006) and Parkinson's, have grown significantly (Tiemeier et al. 2003). Epidemiological studies indicate a close association between the incidence and development of these illnesses and factors such as: age, physical activity, family history and poor diets, such as a lack or low levels of ω-3 (Tiemeier et al. 2003; Sontrop/Campbell 2006; Shinto et al. 2009).

A decrease in omega-3 levels in the blood is associated with neurological disorders (McCann/Ames 2005; McNamara/Carlson 2006), such as cognitive impairment (loss of functions, such as memory, attention and information processing), dementia (Ikemoto et al. 2001; Morris et al. 2003), migraines (Tanskanen et al. 2001a, b; Shinto et al. 2009) and post-partum depression (Valenzuela et al. 2009). Omega-3 deficiency has also been related to behavioural disorders, such as bipolarity, symptoms of depression and aggression (Zanarini/Frankenburg 2003).

Other studies propose that omega-3 should be used to treat illnesses such as depression, schizophrenia (Hui-Min 2010), bipolar disorders (McNamara/Carlson 2006), dementia and psychiatric disorders prevalent during pregnancy and breast-feeding (Carlson 2001). Furthermore, it has been observed that neurodegenerative diseases, such as Parkinson's and Alzheimer's, are associated with a major loss of DHA in the neuronal membranes. With regard to Alzheimer's, the ω-3 in the membranes is replaced by saturated fatty acids (Sanhueza et al. 2004).

As such, research conducted on mice (Carrié et al. 2000) and humans (Helland et al. 2006), associates higher levels of DHA in brain tissue with a greater capacity for learning, through memorization, intelligence (Helland et al. 2006) and motor development tests. Furthermore, it has been shown that the presence of omega-3 is associated with a neuroprotective state, i.e. one that protects the brain (Bazan 2005; Ikemoto et al. 2001; Takeuchi et al. 2002). Other studies conducted on rats show that the consumption of omega-3 during pregnancy and the first days of breast-feeding produces greater resistance among their offspring to the severity and occurrence of convulsions stemming from increasing their body temperature (experimental hyperthermia) (Flores et al. 2007).

These findings, with regard to the health benefits of omega-3, have led to increased interest among the scientific community in studying what the benefits of omega-3 intake are and how it helps to improve mental and physical health; however, the significant changes in human diets have created major alterations in metabolic processes and general health. The scales have shifted towards excessive consumption of sugars, flour, saturated fats and omega-6, and reduced consumption, or lack thereof, of whole grains, fruits, vegetables and omega-3. If we add the lack of awareness of the general public regarding the nutritional value of foods and the influence of media outlets in promoting the consumption of low-quality or junk foods, the outcome is discouraging given that these illnesses continue to increase.

The restructuring of our eating habits towards a healthier diet, rich in ω-3, would lead to improved protection, prevention and treatment of chronic degenerative, psychiatric and behavioural disorders. As such, the creation of a range of healthcare and prevention strategies, such as promoting a balanced diet rich in DHA and a healthier lifestyle that encompasses physical activity and appropriate body weight, could lead to a significant reduction in risk factors among the general population.

Furthermore, if the information created surrounding fatty acids is accurate and widely available, people will be able to choose what they eat carefully and take advantage of the benefits that omega-3 offers. It is important to highlight that the food industry, which extracts ω-3 from fish and some algae, offers a wide range of supplement options for the whole family (capsules, oil). Special supplements have also been created for children, using attractive and tasty options (gummy bears), which are easy to find and economical. These fatty acids have also been added to infant formula, and in some countries omega-3 is being used in infant foods, as it is deemed to be a fundamental element in the development of the central nervous system, in addition to promoting optimal mental and visual development among children.

However, learning more about the nutritional composition and the effect that foods have on our bodies is not enough. Creating a culture of prevention and decreasing risk factors to our health will, without a doubt, reduce the incidence of these illnesses and improve the health of future generations. We need to overcome nutritional problems not only through choosing what food we eat or what supplements we take, but also by finding preventive measures. Our first-aid boxes should be found in the food we keep in the kitchen rather than in our medicine cabinets.

References

Alwan, A. (2011). *Global status report on noncommunicable diseases 2010*. World Health Organization.

Arab-Tehrany, E., Jacquot, M., Gaiani, C., Imran, M., Desobry, S., Linder, M. (2012). Beneficial effects and oxidative stability of omega-3 long-chain polyunsaturated fatty acids. *Trends in Food Science & Technology*, 25, 24–33(2012). https://doi.org/10.1016/j.tifs.2011.12.002.

Badui, D.S. (2006). *Lípidos. En Química de los Alimentos* (4a ed). México: Pearson Education.

Ballesteros-Vásquez, M.N., Valenzuela-Calvillo, L.S., Artalejo-Ochoa E., Robles-Sardin, A.E. (2012). Ácidos grasos trans: un análisis del efecto de su consumo en la salud humana, regulación del contenido en alimentos y alternativas para disminuirlos. *Nutrición Hospitalaria. Scielo,* 27(1), 54–64 (2012).

Bazan, N.G. (2005). Lipid signaling in neural plasticity, brain repair, and neuroprotection. *Molecular Neurobiology,* 32, 89–103(2005). https://doi.org/10.1385/mn:32:1:089.

Bell, S., Cooney, J., Packard, C.J., Caslake, M.J. Deighan, C.J. (2012). The effect of omega-3 fatty acids on the atherogenic lipoprotein phenotype in patients with nephrotic range proteinuria. *Clinical Nephrology,* 77(6), 445–453 (2012). https://doi.org/10.5414/cn107450.

Carlson, S.E. (2001). Docosahexaenoic acid and arachidonic acid in infant development. *Seminars in Neonatology,* 6, 437–449 (2001). https://doi.org/10.1053/siny.2001.0093.

Carrero, J.J., Martín-Bautista, E., Baró, L., Fonollá, J., Jiménez, J., Boza, J.J., et al. (2005). Efectos cardiovasculares de los ácidos grasos omega-3 y alternativas para incrementar su ingesta. *Nutrición Hospitalaria Scielo,* 20 (1), 63–69 (2005). doi:10.3305%2Fnutr+hosp.v20in01.3525.

Carrié, I., Guesnet, P., Bourre, J.M., Francès, H. (2000). Diets containing long-chain n-3 polyunsaturated fatty acids affect behaviour differently during development than ageing in mice. *British Journal of Nutrition,* 83, 439–447 (2000).

Das, U.N. (2003). Long-chain polyunsaturated fatty acids in memory formation and consolidation: further evidence and discussion. *Nutrition,* 19, 988–993 (2003). https://doi.org/10.1016/s0899-9007(03)00174-6.

D'Auria, E., Miraglia Del Giudice M., Barberi, S., Mandelli, M., Verduci, E., Leonardi, S., et al. (2014). Omega-3 fatty acids and asthma in children. *Allergy and Asthma Proceedings,* 35(3), 233–240 (2014). https://doi.org/10.2500/aap.2014.35.3736.

De la Peña, J.E. (2012). Enfermedad del corazón principal causa de muerte en México. Instituto Mexicano de Seguro Social (IMSS). http://archivo.eluniversal.com.mx/sociedad/2014/enfermedad-corazon-causa-muerte-mexico–1040565.html. Accessed 14 Feb 2018.

Dutta-Roy, A.K. (2000). Transport mechanism for long-chain polyunsaturated fatty acids in the human placenta. *The American Journal of Clinical Nutrition,* 71, 315S–322S.

Encuesta Nacional de Salud y Nutrición [ENSANUT] (2012). Resultados Nacionales. México: National Institute of Public Health. http://ensanut.insp.mx/doctos/ENSANUT2012_PresentacionOficialCorta_09Nov2012.pdf. Accessed 18 Apr 2017.

Flores, M.L., Hernández, G.M., Guevara, P.M. (2007). Efecto neuroprotector de los ácidos grasos omega-3: hallazgos neurofisiológicos y comportamiento. In M.A. Guevara, M. Hernández, M. Arteaga, M.E. Olvera (Eds.). *Aproximaciones al estudio de la funcionalidad cerebral y el comportamiento* (pp. 509–538). México: University of Guadalajara.

Gil A., & Gil, M. (2004). Funciones de los ácidos grasos poliinsaturados y oleico durante la gestación, la lactación y la infancia. In J. Mataix., & A. Gil. (Eds.). *Libro blanco de los omega-3, los ácidos grasos poliinsaturados Omega 3 y monoinsaturados tipo oleico y su papel en la salud* (pp. 82–96). Spain: Editorial Médica Panamericana.

Haast, R.A.M., & Kiliaan, A.J. (2014). Impact of fatty acids on brain circulation, structure and function. *Prostaglandins Leukotrienes Essent Fatty Acids,* 92, 3–14 (2014). https://doi.org/10.1016/j.plefa.2014.01.002.

Hartweg, J., Farmer, A.J., Perera, R., Holman, R.R., Neil, H.A.W. (2007). Meta-analysis of the effects of n-3 polyunsaturated fatty acids on lipoproteins and other emerging lipid cardiovascular risk markers in patients with type 2 diabetes. *Diabetologia,* 50, 1593–1602 (2007). https://doi.org/10.1007/s00125-007-0695-z.

Haubner, L.Y., Stockard, J.E., Saste, M. D., Benford, V.J., Phelps, C.P., Chen, L.T., et al. (2002). Maternal dietary docosahexanoic acid content affects the rat pup auditory system. *Brain Research Bulletin,* 58(1), 1–5 (2002). https://doi.org/10.1016/s0361-9230(01)00764-x.

Helland, I.B., Saugstad, O.D., Saarem, K., Van Houwelingen, A.C., Nylander, G., Drevon, C.A. (2006). Supplementation of n-3 fatty acids during pregnancy and lactation reduces maternal plasma lipid levels and provides DHA to the infants. *The Journal of Maternal-Fetal & Neonatal Medicine,* 19(7), 397–406 (2006). https://doi.org/10.1080/14767050600738396.

Hui-Min, S. (2010). Mechanisms of n-3 fatty acid-mediated development and maintenance of learning memory performance. *Journal of Nutritional Biochemistry*, 21, 364–373 (2010). https://doi.org/10.1016/j.jnutbio.2009.11.003.

Ikemoto, A., Ohishi, M., Sato, Y., Hata, N., Misawa, Y., Fujii, Y., et al. (2001). Reversibility of n-3 fatty acid deficiency-induced alterations of learning behaviour in the rat: level of n-6 fatty acids as another critical factor. *Journal Lipid Research*, 42(10), 1655–1663.

Innis, S.M. (2008). Dietary omega 3 fatty acids and the developing brain. *Brain Research*, 1237, 35–43 (2008). https://doi.org/10.1016/j.brainres.2008.08.078.

Jalbert, I. (2013). Diet, nutraceuticals and the tear film. *Experimental Eye Research*, 117,138–146 (2013). https://doi.org/10.1016/j.exer.2013.08.016.

Kaushik, M., Mozaffarian, D., Spiegelman, D., Manson, J.E., Willett, W.C., Hu, F.B. (2009). Long-chain omega-3 fatty acids, fish intake, and the risk of type 2 diabetes mellitus. *American Journal of Clinical Nutrition*, 90 (3), 613–620 (2009). https://doi.org/10.3945/ajcn.2008.27424.

Krummel, D. (2001). Nutrición en Enfermedades Cardiovasculares. In L. Kathleen, & S. Escott-Stump (Eds.). *Nutrición y Dietoterapia de Krause* (pp. 525–568). México: Mc Graw-Hill Interamericana (9ª ed.).

Lagercrantz, H., & Ringstedt, T. (2001). Organization of the neuronal circuits in the central nervous system during developing. *Acta Paediatrica*, 90(7), 707–715 (2001). https://doi.org/10.1111/j.1651-2227.2001.tb02792.x.

Lee, H., Lee, I.S., Choue, R. (2013). Obesity, Inflammation and Diet. Pediatric Gastroenterology, *Hepatology and Nutrition*, 16(3), 143–152 (2013). https://doi.org/10.5223/pghn.2013.16.3.143.

Legrand-Poels, S., Esser, N., L'homme, L., Scheen, A., Paquot, N., Piette, J. (2014). Free fatty acids as modulators of the NLRP3 inflammasome in obesity/type 2 diabetes. *Biochemical Pharmacology*, 92(1), 131–141 (2014). https://doi.org/10.1016/j.bcp.2014.08.013.

Levan, B., Radel, J.D., Carlson, S.E. (2004). Decreased brain docosahexaenoic acid during development alters dopamine-related behaviors in adult rats that are differentially affected by dietary remediation. *Behavioural Brain Research*, 152(1), 49–57 (2004). https://doi.org/10.1016/j.bbr.2003.09.029.

Mata, P., Alonso, R., Mata, N. (2004). Los omega 3 y los omega 9 en la enfermedad cardiovascular. In J. Mataix., & A. Gil (Eds). *Libro blanco de los omega-3, los ácidos grasos poliinsaturados Omega 3 y monoinsaturados tipo oleico y su papel en la salud* (pp. 49–63). Spain: Editorial Médica Panamericana.

McCann, J.C., & Ames, B.N. (2005). Is docosahexaenoic acid, an n-3 long-chain polyunsaturated fatty acid, required for development of normal brain function? An overview of evidence from cognitive and behavioral tests in humans and animals. *The American Journal Clinical Nutrition*, 82, 281–295.

McNamara, R.K., & Carlson, S.E. (2006). Role of omega-3 fatty acids in brain development and function: potential implications for the pathogenesis and prevention of psychopathology. *Prostaglandins Leukotrienesis and Essential Fatty Acids*, 75, 329–349 (2006). https://doi.org/10.1016/j.plefa.2006.07.010.

Merendino, N. Costantini, L., Manzi, L., Molinari, R., D'Eliseo, D., Velotti, F. (2013) Dietary ω-3 Polyunsaturated Fatty Acid DHA: A Potential Adjuvant in the Treatment of Cancer. *Biochemical Medical Research International*, 11, 1–11 (2013). https://doi.org/10.1155/2013/310186.

Montgomery, C., Speake, B.K., Cameron, A., Sattar, N., Weaver, L.T. (2003). Maternal docosahexaenoic acid supplementation and fetal accretion. *The British Journal of Nutrition*, 90, 135–145 (2003). https://doi.org/10.1079/bjn2003888.

Morris, M.C., Evans, D.A., Bienias, J.L., Tangney, C.C., Bennett, D.A., Wilson, R.S., et al. (2003). Consumption of fish and n-3 fatty acids and risk of incident Alzheimer disease. *Jama Neurology Formerly Archives of Neurology*, 60 (7), 940–946 (2003). https://doi.org/10.1001/archneur.60.7.940.

Muriana, F. (2004). Metabolismo de los ácidos grasos. In J. Mataix., & A. Gil. (Eds.). *Libro blanco de los omega-3, los ácidos grasos poliinsaturados Omega 3 y monoinsaturados tipo oleico y su papel en la salud* (pp. 35–47). Spain: Editorial Médica Panamericana.

Nussbaum, R.L., & Ellis, C.E. (2003). Alzheimer's disease and Parkinson's disease. *The New England Journal of Medicine*, 348, 1356–1364 (2003). https://doi.org/10.1056/nejm2003ra 020003.

Organización de las Naciones Unidas para la Agricultura y la Alimentación [FAO] (2009). Visión general del sector pesquero nacional. Japan: FAO. http://www.fao.org/fishery/countrysector/naso_japan/es. Accessed 19 Apr 2017.

Pasinettia, G.M., Wanga, J., Hoa L., Zhao, W., Dubner, L. (2014). Roles of resveratrol and other grape-derived polyphenols in Alzheimer's disease prevention and treatment. Biochimica et Biophysica Acta (BBA) – *Molecular Basis of Disease*, 1852, 1202 (2014). https://doi.org/10.1016/j.bbadis.2014.10.006.

Pirillo, A., & Catapano, A.L. (2013). Omega-3 polyunsaturated fatty acids in the treatment of hypertriglyceridaemia. *International Journal of Cardiology*, 170, S16–S20 (2013). https://doi.org/10.1016/j.ijcard.2013.06.040.

Porqueta, D., Griñán-Ferré, C., Ferrerd, I., Caminsa, A., Sanfeliub, C., Del Valle, J., Pallása, M. (2014). Neuroprotective role of trans-resveratrol in a murine model of familial alzheimer's disease. *Journal of Alzheimer's Disease*, 42, 1209–1220 (2014). https://doi.org/10.3233/jad-140444.

Rahman, M., Beg, S., Ahmad, M.Z., Kazmi, I., Ahmed, A., Rahman, Z., Akhter, S. (2013). Omega-3 fatty acids as pharmacotherapeutics in psoriasis: current status and scope of nanomedicine in its effective delivery. *Current Drug Targets*, 14(6), 708–722 (2013). https://doi.org/10.2174/1389450111314060011.

Sanhueza, J., Nieto, S., Valenzuela, A. (2004). Docosahexaenoic acid (DHA), brain development, memory and learning: the importance of perinatal supplementation. *Revista Chilena de Nutrición*, 31, 138–11 (2004). https://doi.org/10.4067/s0717-75182004000200002.

Shinto, L., Marracci, G., Baldauf-Wagner, S., Strehlow, A., Yadav, V., Stuber, L., Bourdette, D. (2009). Omega-3 fatty acid supplementation decreases matrix metallopro-teinase-9 production in relapsing-remitting multiple sclerosis. *Prostaglandins Leukot Essent Fatty Acids*, 80(2–3), 13–136 (2009). https://doi.org/10.1016/j.plefa.2008.12.001.

Sontrop, J., & Campbell, M.K. (2006). Omega-3 polyunsaturated fatty acids and depression: a review of the evidence and a methodological critique. *Preventive Medicine*, 42, 4–13 (2006). https://doi.org/10.1016/j.ypmed.2005.11.005.

Swan, K., & Allen, P.J. (2013). Omega-3 fatty acid for the treatment and remission of Crohn's disease. Journal *Complementary and Integrative Medicine*, 10, 221–228 (2013). https://doi.org/10.1515/jcim-2012-0010.

Takeuchi, T., Fukumoto, Y., Harada, E. (2002). Influence of a dietary n-3 fatty acid deficiency on the cerebral catecholamine contents, EEG and learning ability in rat. *Behavioural Brain Research*, 131(1–2), 193–203 (2002). https://doi.org/10.1016/s0166-4328(01)00392-8.

Tanskanen, A., Hibbeln, J.R., Hintikka, J., Haatainen, K., Honkalampi, K., Viinamäki, H. (2001a). Fish consumption, depression, and suicidality in a general population. *Jama Psychiatry Formely Archives of General Psychiatry*, 58(5), 512–513.

Tanskanen, A., Hibbeln, J.R., Tuomilehto, J., Uutela, A., Haukkala, A., Viinamäki, H. et al. (2001b). Fish consumption and depressive symptoms in the general population in Finland. *Psychiatric Services*, 52(4), 529–31 (2001). https://doi.org/10.1176/appi.ps.52.4.529.

Tiemeier, H., Van, H.R., Hofman, A., Kiliaan, A.J., Breteler, M.M. (2003). Plasma fatty acid composition and depression are associated in the elderly: The Rotterdam Study. *American Journal of Clinical Nutrition*, 78(1), 40–46.

Valenzuela, B.A. (2005). El Salmon: Un Banquete De Salud. *Revista Chilena de Nutrición*. 32 (1), 8–17 (2005). https://doi.org/10.4067/s0717-75182005000100001.

Valenzuela, B.R., Tapia, O.G., González E.M., Valenzuela, B.A. (2011a). Omega-3 fatty acids (EPA and DHA) and its application in diverse clinical situations. *Revista Chilena de Nutrición*, 38 (3), 383–390 (2011). https://doi.org/10.4067/s0717-75182011000300011.

Valenzuela, B.R., Bascuñan G.K., Chamorro M.R., Valenzuela B.A. (2011b). Ácidos grasos omega-3 y cáncer, una alternativa nutricional para su prevención y tratamiento. *Revista Chilena de Nutrición*, 38 (2), 219–226 (2011). https://doi.org/10.4067/s0717-75182011000200012.

Valenzuela, B.R., Bascuñan, G.K., Valenzuela, B.A., Chamorro, M.R. (2009). Omega-3 Fatty Acids, Neurodegenerative and Psychiatric Diseases: A New Preventive and Therapeutic Approach. *Revista Chilena de Nutrición*, 36 (4), 1120–1128 (2009). https://doi.org/10.4067/s0717-75182009000400009.

Vasudevan, A., Yu, Y., Banerjee, S., Woods, J., Farhana, L., Rajendra, S.G., et al. (2014). Omega-3 fatty acid is a potential preventive agent for recurrent colon cancer. *Cancer Prevention Research*, 7(11), 1138–1148 (2014). https://doi.org/10.1158/1940-6207.capr-14-0177.

Wu, J.H., Micha, R., Imamura, F., Pan, A., Biggs, M.L., Ajaz, et al. (2012). Omega-3 fatty acids and incident type 2 diabetes: a systematic review and meta-analysis. *British Journal of Nutrition*, 107 (2), S214–S227 (2012). https://doi.org/10.1017/s0007114512001602.

Zanarini, M.C., & Frankenburg, F.R. (2003). Omega-3 Fatty acid treatment of women with borderline personality disorder: a double-blind, placebo-controlled pilot study. *The American Journal of Psychiatry*, 160(1), 167–169 (2003). https://doi.org/10.1176/appi.ajp.160.1.167.

Zhou, Y., Tian, C., Jia, C. (2012). Association of fish and n-3 fatty acid intake with the risk of type 2 diabetes: a meta-analysis of prospective studies. *British Journal of Nutrition*, 108(3), 408–417 (2012). https://doi.org/10.1017/s0007114512002036.

Chapter 6
Risk Behaviour of Field Workers with Common Infections

Ninfa Ramírez Durán, Arturo G. Rillo and Horacio Sandoval Trujillo

Abstract With the purpose of understanding the social dimension of infectious diseases caused by actinomycetes, we present the complexity of the triad balance of agent-environment-host through the structural genetic model, highlighting reality levels, structural causes, health risks and risk behaviours. We discuss the possibility of recovering the subjective experience of the patient as a member of a community of sufferers, who present levels of extreme poverty, and whose risk behaviours are determined by networks of meanings with which they interpret the experience of being alive, appropriating and reconstructing the traditions and beliefs that have allowed them to live in a world that is adverse to them.

Keywords Actinomycete infection · Ecological model · Mycetoma
Risk behaviours · Structural genetic model

6.1 Introduction

Humanity still has not forgotten the pandemics, endemics and epidemics of infectious diseases that have devastated the world's population, and now it tries to understand the pandemics of chronic diseases. Whether they are infectious, degenerative or chronic, in each case there have been multiple health care actions that are structured in ideologically sustainable social responses.

The socially organized response is aimed at reducing the risks of disease at various levels involving both health education and protection against specific risks of disease with the intention of addressing health problems from different horizons of understanding that explain the transition between health and disease.

Dr. Ninfa Ramírez Durán, Teacher-Researcher, Faculty of Medicine, Autonomous Mexico State University. Email: nramirezd@uaemex.mx.

Dr. Arturo G. Rillo, Teacher-Researcher, Faculty of Medicine; Autonomous Mexico State University.

Dr. Horacio Sandoval Trujillo, Teacher-Researcher, Department of Biological Systems, Autonomous University Metropolitan-Xochimilco.

© Springer International Publishing AG, part of Springer Nature 2018
Ma. L. Marván and E. López-Vázquez (eds.), *Preventing Health and Environmental Risks in Latin America*, The Anthropocene: Politik—Economics—Society—Science 23, https://doi.org/10.1007/978-3-319-73799-7_6

From magical and religious beliefs to the use of telemedicine, medical robotics, genomics or nanotechnology, organized social response to health has gradually determined the transformation of health care models and health systems, as well as the logic of scientific research in the generation of medical knowledge and its technological application.

Whether they are infectious or chronic diseases, they have an ideological and cultural content; so that we can speak today of diseases derived from working conditions of high social impact, tropical diseases, or diseases linked to eating habits and lifestyles.

Infectious diseases and chronic diseases have represented in the cultural history of the disease an ideological stance (Viniegra-Velázquez 2008), identifying the first group as diseases of poverty and underdevelopment; while diseases of the second group, associated with economic development, were recognized as diseases of countries with a high level of technological development.

In the last 25 years of the twentieth century, different social, economic, political, demographic, and access factors of health care services showed the coexistence of infectious diseases and chronic diseases, a phenomenon that was called epidemiological transition (Rillo 1996). The impact of this transition was reflected in the natural history of the disease, prevention levels and the primary health care strategy, facilitating the identification of risk factors, health determinants and risk behaviours to promote the reduction of morbidity and mortality of disease states.

In this scenario, the explanation and understanding of the health-disease process has moved from uni-cause models towards complex multifactorial models that articulate risk factors, health determinants, risk behaviours and health-orientated behaviours.

Risk factors have been defined as those social, economic, biological, behavioural or environmental conditions that are associated with increased susceptibility of the human being to acquire a disease, suffer an injury or gradually decrease and deteriorate the state of health; while the determinants of health are understood as the set of personal, social, economic and environmental factors that determine the health status of individuals or populations (World Health Organization 1998).

In this sense, the notion of risk behaviours is a conceptual category used to refer to specific forms of human behaviour; of which the relation between this behaviour is known and the increase of the subject's susceptibility to develop a specific disease or to deteriorate his state of health; so that health-orientated behaviours include any activity of the subject that is intended to promote, protect or maintain health status. Health-orientated behaviours may or may not be objectively effective in improving or preserving health status, and are generated independently of the perception of the subject's health status (World Health Organization 1998).

Whether they are risk factors or health determinants, the causal conception of the disease continues to be based on the mechanistic and deterministic paradigm, which, in cultural and social terms, limits understanding of the health-disease process to linear causal relationships. On the other hand, the conceptualization of the intentional participation of subject behaviours related to the state of health or illness approximates an approach to the multifactorial complexity of the health-disease process, although it is limited to a teleonomic perspective; so each of these components that are orientated to the promotion of health is embedded both empirically and epistemologically.

Facing the construction of the global health approach at the end of the 20th century and the beginning of the 21st century, which aims to delimit a field of practices and interdisciplinary interventions in health that derive from the investigation of phenomena, links, interactions and actions in the health-disease process that is generated in the international community, priorities have been established to define policies for health care, both for health problems of greater global incidence and for emerging diseases, whether chronic or infectious (Franco-Giraldo/ Álvarez-Dardet 2009; Garret 2013). In this way, relatively low-frequency diseases attracted attention because they were closely linked to problems of endemic poverty (World Health Organization 2010). This is the case with infections caused by actinomycetes, so the following questions arise: What is the significance of infections caused by actinomycetes? What is the impact of actinomycete infections on public health?

In order to think about infections caused by actinomycetes in terms of their impact on population health at local, regional and global levels, it is necessary to consider the advances in clinical and biomedical research. This makes it easier to understand the social dimension of these diseases in a context of complexity that articulates different levels of appropriation of reality carried out by the health sciences.

6.2 Actinomycete Infection

Actinomycetes are microorganisms that are distributed in natural ecosystems participating in the degradation of organic matter. They were originally classified as fungi but are now recognized as true aerobic bacteria. Most actinomycetes are gram positive, filamentous, partially acid-alcohol resistant, branched. Phylogenetically, they have been grouped into seven genera: *actinomycetes*, *mycobacteria*, nitrogen-fixing actinomycetes, *actinoplanes*, *dermatophilus*, *nocardia*, *streptomycetes* and *micromonospores*.

Actinomycetes are useful to humans; for example, *streptomycetes* produce antibiotics such as streptomycin, neomycin, tetracycline, clindamycin, amphotericin B, erythromycin, nystatin, and chloramphenicol; others, such as the actinomycetes of the genus *Frankia* (nitrogen fixers) are of agricultural interest because of their participation in the biofertilization of soils or in the development and growth of plants (Franco-Correa 2009). More recently, the study of halophilic actinomycetes of industrial interest has been deepened by their extremophile ability to grow in saline environments (Ramírez et al. 2006).

Actinomycetes of medical importance are associated with diseases such as tuberculosis, leprosy, granulomatous, actinomycosis, epidermal infections, nocardiosis and mycetoma. Of these diseases, the clinical and biomedical aspects of mycetoma have been gradually deepened, but there are no epidemiological studies showing the evolution and endemic characteristics of the disease in Mexico.

Recently mycetoma has been included in the list of the World Health Organization (WHO) of Neglected Tropical Diseases (van de Sande et al. 2014), which includes diseases caused by protozoa, bacteria, helminths and viruses, which are mainly found in the 149 poorest countries of the world, affecting more than 100 million inhabitants; thus providing an economic cost of billions of dollars to developing economies (World Health Organization 2010).

In this context, the significance of actinomycete infections is that they are diseases that, following Enrique Dussel's line of thought, can be considered endemic diseases attributed to marginalization, exclusion and social segregation. This is why the study is focuses solely on mycetoma for the analysis of risk behaviours in actinomycete infections. This disease is representative of the social, economic and cultural implications of infectious diseases caused by actinomycetes in the context of the conception of global health, showing through it the ideological and political burden of the health-disease process that affects countries located in the tropical and subtropical strip of the world.

6.3 Conceptualization of Mycetoma

Mycetoma is a very old disease. In the *Atharva Veda* it is called *padvalmicum* (meaning 'anthill of the foot'); Its presence is also reported in the pre-Hispanic culture of Tlatilco (Mansilla-Lory/Contreras-López 2009). In 1842, John Gill recognized the disease in tropical regions of India, and in 1884 the first cases were reported in Africa (Lynch 1964). From 1909 cases in Latin America are noted, and in 1912 the first case in Mexico was reported by Cicero (López-Martínez et al. 2013). The history of the discovery of mycetoma in countries with a colonial heritage exposes the marginality of the disease and contributes to its configuration as a silent disease, because those who are at risk or acquire mycetoma lack a political voice.

This silence is reflected even in the disagreement among the experts in naming it nosologically as a syndrome, an infection or a mycosis, which implies identifying different risk behaviours and prevention measures. Therefore, the starting point is to understand the nature of mycetoma.

A syndrome is a set of symptoms and signs that define an altered function and that are related to each other by means of some anatomical, physiological or biochemical trait associated with multiple diagnosable causes (Jablonsky 1992).

Considering the above, mycetoma is a clinical syndrome with chronic evolution, characterized by being a granulomatous, deforming and suppurative disease, located in the skin, aponeurosis, subcutaneous tissue and bone. A pathogenic process involving a variety of actinomycetes and aerobic eumycetes underlies the lesions.

It is at this point that the differentiation of mycetoma is established as a bacterial infection and a mycosis, thus delimiting the understanding of the natural history of the disease as well as its determinants of health and risk behaviours. The mycetoma produced by various types of fungi is eumycetoma, a deep mycosis produced by *Madurella mycetomatis* in 70% of cases and occurring predominantly in Africa and Asia (Queiroz-Telles et al. 2011).

The type of mycetoma produced by various species of aerobic actinomycetes and predominant in Latin America is actinomycetoma, a chronic infectious granulomatous disease affecting skin, subcutaneous tissue, fascia and bone. It may spread through the thoracic, abdominal, or other regions of the body. It is usually accompanied by secondary infections, is destructive and invades the bones from very early stages of infection. The inflammation observed in actinomycetoma is progressive, with increased volume and deformation of the area where it is located. It is generally painless and manifests itself as abscesses, ulcers and fistulas, from which drains a serous fluid with 'grains', containing the etiological agent (Licón-Trillo et al. 2003).

To improve understanding of the risk behaviours associated with increased susceptibility of the subject to be infected by actinomycetes and develop actinomycetoma, it is necessary to consider the etiological causal agent and its interrelation with the environment and the human being. Therefore we now proceed to the analysis of the agent-environment-host relationship from the ecological model of the health-disease process.

6.4   Ecological Understanding of Actinomycetoma

The ecological model of the health-disease process is determined by the homeostatic interaction between two ecological domains: the internal environment of the human subject and the multiple environments where he lives. So the dialectic of the ecological triad is the nodal point in understanding the subject's transit through the health-illness continuum, by demonstrating the dynamics of the triadic relationship of the agent-environment-host.

6.4.1   Agent

More than 33 species of bacteria and fungi are the etiological agents of mycetoma; in Latin America, 60% of cases are due to bacteria and 40% to fungi. Globally, the etiologic agent of actinomycetoma comprises the genera: *Nocardia*, *Actinomadura*, *Streptomyces*, *Rhodococcus* and *Nocardiopsis*. Regional distribution depends on climatic and ecological factors.

In Mexico, the main causative agent is *Nocardia brasiliensis* (85%), followed by *Actinomadura madurae* (10%), *Streptomyces somaliensis*, *Actinomadura pelletieri*,

Nocardia asteroides and *Nocardia otitidiscaviarum* (Bonifaz et al. 2007; Serrano/ Sandoval 2003).

The reservoir of the causative agents of actinomycetoma is the soil. Other ecological niches include plants and cow manure. The vectors described include insects, snakes and the yellow wasp. The incubation period is from 3 months to 9 years. The pathogenicity is low and the virulence of the strains is greater in the exponential phase of growth, although not all are pathogenic (Salinas Carmona et al. 2002). The characteristics that determine the pathogenicity of actinomycete are unknown, but it has the ability to evade host defences through adaptations such as thickening of the cell wall, melanin production and the production of substances with antifungal and/or antibacterial activity (Iffat/Abid 2011).

The presence of actinomycete is the basic, necessary but not sufficient cause for initiating the disease process in the susceptible human host under favourable environmental and social conditions. The characteristics described help to explain the chronicity of the disease, the economic impact of the pharmacological treatment of the disease, as well as the social impact attributed to the permanent sequelae that incapacitate the patient, making it harder to carry out his activities of daily life.

6.4.2 Host

In the host, the predisposing causes are set up in terms of susceptibility or resistance to an agent that, when in effective contact with it and in favourable conditions, initiates a disease process or only runs the risk of suffering it. The reports of cases of mycetoma by actinomycetes have increased in Mexico, so the question arises: Are there modifications to the characteristics of populations at risk of contracting this disease?

Actinomycetoma affects all age groups, predominating in people aged between 20 and 40 years. It is more common in men than in women, in a ratio of 4:1 worldwide and 3:1 in Mexico, where 57.32% of cases are concentrated between 16 and 40 years of age (López-Martínez et al. 2013).

The earliest reports of mycetoma indicate that it affected peasants and farmers and, later, day labourers, gardeners and herders. In Mexico, the predominant occupation in 58.41% of cases is agricultural labourers in fields, followed by housewives (21.79%), labourers (5.74%), students (4.6%), bricklayers (3.13%); employees (2.02%), traders (1.53%), cleaning workers (1.28%), drivers (0.83%), professionals (0.32%), and other unspecified work activities (0.89%) (López-Martínez et al. 2013).

This frequency is consistent with that reported in countries in Asia, Africa, Central America and South America; this is a disease that develops in rural areas where the poverty conditions of the population are linked to poor hygiene, low socio-economic status, inadequate nutritional status and inadequate work practices; for example, day labourers habitually carry sugar cane, straw, wood or agricultural products, generally on their back (Serrano/Sandoval 2003; Welsh et al. 2008).

All these risk factors are closely linked to the susceptibility-resistance binomial determined by the effectiveness of the patient's immune response. In the case of actinomycetoma, innate immunity eliminates the bacterium when invading the tissues, which is why many infected people do not become ill. In infection caused by *Nocardia brasiliensis*, antibodies of the IgM type have an inhibitory and protective effect against infection in the first month of contamination (Salinas-Carmona/ Pérez-Rivera 2004). IgM antibodies increase over time, but do not have a protective effect on experimental infections (Welsh et al. 2008).

In *Nocardia brasiliensis* infections, there is an intense response characterized by the appearance of an inflammatory infiltrate of polymorphonuclear neutrophils that forms microabscess in lesions after several years of evolution. The inflammatory response is associated with depression of cell-mediated immunity, facilitating the pathogenesis and chronicity of the disease; it has also been linked to the failure of conventional treatment.

The most frequent topography of the disease is the lower limb (75%), of which the foot is the most affected (44%); followed by the trunk in the anterior aspect of the thorax, deltoid region and lumbar region (10%) and upper arms and forearm (10%). Multiple conditions involving feet and legs have been reported in 4% of cases; back and neck; thorax and arm; arm and elbow, elbow and foot (Rodríguez Acar et al. 2002).

In endemic areas, the head, cheek, neck, back of the thorax, abdomen, perineum, thigh, knee and leg may also be affected.

Perianal actinomycetoma is associated with personal hygiene during defecation, since most patients report the habit of cleaning the anus with leaves of plants, causing small lesions in which the microorganism is inoculated. In addition, in homosexual patients it can be associated with sexual practices of risk, such as the introduction of objects in the anus like carrots, cucumbers and radishes, since they are tubers that grow underground and, therefore, different actinomycetes and fungi may be present (Chávez et al. 2002).

The clinical diagnosis is usually made late, since the time between the manifestation of the disease and the first medical visit is, on average, five years. This significantly alters the evolution and prognosis of actinomycetoma, since, in addition, to being determined by the causal agent, the clinical location, the spread and extension of the disease (Bonifaz et al. 2007), the accessibility of health services and the perception of the disease are involved. Consequently, it is constituted as a public health problem not because of its magnitude at population level, but because of the difficulty of controlling it and the recurrent infection.

The extent and prognosis of the disease is associated with the immunological state of the patient, because in immunocompromised patients, such as those affected by AIDS, cancer, lung disease, or immunosuppressive treatment, skin infections occur and it rapidly spreads into the organism (Hogade et al. 2011). This exposes the presence of mixed factors, since in the development of the disease, the influence of environmental and behavioural factors is performed by acting on an endogenous predisposition of genetic origin (Salleras 1992).

Currently, there has been an increase in the number of cases in travellers and immigrants from endemic areas (Hernández-Bel et al. 2010), and there have even been reports of people developing mycetoma after suffering a car accident (Welsh et al. 2011).

6.4.3 Environment

The factors which influence the way in which human beings develop can be divided into different categories. Endogenous factors (those which have an internal cause) are associated with phylogenetic (evolutionary) and ontogenetic (individually developed) characteristics; behavioural factors can be associated with risky practices and activities; and environmental factors are biological, physical or chemical in nature (Salleras 1992).

In the conception of the ecological triad, contributing causes and triggers are located in the environment, so it is defined as the set of external conditions and influences provided by nature but also encompassing those that man himself has created as a group. Therefore the patterns of infection are expected to be modified in relation to the changes caused by environmental risk factors. Does this also happen in actinomycetoma? The environmental factors contributed by nature are physical and biological. In the case of actinomycetoma, the distribution of the causal agent has wide variations that depend on climatic factors, mainly of the precipitation pluvial.

Actinomycetoma is endemic in many tropical and subtropical regions. The world distribution of this disease is located in the Tropic of Cancer and the Tropic of Capricorn, the latitude between 15° South and 30° North, which is also known as the 'mycetoma belt' and is comprised of the following countries that present the highest prevalence of the disease: Sudan, Somalia, India, Yemen, Mexico, Venezuela, Colombia, Argentina and Brazil.

It is also present at an altitude of 1,500–2,000 metres above sea level, and in very dry areas where the flora with greater presence and distribution are several species of Acacia, as well as thorny trees and shrubs (Fahal 2006).

In African countries located in the 'mycetoma belt', where the most frequent causative agent is *Madurella mycetomatis*, the areas of highest mycetoma prevalence are relatively arid areas; but in Mexico it occurs in areas with high precipitation, since *Nocardia brasiliensis* is found in humid climates, with rainfall of 600–2,000 mm per year, affecting all the states of the republic except Baja California, Baja California Sur, Sonora, Chihuahua, Durango, Aguascalientes, Tlaxcala and Quintana Roo (López-Martínez et al. 1992, 2013).

6.5   Complexity of Triadic Balance in the Genesis of Actinomycetoma

The characteristics of the elements that make up the ecological triad have been described and it is recognized that actinomycetoma is not an infection that is transmitted from animal to human or from human to human; Therefore, the question arises: How do these elements get together to produce actinomycetoma?

The equilibrium of the agent-environment-host relationship begins upon contacting the agent and the host in a favourable environment for producing the disease.

The main route of infection is not yet clearly identified; but because actinomycetes are present in soil, it is accepted that the agent enters the organism when an injury caused by, for example, thorns, wood chips, nails, stones or agricultural tools, propitiates an entry point. Pre-existing abrasive lesions may also enable actinomycetes to enter the body (Agarwal et al. 2014). Although only half of the patients can remember a history of previous pre-disease injury, this hypothesis is supported by the fact that cases are common among field workers or people walking barefoot. In addition, the main areas of the body affected are feet and hands.

Subsequently, the patient may feel pain or discomfort at the site of the injury; developing abscesses, granulomas and subcutaneous painful nodules that develop and diffuse in the affected area, slowly and producing fistulas. The nodules increase in size and quantity, with accompanying drainage of serous fluid, blood or pus.

The lesions will cause deformity of the affected area and ankylosis. Although actinomycetoma is rarely fatal, it is a disabling disease. Neglect of the disease is associated with the need for amputation of the affected limb. In chronic cases, lymphatic obstruction and fibrosis can occur, causing lymphedema, as well as varicose veins. Prolonged treatment may produce toxicity from the drugs used.

6.6   Interpretation of Actinomycetoma Through the Structural Genetic Model

The structural genetic causal model was first proposed by Sagatovski/Antipov (1966) to analyse the relationship between social structure and the health-disease process (Yazlle-Rocha 1978). Subsequently, it was modified by Laurell de Leal/Blanco-Gil (1975) to express the historical determination of social processes in the genesis of health and disease. Finally, it was adapted by García de Alba/Salcedo-Rocha (2013), establishing synchronic and diachronic articulations between different levels of reality and the structural processes involved in the development of the health-disease process.

In this model, the phenomenon that precedes another and that is necessary for its appearance is called genetic cause, whereas the set of necessary and sufficient interacting conditions for any phenomenon to occur is called a structural cause;

so that when applying this model to move from the identification of risk factors and health determinants towards the understanding of risk behaviours in infectious diseases by actinomycetes, the social structure is articulated with the agent-environment-host triad in the genesis of the disease.

The social structure in the structural genetic model is interpreted in terms of environmental factors that are provided by man as a social group, defined as socio-economic and cultural factors. These include dominant modes of production, occupation, education, health policies, religious norms, cultural patterns, temporary and permanent distribution of the population (agglomerations, overcrowding, migration), composition of the population, roads, and environmental sanitation.

In this context, the interpretation of actinomycetoma is performed in three levels of reality: general, particular and individual. The general level is linked to the socio-economic processes determined by the mode of production of the social and economic formation in which the actinomycetoma develops and that corresponds to developing countries that have adopted neoliberalism as a model of development.

In recent decades, the epidemiological profile of mycetoma in general and actinomycetoma in particular has been verified and attributed to the geo-ecological change produced by the development of industrial activities and the transformation of agricultural practices.

The particular level refers to the interpretation of the disease that is carried out through the social group to which the patients who have fallen ill from actinomycetoma belong. As García de Alba/Salcedo-Rocha (2013) indicate, it represents an element that vertically articulates the level of general reality with the individual, and considers the development of social and cultural conditions of a group nature related to work and consumption levels. It has been previously reported that actinomycetoma mainly affects rural workers and rural housewives, which constitutes diachronic social and cultural processes that increase the risk of actinomycetoma.

The level of consumption articulates synchronous processes related to the satisfaction of basic needs (simple consumption), among which food, housing, clothing and education stand out. On the other hand, it also includes the satisfaction of social needs (expanded consumption), in addition to recreation and habits harmful to health. In the case of actinomycetoma, the simple consumption of the population is characterized by inadequate clothing for field activities (mainly lack of footwear), inadequate feeding and low educational level; while expanded consumption refers to inappropriate use of leisure time, alcoholism and insufficient health education.

Even though farmers are at greater risk of injury from actinomycete, trauma is not the only factor in the development of the disease, since the viability of the microorganisms and the local conditions of the wound are also involved in the evolution of the infection. Thus, actinomycetes that have the possibility of colonizing superficial wounds in the short term will develop infection, and will be linked with the bacterial flora of the affected anatomic region.

The individual level refers to the state of health and illness of the particular subject. The state of health of the individual is linked to the perception that he has of health; this implies feeling good, which sometimes conflicts with the actual state of health, considering above all the conception of normality held by the individual and the prolonged period of the actinomycete infection.

The disease state expresses the cultural and ideological burden of the disease, because even when an inflammatory lesion is perceived in some extremity, the absence of pain enables the patient to perform his daily activities; it will not be until the actinomycetoma generates discomfort, pain or incapacity that the patient attends medical consultation. The perception of the state of health and the cultural load of the disease determine different risk factors, such as the decrease in the use of protective clothing, mainly footwear, due to the warmer conditions of the climate, but especially to the endemic conditions of poverty, which are also accompanied by a deteriorated general health status, malnutrition or diabetes mellitus (Agarwal et al. 2014).

Table 6.1 shows the temporo-spatial aggregate matrix to derive risk behaviours in actinomycete infection, exemplified by actinomycetoma, performed through the application of the structural genetic model to interpret the health-disease process. These types of matrices link three axes of coordinates: axis of aggregation of the target population, time axis with reference to the historical nature of the vulnerability of the population; and spatial axis that recovers the inequality in the geographical distribution of the disease (Pérez 2003).

In the axis of aggregation of the target population, it was considered that it affects a specific nucleus of the population that is represented mainly by field workers, which generates the following questions: What is the impact of the changes experienced in agricultural areas as a result of agricultural policies and technological innovation, after 100 years of having identified mycetoma in Mexico? What is the impact of socio-demographic changes on the incidence of the disease?

Table 6.1 Temporal-spatial aggregate matrix to articulate reality levels with risk behaviours for actinomycetoma (own elaboration)

Level of reality	Structural causes		Health risks	Risk behaviours
	Diachronic processes	Synchronous processes		
General	Neoliberal economic model	Rural areas of tropical and subtropical climate	Low socio-economic level	Agricultural practices
Particular	Farm workers, housewives	Lack of footwear Inadequate feeding Housing with poor hygiene Low educational level Misuse of free time Alcoholism	Do not use protective equipment	Walking barefoot Loading agricultural products in the back Performing labour activities lying on the ground
Individual	Perceived and actual health status	Cultural and ideological burden of the disease state	Inadequate work practices	Inappropriate use of footwear

Source The authors

In the temporal axis, the historical nature of the vulnerability of the population that focuses on the economically active population stands out. Ever since the first cases reported in 1912, it continued to mainly affect people aged between 20 and 40 years old. Although there are currently reports of cases in children and people over 60 years, this does not mean that the population window of those affected by actinomycetoma is increased. On the contrary, it only represents the incorporation of new strategies for medical diagnosis and an increase in access to health services, as the cases are reported by third-level hospitals, such as the Pascua Dermatological Centre, the National Medical Centre of the Mexican Social Security Institute and even the General Hospital of Mexico, where mycetoma cases throughout the country are concentrated.

In the spatial axis from which the inequality in the geographical distribution of the disease is recovered, the analysis was located in the magnitude of the disease as a public health problem. The magnitude of actinomycetoma in epidemiological and public health terms in Mexico is not known with certainty. There are studies that report isolated cases but, when comparing them with those reported by López-Martinez et al. (2013), it is verified that each study starts from different measurements, which makes definitive comparisons impossible. However, in general terms, an average of 73 new cases per year are reported, without differentiating the number of cases already existing in the same period to distinguish the prevalence and incidence of new cases of the disease.

6.7   Epilogue

Over one hundred years have passed since the first description of mycetoma in Mexico. Undoubtedly, Mexico in 1912 is not the same as in 2017. However, 105 years of history and social, economic and political development in all areas, mainly in health, have not contributed to the fact that the patients described by Cicero are different from those described by the group of López-Martínez et al. (2013). Prerevolutionary poverty has roots so deep that it continues to plague 45.5% of the Mexican population. In this sense, conclusions can't be drawn; on the contrary, there have been ways of conducting studies, analyses, reflections, interpretations, projections, for which many questions remain unanswered, among which the following stand out:

- What has been the effect of the use of antibiotics in livestock on the prevalence of etiological agents, and that of etiological agents, in turn, on the incidence of the disease?
- What is the prediction of the incidence and distribution of the disease in relation to climate change and the redistribution of specific ecosystem ranges?
- What public health implications are derived from these changes?
- What are the indicators of health economics of actinomycete infectious diseases?

The analysis made and discussed above delimits a horizon of understanding by pointing out some scales to explore the above issues. For example, it is a fact that the historical background of mycetoma study supports the concept of it being a disease of deep colonial heritage that represents the neglect of the population that has been marginalized, excluded and segregated in contemporary Mexico.

Actinomycetoma is not an occasional disease, because together with eumycetoma it is widely distributed with a prevalence that is increasing in the 'mycetoma belt', representing a very wide area around the world.

Because it mainly affects people who work in fields or are frequently exposed to contact with the soil, it is an occupational disease.

Since 1970 it has been recognized that patients lead a desperately resigned life, as the lack of effective medical services and pharmacological treatment faces them with the inevitable and irreparable loss of the limbs affected, facing them with a bleak future. This community presents levels of existence located in poverty. Their risk behaviours are determined by the traditions and beliefs they use to interpret the experience of living in a world that is unfavourable and adverse to them.

Understanding actinomycetoma to delineate horizons that define risk behaviours for actinomycete infectious diseases has made it possible to rehabilitate a conceptual framework, articulating causal models that have been used to understand the relationship between health and disease in the context of social processes. This means that risk behaviour can be considered an organized social response to dealing with adverse living conditions. In this conception, response strategies should include the development of life skills and the creation of healthy environments that support preventive and health education actions.

Finally, as Viniegra-Velázquez (2008: 527) points out: 'caring for health is only a component of something superior: caring for life … Caring for life is to seek dignity, satisfaction, serenity, fraternity; is to seek a pluralistic, inclusive, egalitarian, fair and supportive social environment'.

References

Agarwal, P., Goel, K., Relhan, V., Garg, V.J. (2014). Mycetoma: a review. *Tropical Clinics of Dermatology,* 1(2), 1–11.

Bonifaz, A., Flores, P., Saul, A., Carrasco-Gerard, E., Ponce, R.M. (2007). Treatment of Actinomycetoma due Nocardia spp. with amoxicilin-clavulanato. *British Journal of Dermatology,* 156, 308. https://doi.org/10.1111/j.1365-2133.2006.07557.x.

Chávez, G., Estrada, R., Bonifaz, A. (2002). Perianal actinomycetoma experience of 20 cases. *International Journal of Dermatology,* 41, 491–493. https://doi.org/10.1046/j.1365-4362.2002. 01550.x.

Fahal, A. (2006). *Mycetoma: clinico-pathological monograph.* Khartoum: University Press.

Franco-Correa, M. (2009). Utilización de los actinomicetos en procesos de biofertilización. *Revista Peruana de Biología,* 16, 239–242. https://doi.org/10.15381/rpb.v16i2.213.

Franco-Giraldo, A. & Álvarez-Dardet, C. (2009). Salud pública global: un desafío a los límites de la salud internacional a propósito de la epidemia de influenza humana A. *Revista Panamericana de Salud Pública,* 25, 540–547. https://doi.org/10.1590/s1020-49892009000600011.

García de Alba, G.J.E., & Salcedo-Rocha, A.L. (2013). Historia natural de la enfermedad. En: R. Martínez y Martínez, *La salud del niño y del adolescente* (pp. 14–34). México: Manual Moderno.

Garret, L. (2013). *Existential challenges to global health.* New York: New York University.

Hernández-Bel, P., Mayorga, J., Pérez, M.E. (2010). Actinomicetoma por nocardia brasiliensis. *Anales de Pediatría (Barcelona),* 73, 213–214. http://www.analesdepediatria.org/es/actinomicetoma-por-nocardia-brasiliensis/articulo/S1695403310002018/.

Hogade, S., Metgud, S.C., Swoorooparani. (2011). Actinomycetes mycetoma. *Journal of Laboratory Physicians,* 3, 43–45. https://doi.org/10.4103/0974-2727.78564.

Iffat, H., Abid, K. (2011). Mycetoma revisited. *Our Dermatol on line,* 2(3), 147–150.

Jablonsky, S. (1992). Syndrome—a changing concept. *Bulletin of Medical Library Association,* 80 (4), 323–327.

Laurell de Leal, A.C. & Blanco-Gil, J. (1975). Morbilidad, ambiente y organización social. *Salud Pública de México,* 17(4), 471–478.

Licón-Trillo, A., Castro-Corona, M.A., Salinas-Carmona, M.C. (2003). Immunogenicity and biophysical properties of a Nocardia brasiliensis protease involved in pathogenesis of mycetoma. *FEMS Immunology & Medical Microbiology,* 37, 37–44. https://doi.org/10.1016/s0928-8244(03)00102-0.

López-Martínez, R., Méndez-Tovar, L.J., Bonifaz, A., Arenas, R., Mayorga, J., Welsh, O., et al. (2013). Actualización de la epidemiología del micetoma en México. Revisión de 3,933 casos. *Gaceta Médica de México,* 149(5), 586–592.

López-Martínez, R., Méndez Tovar, L.J., Lavalle, P., Welsh, O., Saul, A., Macotela Ruiz, E. (1992). Epidemiología del micetoma en México: estudio de 2105 casos. *Gaceta Médica de México,* 128(4), 474–481.

Lynch, J.B. (1964). Mycetoma in the Sudan. *Annals of the Royal College of Surgeons of England,* 35(6), 319–340.

Mansilla-Lory, J., Contreras-López, E.A. (2009). Micetoma en México prehispánico. Estudio en la colección esquelética de la cultura de Tlatilco. *Revista Médica del Instituto Mexicano del Seguro Social,* 47(3), 237–242.

Pérez, R. (2003). Las matrices de agregados, temporales y espaciales de la promoción de la salud. *Revista PsicologíaCientífica.com,* 5(6), http://www.psicologiacientifica.com/matrices-promocion-de-salud/ Accessed: 17 May 2017.

Queiroz-Telles, F., Nucci, M., Colombo, A.L., Tobon, A., Restrepo, A. (2011). Mycoses of implantation in Latin America: an overview of epidemiology, clinical manifestations, diagnosis and treatment. *Medical Mycology,* 49, 225–236. https://doi.org/10.3109/13693786.2010.539631.

Ramírez, N., Serrano, J.A., Sandoval, H. (2006). Microorganismos extremófilos. Actinomicetos halófilos en México. *Revista Mexicana de Ciencias Farmacéuticas,* 37(3), 56–71.

Rillo, A.G. (1996). Actitudes culturales y salud. *Convergencia,* 4(12–13), 119–141.

Rodríguez Acar, M., Alvarado, D.A., Padilla, M.C., Ramos-Garibay, A. (2002). Micetoma de inoculación múltiple por Nocardia brasiliensis. Reporte de un caso. *Revista del Centro Dermatológico Pascua,* 11(3), 126–130.

Sagatovsky, V.N. & Antipov, I.G. (1966). Acerca de la correlación de los conceptos 'causa', 'condición', 'etiología' y 'patogénesis'. *Vestnick Rossiskoi Akademii Meditsinskikh Nauk SSSR,* 21(1), 34–40.

Salleras, L. (1992). La medicina clínica preventiva: el futuro de la prevención. *Medicina Clínica (Barcelona),* 102(1), 5–12.

Salinas Carmona, M.C., Casto Corona, M.A., Licón Trillo, A. (2002). Mecanismos de patogenicidad de Nocardia brasiliensis y Nocardia asteroides. *Medicina Universitaria,* 4(15), 97–101.

Salinas-Carmona, M.C., Pérez-Rivera, I. (2004). Humoral immunity through immunoglobulin M protects mice from an experimental actinomycetoma infection by Nocardia brasiliensis. *Infection and Immunity,* 72, 5597–5604. https://doi.org/10.1128/iai.72.10.5597-5604.2004.

Serrano, J.A., Sandoval, A.H. (2003). El Micetoma. Revisión. *Revista de la Sociedad Venezolana de Microbiología,* 23(1), 70–79.

van de Sande, W.W.J., Maghoub, E.S., Fahal, A.H., Goodfellow, M., Welsh, O., Zijlstra, E. (2014). The mycetoma knowledge gap: identification of research priorities. *PLoS Neglected Tropical Disease*, 8, e2667. https://doi.org/10.1371/journal.pntd.0002667.

Viniegra-Velázquez, L. (2008). La historia cultural de la enfermedad. *Revista de Investigación Clínica,* 60(6), 527–544.

Yazlle Rocha, J.S. (1978). Salud enfermedad y estructura social. *Salud Problema UAM-X,* 3, 5–16.

Welsh, O., Morales-Toquero, A., Cabrera-Vera, L., Vázquez-Martínez, O., Gómez-Flores, M., Ocampo-Candiani, J. (2011). Actinomycetoma of the scalp after a car accident. *International Journal of Dermatology, 50,* 854–857. https://doi.org/10.1111/j.1365-4632.2011.04874.x.

Welsh, O, Salinas-Carmona, M.C., Vera-Cabrera, L. (2008). Los Micetomas en México. *Gaceta Médica de México,* 144(2), 125–127.

World Health Organization. (1998). *Health promotion. Glossary.* Geneva: World Health Organization Press.

World Health Organization. (2010). *Working to overcome the global impact of neglected tropical diseases.* France: World Health Organization Press.

Chapter 7
Sedentary Behaviour in Adolescents: A Risky Conduct for Health

Roseane de Fátima Guimarães Czelusniak

Abstract This chapter is about sedentary behaviour in adolescents, as well as suggestions to induce the adoption of a healthy lifestyle in this phase of life. Sedentary behaviour among adolescents is a public health problem, since it leads to an increase in obesity and morbidity in adult age. A significant association has been shown between the time spent in front of a screen and obesity in adolescents, and, besides that, sedentary behaviour also leads to a decrease of physical activity levels and to unhealthy eating habits. According to this evidence, we can conclude that regular physical activity, healthy eating habits and low sedentary behaviour can prevent non-communicable chronic diseases in adult age. These findings could be helpful for public health professionals seeking to formulate strategies to improve the lifestyle of adolescents, using both comprehensive campaigns and intervention programmes for this population.

Keywords Sedentary behaviour · Physical activity · Eating habits
Obesity · Adolescents · Healthy lifestyle · Intervention

7.1 Introduction

Childhood and adolescence are phases of extreme importance for the development of a healthy lifestyle, since habits acquired in these phases of life are perpetuated into adulthood (Souza et al. 2011). These days, adolescents are much more exposed to behaviours that pose health risks, such as sedentary activities, low physical activity levels, and bad eating habits (Hallal et al. 2006; Ekelund et al. 2007; Freedman et al. 2012; Guimarães et al. 2013a, b; 2014).

Different authors demonstrate the presence of several cardio metabolic risk factors during childhood and adolescence (Ekelund et al. 2007; Martinez Gómez

Dr. Roseane de Fátima Guimarães Czelusniak, Teacher- Researcher, Head of the Department of Physical Activity, Salesian University Center of Sao Paulo (UNISAL). Email: roseanefguimaraes@gmail.com.

© Springer International Publishing AG, part of Springer Nature 2018   87
Ma. L. Marván and E. López-Vázquez (eds.), *Preventing Health and Environmental Risks in Latin America*, The Anthropocene: Politik—Economics—Society—Science 23, https://doi.org/10.1007/978-3-319-73799-7_7

et al. 2010; Silva et al. 2010; May et al. 2012; Freedman et al. 2012; Bozza 2013; Guimarães et al. 2013a, b; 2014).

Andersen et al. (2006) state that in the paediatric ages, it is very important to decrease sedentary behaviours and increase physical activity, and intervention programmes such as the practice of sports in school can offer opportunities for all students to be physically active, as it is known that lack of physical activity significantly contributes to excess body weight (Strong et al. 2005).

7.2 Lack of Physical Activity and Sedentary Behaviour in Adolescence

Physical activity consists of any voluntary human movement which results in energy expenditure above resting levels, characterised by daily activities and physical exercises (Federal Council of Physical Education—CONFEF 2010).

Hallal et al. (2006) state that physical activity has several health benefits for adolescents, both short and long term, including a higher probability of continuing to be physically active into adulthood (Azevedo et al. 2007). Physical activity of any intensity in adolescence, as long as done regularly, has benefits over control of cardiovascular risk factors, such as obesity, dyslipidaemia, diabetes mellitus, smoking, and hypertension, as well as over aerobic functional capacity, osteoporosis prevention and the psychological health of those who practise physical exercise (Brazilian Society of Cardiology—SBC 2005).

The World Health Organization (WHO 2002) estimates that 22% of cardiac diseases and 10–16% of type II diabetes cases and breast, colon and rectal cancer cases could be avoided with the practice of physical activity within specific recommendations for each age range.

In addition to these benefits, it is known that good physical activity levels are a determining factor in bodily characteristics in adolescents. Thus, lack of physical activities commingled with other factors, such as inadequate eating habits and sedentary behaviours, lead to bodily alterations, and consequently, to increased body weight and/or obesity (Dias et al. 2014).

Sedentary behaviour is characterised by the inexistence of, or decrease in, physical activity in leisure time, at work, on commute, and in household activities (Malta et al. 2009). It is determined by an increase in activities which do not generate energy expenditure substantially above a resting state, e.g. sitting, watching television, playing video games, being in front of a computer, among others (Pate et al. 2008).

The American Academy of Paediatrics (AAP 2001) characterises sedentary behaviour in adolescents as over two hours daily spent on communication media and electronic entertainment.

7.3 How Sedentary Behaviour in Adolescents Is Detected

Nowadays, with advances in technology and modernisation, the adoption of a sedentary lifestyle is on the rise; in light of this, researchers are turning their focus to the analysis of prevalence as well as of factors associated with sedentary behaviours in adolescents.

To detect such sedentary behaviours, there are several instruments, subjective (interviews, questionnaires, records) and objective (accelerometers). The subjective instruments focus on screen time. Among these is a comprehensive questionnaire of adolescent sedentary behaviours, which involves educational, commuting, leisure, cultural, and social aspects (Hardy et al. 2007), and which has also been validated for Brazilian adolescents (Guimarães et al. 2012).

Among the objective instruments, accelerometers are the most accurate to estimate time spent with sedentary activities, since the value obtained (counts) is turned into the intensity of the activity, classifying it as sedentary, light, moderate and vigorous (Rosenberger et al. 2013). Studies involving adolescents, such as the one by Sardinha et al. (2008), used this valid and objective method to evaluate sedentary behaviour, together with physical activity levels, in order to assess the association of such behaviours with health risk factors.

7.4 Sedentary Behaviour and Health Risk Factors in Adolescents

In a previous study by Guimarães et al. (2013a, b), conducted with 572 adolescents from Curitiba (Brazil), a significant positive association between total cholesterol and increased sedentary screen time was established, i.e., the subjects who related spending more time in front of the television, video games, computers, among other electronic devices, had a higher chance of having total cholesterol above desirable levels when compared to the other adolescents. In addition to total cholesterol, a significant positive association between LDL-c and increased screen time was observed.

In the study by Dias et al. (2014), an association was found between sedentary behaviour and other behaviours considered to be of risk, such as consumption of alcohol and tobacco, as well as bad eating habits.

In other studies, leisure time in front of television has been shown to be associated with cardio metabolic risk factors, independently of time spent with physical activities; the predominant factors are overweight status and obesity, which can continue on into adulthood (Ekelund et al. 2006; Sardinha et al. 2008; Hardy et al. 2010). Thus, it is observed that sedentary lifestyle and the practice of physical activity are different types of behaviours, and that such behaviours are potentially associated with different metabolic changes related to health.

According to Dias et al. (2014), high amounts of screen time influence adolescents' food choices, as foods advertised on media are usually not healthy and possess high calorie values, contributing to an increase in obesity among this population.

7.5   Eating Habits in Adolescence

Eating habits are what people are accustomed to eating and are influenced by several factors, among which are cultural, economic, psychological, and physiological factors (Lemos/Dallacosta 2005). They develop in childhood, from the age of six months old, a time when children, in general, initiate complementary feeding. In this phase children tend to accept any kind of food given by the people to whom they are close and with whom there is affection, transforming these feeding experiences into habits (Mintz 2001).

According to Lemos/Dallacosta (2005), in the past the attitude towards eating was different from today's, since the main idea was the importance of eating to live, and today people are encouraged to live to eat. In general, people tend to adopt eating habits that are in harmony with the social group to which they belong. Family also has a great influence on the individual's diet, which, once consolidated, will be difficult to change.

Adolescents can easily change their eating habits in order to conform to a social group, be it a group of friends, peers, or others, as some theories suggest that it is through one's eating habits that an individual integrates into any given group (Mezomo 2002). Caldeira (2000) interviewed a number of adolescents who cited friends, television and family as influencing factors in the formation of eating habits. These were the same results yielded by Lemos/Dallacosta's research (2005).

This influence of television or any other media takes the form of encouraging consumption of frozen foods and fast food, which adolescents who don't have time for a healthy diet love; they come in attractive packaging and are heavily advertised, making such individuals very keen on unhealthy foods which have no nutritional value but are tasty and practical (Santos et al. 2012).

Another influencing factor, mentioned by Fisberg et al. (2002), is the fact that adolescents tend to be 'lazy' and therefore favour foods that are ready for consumption. Hanley et al. (2000) claim that these changes in eating habits, especially for adolescents, can have negative consequences, and that the association between such habits and the occurrence of diseases is being studied widely (Lemos/Dallacosta 2005; Ilias 2006).

Many diseases could be avoided through regular physical exercise and changes in eating habits (Lemos/Dallacosta 2005). According to the findings of Sichieri et al. (2003), adolescents who have bad eating habits have a greatly increased risk of becoming obese adults and developing chronic non-communicable diseases, such as cardiovascular diseases, cancer, diabetes, hypertension, obesity, and dyslipidemias.

7.6 Overweight Status and Obesity in Adolescence

Overweight status and obesity are defined as an excessive accumulation of fat which can pose health threats; both are considered the biggest risk factors for the development of chronic illnesses, such as diabetes, cardiovascular diseases and cancers (WHO 2011).

There are many factors involved in the development of overweight status and obesity in adolescents. Daily physical activity levels and sedentary behaviour have been demonstrated to be directly linked to the prevalence of overweight status and obesity (Jenovesi et al. 2003; Suñe et al. 2007), just like bad eating habits—including being empty stomached—skipping meals and the excessive fat present in the foods eaten.

According to Guedes et al. (2006), many authors are looking for answers for the increased prevalence of health risk factors in adolescents, not just through biological indicators, such as body mass index (BMI), waist circumference, and blood pressure, but also through health-damaging behavioural indicators, such as sedentary behaviours, high fat, cholesterol, and sugars intake, and physical activity levels below what is recommended.

Excessive sedentary behaviour, inadequate eating habits, and low physical activity levels, factors which are highly related to the development of overweight status and obesity (Fonseca et al. 1998), may lead to dysfunctional levels of plasmatic lipids-lipoproteins (HDL-C e LDL-C), total cholesterol, triglycerides, blood glucose, and insulin (Martinez-Gómez et al. 2010; Guimarães et al. 2014).

Cases of overweight status and obesity can be prevented and even reversed with a few changes in daily behaviours, improved eating habits, and increased energy expenditure. These preventive measures are important from early childhood to stop health problems developing in childhood and adolescence and reduce the risk of morbidities and premature mortality occurring in adulthood (Mello et al. 2004).

Santos et al. (2012) state that intervention programmes aimed at preventing or reversing cases of unhealthy anthropometric changes must encourage the practice of daily physical activity in order to motivate children and adolescents, together with their parents/legal guardians, to incorporate it into their daily lives.

Eating habits also play an important part in the prevention of these conditions. Mello et al. (2004) state that not having breakfast, eating high-calorie dinners, consuming a limited variety of foods, ingesting high-calorie drinks excessively—i.e., having inadequate eating habits, including those stemming from bad examples set by parents/legal guardians—constitute unhealthy practices that lead to obesity.

The main complications that obesity can bring about in the lives of children and adolescents are the following:

- Of the joints: Greater predisposition towards arthrosis, osteoarthritis, Epiphysiolysis of the upper femur, and genu valgum.
- Cardiovascular complications: Arterial systemic hypertension, cardiac hypertrophy.
- Surgical complications: Increased surgical risk.

- Growth-related complications: Advanced bone age, height increase, and premature menarche.
- Cutaneous complications: Greater predisposition towards ringworm, dermatitis, and pyoderma.
- Endocrine/metabolic complications: Insulin resistance with a greater predisposition towards diabetes, hypertriglyceridemia, and hypercholesterolemia.
- Gastrointestinal complications: Increase in the frequency of biliary lithiasis, hepatic steatosis, and steatohepatitis.
- Neoplastic complications: Higher incidence of cancer of endometrium, breast, gall bladder, colon/rectum, and prostate.
- Respiratory complications: A tendency towards hypoxia due to increase in ventilatory demand, increase in respiratory effort, decrease in muscular efficiency, decrease in functional reserve, microectasia, sleep apnoea, and pickwickian syndrome, infections, and asthma.
- Mortality: Increase in the risk of mortality.
- Psychosocial complications: Social discrimination and isolation, withdrawal from social activities, and difficulty in expressing one's feelings (Mello et al. 2004).

Figure 7.1 shows which aspects are fundamental to changing one's lifestyle and preventing the development of overweight status and obesity in this age range.

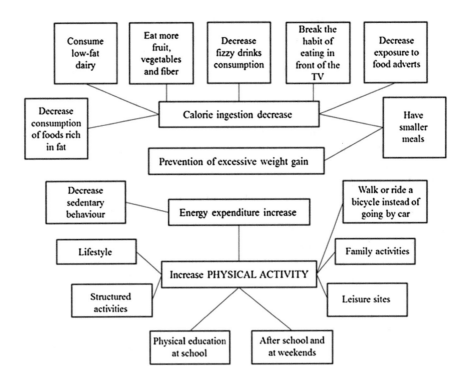

Fig. 7.1 Important aspects for the prevention of overweight status and obesity. *Source* Adapted from Mello, E. D., Luft, V. C., Meyer, F. (2004). Obesidad infantil: como podemos ser eficazes? Jornal de Pediatria, 80(3), 173–82

A meta-analysis study by Young et al. (2007) has shown that interventions with a family conduct component produced more significant effects than alternative treatment groups without monitoring and motivational behaviours by the subject's family, demonstrating that inclusion of one's family in the treatment programme must be encouraged, together with preventive measures with a family focus.

7.7 Dysfunctional Levels of Lipids and Blood Glucose in Adolescence

The Brazilian Cardiology Society (SBC 2001) states that cholesterol, triglycerides, fatty acids, and phospholipids are molecular species of lipids present in plasma, and important from a physiological and clinical standpoint; however, dysfunctional levels of lipids constitute risk factors which create an environment conducive to the emergence and progression of the atherosclerotic process, in this phase in life and in adulthood (SBC 2001).

Blood glucose is the concentration level of glucose in blood plasma. Blood glucose levels go up as a result of carbohydrate-rich diets; carbohydrates turn into glucose after ingestion, to be used as energy by the cells (Watkins 2003). Diabetes occurs through factors which interfere with the role of insulin, a hormone produced by the pancreas, resulting in the increase of blood glucose levels (hyperglycaemia) (Watkins 2003).

Studies such as the one by Martínez-Gómez et al. (2010) show that adolescents with excessive sedentary behaviours have higher levels of systolic blood pressure, triglycerides, and blood glucose. These findings corroborate the findings of Ekelund et al. (2007), which, in research conducted with 1,709 children and adolescents, showed a significant association between time spent watching TV and cardiovascular disease risk factors, among which are elevated levels of systolic and diastolic blood pressure, and dysfunctional levels of triglycerides, blood glucose, and insulin.

The results of this study show that sedentary behaviour, as well as low physical activity levels and inadequate eating habits, can lead to the development of diabetes and dyslipidemias, and, consequently, to cardiovascular diseases in adolescence (Martínez-Gómez et al. 2010).

Data resulting from The Bogalusa Heart Study reveal that the early stages of the development of atherosclerosis are related to serum levels of lipoproteins in childhood, reinforcing the idea that dysfunctional levels in lipid profile in this phase can be predictive of coronary artery disease in adulthood.

Additionally, results from this same study have also shown that excess body weight has significant explanatory power in dysfunctional levels in lipid profile, demonstrating, through multivariate analysis, association between adiposity and low levels of HDL-c as well as elevated levels of triglycerides, which increase more with weight gain than with elevated body weight per se.

According to Gabbay et al. (2003), most obese individuals also suffer from type II diabetes mellitus—approximately 70–90% of children or adolescents with type II diabetes are obese, 38% being morbidly obese. A study conducted by Weiss et al. (2004) found a prevalence of 27% of the obese subjects for type II diabetes.

Since overweight status and obesity influence the development of dyslipidemias and hyperglycaemia—factors which can lead to the emergence of cardiovascular diseases in adulthood—this state must be avoided or reversed. Thus, the prevention of cardiovascular diseases must start early in life, correcting excess weight and improving life habits through implementation of strategies such as preventing the development of dyslipidemias in childhood and adolescence.

7.8 Empirical Data About Health Risk Factors and Behaviour in Adolescents

In a previous study by Guimarães et al. (2013a, b), conducted with 572 adolescents, it was observed that 30.5% of the boys and 27% of the girls were overweight or obese. When eating habits were assessed, 57.7% of the male participants and 59.8% of the female ones were shown to have cardiovascular risk factors, with precisely half the female participants in the sample (50%) having elevated total cholesterol, and 66% of them having dysfunctional HDL-c levels. As for the male subjects, 67.5% had dysfunctional HDL-c levels.

Some health risk factors, such as excess bodily adiposity and dysfunctional blood metabolic levels, can be reduced with regular physical activity and improved eating habits (Oliveira 2005; Paterson/Warburton 2010; Guimarães et al. 2014).

Transversal and longitudinal observational studies have shown that adolescents, both male and female, who have high levels of physical activity have lower adiposity values when compared with the less physically active ones (Paterson/ Warburton 2010). In addition to the influence over adiposity, a study which evaluated lipid values in a representative sample of American adolescents (n = 3.110) showed that both boys and girls with low cardiorespiratory capacity had a greater probability of having hypercholesterolemia and dysfunctional levels of HDL-c than those with moderate or high cardiorespiratory capacity (Carnethon et al. 2005).

There is also evidence of improvement in blood pressure levels with the practice of physical activity. A meta-analysis by Kelley/Kelley (2003) and a study by Barros et al. (2013) indicate a clear association between physical activity and reduction of blood pressure in adolescents; additionally, the American College of Sports Medicine (2010) suggests that the practice of physical activity reduces the risk of arterial hypertension, since sedentary individuals have a 30–50% higher chance of developing the condition than their physically active peers.

Moreover, moderate intensity physical exercise programmes, with a duration of 30–60 min, 3–7 days a week, can reduce visceral adiposity and body mass in children and adolescents with excess body weight (Strong et al. 2005). What is more, several studies demonstrate an association between physical activity and improvement in the metabolic syndrome components in obese and non-obese young people (Janssen 2007).

Physical activity recommendations can be met in a cumulative manner at school, during physical education classes, breaks, and extracurricular sports, among others. The motivation for adolescents to participate in physical activities is readily available at home and at school, as well as in community settings (Strong et al. 2005).

Silva et al. (2012) have shown, through an intervention study, that changes in everyday behaviours regarding diet, physical activity levels and healthy living lead to significant positive reductions in the values of body mass index (BMI) and waist circumference, as well as to improved body composition in general and improved physical capacity in relation to health, with an increase in levels of VO_2 max, flexibility, and abdominal strength/stamina. Hallal et al. (2006) also provided evidence that the practice of regular physical activity is important to combat overweight status and obesity, with the added benefit of maintaining adequate levels of blood lipids.

In the systematic review by Souza et al. (2011), the work they analysed which promoted the practice of regular physical activity with intervention activities evidenced improvement in body composition and in the weekly practice of physical activities among the students. Along the same lines, Bozza (2013) argues that prevention strategies implemented as early as possible have greater potential to avoid the precursors that lead to elevated blood pressure, and that behavioural changes are necessary for young individuals who are pre-hypertensive or who suffer from arterial hypertension.

7.9 Healthy Lifestyle in Adolescence

Based on the concepts and findings related above, we—the public health professionals and other related professionals—ought to direct our focus to the development of programmes capable of generating positive attitudes in young people, especially within schools, in physical education classes, where school-age adolescents have opportunities to change and consolidate their attitudes towards the practice of habitual physical activity. Thus, creating good school physical education programmes is extremely important to establish the habitual lifelong practice of incorporating sufficient physical activity into every individual's daily routine.

To place more emphasis on sedentary lifestyles, there is ample evidence about the maximum amount of time that should be spent on sedentary activities, especially screen time, which should be limited for both children and adolescents (Pate et al. 2008; Martínez-Gómez et al. 2010).

Dias et al. (2014) demonstrated an inverse association between the practice of physical activity and sedentary behaviour, which may suggest that there is a need for intervention strategies for this population, such as educational actions on the part of family members, school, and health professionals. These interventions, especially with regard to encouraging the practice of activities with energy expenditure, reduce sedentary behaviours and, consequently, excess body weight among young people.

It is known that adolescents are constantly influenced by the sociocultural environment in which they find themselves, and nowadays encouragement to exercise is neither sufficient nor adequate, increasing sedentary activity. A typical example is the long time spent in front of electronic devices; this occurs because adolescents are not presented with other options for leisure. Therefore, motivation for the practice of physical activity, in this age range, is related to the pleasurable aspects of doing such activities, evolving over the years to being related to the health and welfare aspects of physical activity (Guedes et al. 2001).

To develop intervention programmes that lead adolescents to practise more physical activity, ensuring commitment and a higher frequency, this population's interests must be taken into account. Knowing that young people favour taking part in group activities, one idea is to involve them in the regular practice of group sports or physical exercises done with friends.

Moreover, considering the influence physical education teachers have over adolescents with regard to the practice of physical activity, the former could be thought of as role models of extreme importance for young people in the acquisition and perpetuation of physical activity habits.

Thus, for adolescents, knowing how to be physically active may be more important than knowing why they should be physically active. Sallis et al. (2000) present four reasons that favour adolescents' participation in physical activity: (1) practical knowledge about the regular practice of physical activities; (2) physical space availability; (3) intention to practise a physical activity; and (4) perception, on the part of the young person, that they are capable of doing the physical activities. What is more, it should be explained to adolescents how the practice of physical activity can be greatly beneficial to one's health and why we should constantly practise it.

7.10 Final Reflection

Physical inactivity in adolescence has been considered a public health problem, since it leads to an increase in the incidence of obesity and morbidity in adulthood, and also to a reduction in physical activities and to bad eating habits.

With the evidence presented in this chapter as a starting point, professionals of public health and related areas ought to devise strategies aimed at improving the lifestyle of adolescents, decreasing time spent on sedentary activities, changing their eating habits and increasing the level of physical activity, thus preventing possible non-communicable chronic diseases in adulthood, which can be caused by dysfunctional values in the anthropometric, lipid, and glucose profiles, as evidenced by the findings shown here.

An active lifestyle must be encouraged and campaigned for, especially in the school setting, which is fertile soil for a good intervention aimed at promoting health, as school is where most of the adolescents of a country spend much of their time.

References

American Academy of Paediatrics. (2001). Committee on Nutrition. Cholesterol in childhood. *Paediatrics*, 101:141–147.

American College of Sports Medicine—ACSM's (2010). *Guidelines for Exercise Testing and Prescription*. USA: Wolters Kluwer.

Andersen, L.B., Harro, M., Sardinha, L.B., Froberg, K., Ekelund, U., Brage, S., et al. (2006). Physical activity and clustered cardiovascular risk in children: a cross-sectional study (The European Youth Heart Study). *Lancet,* 368(9532), 299–304.

Azevedo, M.R., Araujo, C.L., da Silva, M.C., Hallal, P.C. (2007). Tracking of physical activity from adolescence to adulthood: a population-based study. *Revista de Saúde Publica,* 41(1), 69–75.

Barros M.V.G., Ritti-Dias R.M., Barros S.S.H., Mota J., Andersen L.B. (2013). Does self-reported physical activity associate with high blood pressure in adolescents when adiposity is adjusted for? *Journal of Sports Sciences,* 31(4).

Bozza, R. (2013). *Pressão arterial elevada em adolescentes: prevalência e fatores associados*. (Ph.D. thesis. Programa de Pós-graduação em Educação Física). Curitiba, PR, Brasil: Universidade Federal do Paraná.

Brazilian Society of Cardiology—SBC. (2001). III Diretrizes Brasileiras Sobre Dislipidemias e Diretriz de Prevenção da Aterosclerose do Departamento de Aterosclerose da Sociedade Brasileira de Cardiologia. *Arquivos Brasileiros de Cardiologia,* 77(3).

Brazilian Society of Cardiology—SBC. (2005). I Diretriz de Prevenção da Aterosclerose na Infância e na Adolescência. *Arquivos Brasileiros de Cardiologia,* 85(6).

Caldeira G.V. (2000). Fatores que influenciam a formação dos hábitos alimentares de crianças. *Anais do Simpósio Sul Brasileiro de Alimentação e Nutrição: História, Ciência e Arte,* Universidade Federal de Santa Catarina, Florianópolis, SC, Brasil, 181–185.

Carnethon, M.R., Gulati, M., Greenland, P. (2005). Prevalence and cardiovascular disease correlates of low cardiorespiratory fitness in adolescents and adults. *JAMA,* 294(23), 2981–2988.

Federal Council of Physical Education in Brazil—CONFEF. (2010). Estatuto do conselho federal de educação física. Publicado no DO. n. 237, Seção 1, 137–143, 13/12/2010. http://www.confef.org.br. Accessed 13 Dec 2014.

Dias P.J.P., Domingos, I.P., Ferreira, M.G., Muraro, A.P., Sichieri R., Gonçalves-Silva, M.R.V. (2014). Prevalência e fatores associados aos comportamentos sedentários em adolescentes. *Revista de Saúde Pública,* 48(2), 266–274.

Ekelund, U., Anderssen, S.A., Froberg, K., Sardinha, L.B., Andersen, L.B., Brage, S. (2007). Independent associations of physical activity and cardiorespiratory fitness with metabolic risk factors in children: the European youth heart study. *Diabetologia,* 50, 1832–1840.

Ekelund, U., Brage, S., Froberg, K. (2006). TV viewing and physical activity are independently associated with metabolic risk in children: the European Youth Heart Study. *PLoS Med,* 3(12), 488.

Ferreira, M.S., Najar, A.L. (2005). Programas e Campanhas de Promoção da Atividade física Ciência. *Ciência e Saúde Coletiva,* 10, 207–219.

Fisberg, M., Priore, M.E., Vieira, V.C.R. (2002). Hábitos alimentares na adolescência. *In: Atualização Científica em Nutrição: nutrição da criança e do adolescente* (pp. 66–93). Porto Alegre: Atheneu.

Freedman, D.S., Goodman, A., Contreras, O.A., Dasmahapatra, P., Srinivasan, S.R., Berenson, G.S. (2012). Secular Trends in BMI and Blood Pressure Among Children and Adolescents: The Bogalusa Heart Study. *Pediatrics,* 130, 159–166.

Gabbay, M., Júdice, P.R.C., Dib, S.A. (2003). Diabetes melito do tipo 2 na infância e adolescência: revisão da literatura. *Jornal de Pediatria,* 79(3), 201–208.

Guedes, D.P., Guedes, J.E.R.P., Barbosa, D.S., Oliveira, J.A. (2001). Níveis de prática de atividade física habitual em adolescentes. *Revista Brasileira de Medicina do Esporte,* 7(6), 187–199.

Guedes, D.P., Guedes, J.E.R.P., Barbosa, D.S., Oliveira, J.A. Stanganelli L.C.R. (2006). Fatores de risco cardiovasculares em adolescentes: indicadores biológicos e comportamentais. *Arquivos Brasileiros de Cardiologia,* 86(6), 439–50.

Guimarães, R.F., Silva. M.P., Legnani, E., Mazzardo, O., Campos, W. (2013a). Reproducibility of adolescent sedentary activity questionnaire (ASAQ) in Brazilian adolescents. *Revista Brasileira de Cineantropometria e Desempenho Humano,* 15(3), 276–285.

Guimarães, R.F., Silva, M.P., Mazzardo, O., Martins, R.V., Campos, W. (2013b). Associação entre comportamento sedentário e perfil antropométrico e metabólico entre adolescentes. *Motriz: Revista da educação física,* 19(4), 753–762.

Guimarães, R.F., Silva. M.P., Mazzardo, O., Martins, R.V., Campos, W. (2014). Atividade física e alimentação associadas aos perfis antropométrico e lipídico em adolescentes. *ConScientiae Saúde,* 13(3), 340–348.

Hallal, P.C., Victora, C.G., Azevedo, M.R., Wells, J.C. (2006). Adolescent physical activity and health: a systematic review. *Sports Medicine,* 36(12), 1019–1030.

Hanley, A.J., Harris, S.B., Gittelsohn, J., Wolever, T.M., Saksvig, B., & Zinman, B. (2000). Overweight among children and adolescents in a Native Canadian community: prevalence and associated factors. *American Journal of Clinical Nutrition,* 71(3), 693–700.

Hardy, L.L., Booth, M.L., Okely, A.D. (2007). The reliability of the Adolescent Sedentary Activity Questionnaire (ASAQ). *Preventive Medicine,* 45(1), 71–74.

Hardy, L.L., Denney-Wilson, E., Thrift, A.P., Okely, A.D., Baur, L.A. (2010). Screen time and metabolic risk factors among adolescents. *Archives of Pediatric & Adolescent Medicine,* 164(7), 643–649.

Ilias, E.J. (2006). Hábitos alimentares e câncer digestivo. *Revista da Associação Médica Brasileira,* 52(5), 281–91.

Janssen, I. (2007). Physical activity guidelines for children and youth. *Canadian Journal of Public Health,* 98(2), 109–121.

Jenovesi, J.F., Bracco, M.M., Colugnati, F.A.B., Taddei, J.A.A.C. (2003). Perfil de atividade física em escolares da rede pública de diferentes estados nutricionais. *Revista Brasileira de Ciência e Movimento,* 11, 57–62.

Kelley G.A, Kelley K.S. (2003). Exercise and resting blood pressure in children and adolescents: a meta-analysis. *Pediatric Exercise Science,* 15, 83–97.

Lemos, M.C.M., Dallacosta, M.C. (2005). Hábitos alimentares de adolescentes: Conceitos e práticas. *Arquivos de Ciências da Saúde da Unipar,* 9(1), 3–9.

Malta, D.C., Moura, E.C., Castro, A.M., Cruz, D.K.A., Morais Neto, O.L, Monteiro, C.A. (2009). Padrão de atividade física em adultos brasileiros: resultados de um inquérito por entrevistas telefônicas, 2006. *Epidemiologia e Serviços de Saúde,* 18(1), 7–16.

Martínez-Gómez, D., Eisenmannb, J., Gómez-Martínezc, S., Vesesc, A., Marcosc, A., Veiga, O.L. (2010). Sedentary Behavior, Adiposity and Cardiovascular Risk Factors. *Revista Española de Cardiologia,* 63(3), 277–285.

May, A.L., Kuklina, E.V., Yoon, P.W. (2012). Prevalence of Cardiovascular Disease Risk Factors Among US Adolescents, 1999–2008. *Pediatrics,* 129, 1035-1041.

Mello, E.D., Luft, V.C., Meyer, F. (2004). Obesidade infantil: como podemos ser eficazes? *Jornal de pediatria,* 80(3), 173–82.

Mezomo, I.B. (2002). *Os serviços de alimentação: planejamento e administração.* 5ª. Ed. São Paulo: Manole.

Ministério da Saúde—Brasil. (2002). Secretaria de Políticas de Saúde. Programa Nacional de Promoção da Atividade Física 'Agita Brasil': Atividade física e sua contribuição para a qualidade de vida. *Revista Saúde Pública,* 36(2), 254–256.

Mintz, S.W. (2001). Comida e Antropologia: uma breve revisão. *Revista brasileira de Ciências Sociais,* 16(47), 31–41.

Oliveira, R.J. (2005). *Saúde e Atividade Física: Algumas abordagens sobre atividade física relacionada à saúde.* Rio de Janeiro: Shape.

Pate, R.R., O'neil, J.R., Lobelo, F. (2008). The envolving definition of sedentary. *Exercise and Sport Sciences Reviews,* 36, 173–178.

Paterson, D.H., Warburton, D.E.R. (2010). Physical activity and functional limitations in older adults: a systematic review related to Canada's Physical Activity Guidelines. *International Journal of Behavior Nutrition and Physical Activity,* 7, 38–60.

Rosenberger, M.E., Haskell, W.L., Albinali, F., Mota, S., Nawyn, J., Intille, S.S. (2013). Estimating activity and sedentary behavior from an accelerometer on the hip or wrist," *Medicine Science in Sports and Exercise,* 45(5), 964–975.

Sallis, J.F., Prochaska, J.J., Taylor, W.C. (2000). A review of correlates of physical activity of children and adolescents. *Medicine & Science in Sports Exercise,* 32(5), 963–75.

Santos C.C., Stuchi R.A.G., Arreguy-Sena C., Pinto N.A.V.D. (2012). A influência da televisão nos hábitos, costumes e comportamento alimentar. *Cogitare Enferm,* 17(1), 65–71.

Sardinha, L.B., Andersen, L.B., Anderssen, S.A. (2008). Objectively measured time spent sedentary is associated with insulin resistance independent of overall and central body fat in 9-to10-year-old Portuguese children. *Diabetes Care,* 31(3), 569–575.

Sichieri, R., Castro, J.F.G., Moura, A.S. (2003). Fatores associados ao padrão de consumo alimentar da população brasileira urbana. *Cadernos de Saúde Pública,* 19(1), 47–53.

Silva, D.F., De Souza, L.L., Delfino, R.D., Bianchini, J.A., Hintze, L., Nardo Junior, N. (2012). Efeitos de um programa multiprofissional de tratamento da obesidade e de sua cessação sobre a aptidão física relacionada à saúde de adolescentes. *Revista da Educação Física,* 23(3).

Silva, M.P., Gasparotto, G.S., Bozza, R., Stabelini Neto, A., Campos, W. (2010). Tempo gasto em atividades hipocinéticas relacionado a fatores de risco cardiovascular em adolescentes. *Revista da Educação Física,* 21(2), 279–285.

Souza, E.A., Barbosa Filho, V.C., Nogueira, J.A.D., Azevedo Junior, M.R. (2011). Atividade física e alimentação saudável em escolares brasileiros: revisão de programas de intervenção. *Caderno de Saúde Pública,* 27(8), 1459–1471.

Strong, W.B., Malina, R.M., Blimkie, C.J., Daniels, S.R., Dishman, R.K., Gutin, B., et al. (2005). Evidence based physical activity for school-age youth. *Journal of Pediatrics,* 146(6), 732–737.

Suñé, F.R., Dias-Da-Costa, J.F., Olinto, M.T.A., Pattussi, M.P. (2007). Prevalência e fatores associados para sobrepeso e obesidade em escolares de uma cidade no Sul do Brasil. *Caderno de Saúde Pública,* 23(6), 1361–1371.

Young, K.M., Northern, J.J., Lister, K.M., Drummond, J.A., O'brien, W.H. (2007). A meta-analysis of family-behavioral weight-loss treatments for children. *Clinical Psychology Review,* 27, 240–249.

Watkins, P.J. (2003). *ABC of diabetes.* 5ª ed. Spain: BMJ Publishing Group Ltd.

Weiss, R., Dziura, J., Burgert, T.S., Tamborlane, W.V., Taksali, S.E., Yeckel, C.W. (2004). Obesity and the metabolic syndrome in children and adolescents. *The New England Journal of Medicine,* 350, 2362–2374.

World Health Organization [WHO]. (2011). Obesity and overweight. Geneva: WHO. Recuperado en http://www.who.int/en/. Accessed in 13 Dec 2014.

World Health Organization [WHO]. (2002). The world health report 2002*: reducing risks, promoting healthy life.* Geneva: WHO. Recuperado en http://www.who.int/en/. Accessed in 13 Dec 2014.

Chapter 8
International Migration as a Trigger for Risks in Mental Health

Jorge Luis Arellanez Hernández

Abstract International migration has increased as never before. The main reasons to move to another country are related to the search for a better quality of life. One of the factors that has also contributed to the increase in population mobility is related to the different outbreaks of violence that have been occurring globally. This scenario makes millions of people vulnerable, putting them in a condition of emotional fragility in the different phases of the migratory process and consequently exposing them to a series of risks. This condition, which is currently experienced by thousands of people, had not been considered an important phenomenon to study, but now some researchers have explored the impact of male migration on the people who actually migrate and on their partner who stays in the country of origin. This chapter describes the post-traumatic stress associated with the migratory experience in Mexican men who migrate, as well as the stress associated with familiar conflicts. In migrant women it has also been observed that the migratory experience may be a risk factor for the consumption of illegal drugs, especially in the youngest migrants. The study on the consumption of drugs shows that in migrant adolescents the percentage of users of drugs is higher than in adult migrant women and slightly lower than in adult migrant males. While these topics have started to be addressed by some researchers, others still go unnoticed, for example sexuality.

Keywords International migration · Mental health · Psychosocial's risk factors
Acculturation

8.1 Introduction

During the past 20 years, international migration has increased as never before. The main reasons to move to another country are related to the search for a better quality of life for both the migrant and the family that remains in the country of origin (ECLAC 2006; CONAPO 2014).

Dr. Jorge Luis Arellanez Hernández, Researcher, Institute of Psychological Research, Universidad Veracruzana. Email: arellanez@hotmail.com.

© Springer International Publishing AG, part of Springer Nature 2018   101
Ma. L. Marván and E. López-Vázquez (eds.), *Preventing Health and Environmental
Risks in Latin America*, The Anthropocene: Politik—Economics—Society—
Science 23, https://doi.org/10.1007/978-3-319-73799-7_8

One of the factors that has also contributed to the increase in population mobility is related to the different expressions of violence that have been occurring globally. On the one hand, the ineffectiveness of the economic system—meaning globalisation and neoliberalism—has contributed to the mobility of thousands of goods and services, but limited the free transit of people. This is contradictory, since within a situation of supply and demand, only objects have free mobility, but people do not. Additionally, government capacity to offer opportunities for development to citizens is poor in many nations, and this has also been reinforced with multiple economic crises and the increasingly evident fragmentation of institutions that comprise systems of government (Gómez-Tagle/López 2004).

This can be understood as a form of structural violence, since it has not only limited the optimal personal development for each person and his or her family, but has forced thousands of adults, adolescents, children, men and women, even complete families, to move to the most developed countries with the aim of improving their quality of life.

Furthermore, the process of relocating to another place has fostered the emergence of another type of violence: social. Since the majority of people who emigrate do so undocumented by slipping unperceived past border authorities, they have become an easy prey for criminal groups. As a matter of fact, with an extensive network between countries, these criminal groups (human traffickers) have taken advantage of thousands of men and women from all ages, and have exerted violence upon them from their places of origin through their transfer to their migratory destiny, by robbing them, kidnapping them, violating all their individual guarantees, making them objects of human trafficking and even depriving them of their lives if they oppose their mandates (Chiarotti 2003).

This scenario makes millions of people vulnerable, putting them in a condition of emotional fragility in the different phases of the migratory process and consequently exposing them to a series of risks.

This condition, which is currently experienced by thousands of people, has not been considered an important phenomenon of study and attention. At least in Mexico, it is an area of knowledge that has been addressed poorly although there exists a wide migratory tradition towards the United States. Research on the physical and mental health of migrants as well as that of the families that remain in Mexico has not been a topic of interest for specialists on health.

8.2 The Study of Health in International Migration: The Case of Mexico-United States Migration

Some works have explored the occurrence of sexually transmitted diseases, others the presence of HIV/AIDS; in both cases, a high correlation between extramarital sexual practices by migrants and a high level of contagion in their partners that remain in Mexico has been reported (Bettini 1999; Bronfman/Minello 1995;

Bronfman et al. 1998; Hernández-Rosete et al. 2005; Macías 2002; Magis et al. 1995; Organista 2004; Salgado de Snyder 1996, 1998).

Also recently, diseases that apparently had been eradicated in Mexico, such as tuberculosis (Moya/Uribe 2007), have been reported present in migrants; furthermore, a high proportion of coronary and brain vascular diseases, hypertension, diabetes mellitus and obesity has been observed in the Mexican population migrating towards the United States, as well as illnesses related to environmental and occupational risks (Foladori et al. 2004; Secretaría de Salud 2002, 2008).

8.3   The Consequences of International Migration on Mental Health

It was only in the 1970s that the first document was published which studied the migratory phenomenon as a factor that harms the mental health of the migrant (De la Fuente 1979). Almost a decade later, Trigueros/Rodríguez (1988) explored the impact of male migration for the partner who stays in Mexico with the responsibility of maintaining stability and familiar unity. While these changes for women that stay include the strengthening of their own selves, which has been an important source for their own development and for their families, it is also known that confronting the migration of the partner fosters in her a high level of stress, the presence of sentiments of despair, symptoms of depression, anxiety and diverse psychosomatic illnesses (Salgado de Snyder 1990, 1992). In particular, in males who migrate the post-traumatic stress associated with the migratory experience has been studied (Salgado de Snyder et al. 1990), as well as the stress associated with family conflicts and the lack of social networks of support that facilitate the migratory stay in the United States (Salgado de Snyder 1991, 1996).

The studies mentioned have contributed relevant findings to knowledge of some of the emotional characteristics of migrants and those who stay, but it is also true that they were carried out in a historical moment in which Mexican migration to the United States had different characteristics to those that can be observed nowadays. The lack of continuity in a psychological study of this phenomenon has created gaps and limits, but at the same time, it is a niche of opportunity for exploration.

Currently, the profile of the Mexican migrant who goes to the United States has changed drastically and so have the means of transportation. Not only migration from rural zones is registered, but the scarcity of opportunities in labour and development has led to the migration of men, women and adolescents from urban areas. Federative entities that at the start of the 1990s did not register international migration have incorporated it in quite dynamic ways during the past years, for example the cases of Veracruz, Tabasco, Estado de México and Yucatán, among others.

The conditions to cross the border have changed as well. Ever since the North American Free Trade Agreement, the US government has increased the surveillance of its border with Mexico with the purpose of discouraging immigration, creating walls, virtual fences through the use of military technology, and some states have even established immigration laws that criminalise the undocumented migrant (like HB-56 in Alabama and SB1070 in Arizona); more than stopping the migratory flux towards the country, it has increased the risks faced by migrants trying to cross the border. Currently, they penetrate through areas of the desert of Sonora and the more profound parts of the river Bravo to avoid the border patrol. These circumstances have only generated yearly reports on the death of tens of migrants in their attempt to arrive in the United States.

While it has to be recognised that these initial research projects have only explored some of the psychological characteristics, these are a basis for the construction of new perspectives for the study of the migratory phenomenon. Specially in the last decade, apart from encountering some studies that once more include the analysis of stress in migrant population (Arellanez-Hernández 2010), an interest has emerged in exploring the relationship between the migratory experience in the returning Mexican migrant population and the consumption of alcohol, tobacco or other drugs as a way of diminishing the psychological discomfort that this type of experience generates.

Various studies have found that migrants increase their level of alcohol consumption while they stay in the United States, drinking great amounts at the weekend (like the habitual cultural pattern in Mexico); but also, they drink during the week, although lower quantities (Borges/Cherpitel 2001; Borges et al. 2007). This pattern of consumption in the medium term can be a risk factor for the appearance of antisocial behaviours that can lead the migrant to forge links with social networks with antisocial or criminal behaviours, and may even increase the risk of consuming other type of substances, such as illegal drugs.

As a matter of fact, several specialists have demonstrated that the migratory experience itself does not influence male Mexican migrants to start using illegal drugs, but it does carry the risk of them increasing consumption or trying new drugs if they have already tried drugs in Mexico. For example, it has been found that some Mexican migrants who had consumed a drug like marihuana or cocaine in Mexico consumed these substances and even more addictive drugs such as ecstasy and methamphetamines during their stay in the United States. It is worth mentioning that the main reasons for the consumption of these substances are associated with enhancing their performance at work or with the emotional difficulties of adapting to life in the new destination, such as feeling alone and isolated (Sánchez-Huesca et al. 2006a, 2007a; Sánchez-Huesca/Arellanez-Hernández 2011). The supply, as well as the conditions of the context in the United States and their emotional situation, can sometimes boost the use of psychoactive substances.

In other studies, focused on migrant women between 14 and 65 years old, it has also been observed that in some of them the migratory experience can be a risk factor for the consumption of illegal drugs, especially in the youngest who also originate from areas that are predominantly urban. Although the use of substances is more of an experimental character, during the migratory state, along consuming marihuana, they used substances of stimulating character like cocaine and ecstasy.

Some of the reasons for beginning drug consumption are related to the fact that a friend or their sentimental partner offered the drug (Arellanez-Hernández/Sánchez-Huesca 2008; Sánchez-Huesca et al. 2006b, 2007, 2008a, b).

Another population group of Mexican migrants in which the relationship between consumption of psychoactive substances and the migratory experience has been studied is adolescents. In order to understand the ways in which the migratory experience can encourage some psychological modifications in this population group, it must be understood that this life period by itself presents a range of physical and psychological transformations that, combined with a rough change in the context, can complicate the definition of self-identity and membership of a social group. Studies carried out have required the development of sensitive methodologies for this sector of the population, not only because of the ethical-judicial conditions of studying subjects less than 18 years old, but also because of the physical and emotional modifications that some of them show when being returned to national territory by force, often 'violently', by the border patrol.

A common feature that has been observed, especially in adolescents, is the risk-taking to explore new things; migrating to the United States for some of them represents the only way to help their family. Indisputably, their experience of crossing the border differs from that of adults; they are exposed to other types of risks. As well as facing geographic difficulties, the possibility of being ravaged, falling into gangs, being subjected to human trafficking, or being abandoned in the desert by the *pollero*, *patero* or *lanchero* supposed to help them arrive in the United States, they also risk becoming a *burrero*, meaning someone who transports drugs across the border in order to 'pay' for their passage.

Research that has addressed the consumption of drugs in migrant adolescents shows that the percentage of drug users is larger than that registered for adult migrant women and slightly lower than that observed for adult migrant males. This finding suggests that, along with the period in life that makes them more vulnerable, the differences between their old and new psychosocial context can lead to adaptation mechanisms that are not necessarily the healthiest (Sánchez-Huesca et al. 2006, 2007b, 2008a, b, 2009; Sánchez-Huesca/Arellanez-Hernández 2009). Particularly in adolescent males who migrate, drinking large quantities of alcohol in front of peers is a way of showing who can 'withstand more'; experimenting with illegal drugs can also be considered a form of assumed incorporation into the context of their destination. This is not necessarily observed in migrant adolescent women; on the contrary, the family itself tends to include them in activities that help them integrate into the sociocultural context of arrival.

During the last decade, the emotional repercussions that can afflict the children of a father who migrated to the United States have been explored. In interviewing some patients being treated for drug consumption and their mothers, it was found that the temporary absence of the father fosters a family crisis that requires considerable adjustment and leads to the exacerbation of conflicts or the emergence of new problems. A restructuring of traditional family roles is observed, mainly the ones assigned to the woman, who needs to act as mother and father at the same time. This can lead to the disruption of authority, routines and family control, unclear rules and general 'chaoticisation' of family life. These situations have a generalised affective impact of stress and depression in women associated with the temporary loss of the partner. These conditions contribute to children using and abusing alcohol and other drugs (Pérez-Islas et al. 2004).

Clearly, the earliest studies, as well as those undertaken in the first decade of this century, have examined the migratory phenomenon from a perspective in which the process of health-illness establishes a practically direct causal relation between the migratory experience and the presence of stress, depressive symptoms or the development of a psychosomatic disease, where drug consumption can be a palliative mechanism to help those affected face these types of emotional discomforts.

However, this explanatory scheme only takes into account migrants' behaviour The psychological currents also need to be considered. From this point of view, migrating does not only imply the act of moving from one place to another, from one region to another, from one country to another; it also implies leaving behind family, home, place of residency, language, traditions and customs, everything that is emotionally significant; depriving oneself of self-identity (to 'not be from here or there'), and self-citizenship; becoming an 'immigrant', 'undocumented', 'illegal'... an invisible being for whom the violation of their human rights and the exposure to situations of violence are so frequent that the individual is subjected to a constant state of tension and uncertainty.

8.4 What Remains to be Explored in Studying the Impact of Migration on Mental Health

In addition to establishing the emotional situation experienced by migrants—men or women, adolescent or adult—this discussion leads to multiple questions. In those who migrate, for example, the total or partial loss of loved ones, objects, places, customs and traditions is one of the constants that appear independently of being a man or a woman, adult, adolescent or child, documented or undocumented. What implications does the letting go of his or her identity, rights and citizenship have on his or her psyche, meaning, for his or her mental health? What are the effects for an adolescent or an adult, for a male or a female? What are the risks to which they get exposed? The varying absences, losses and affective splits facilitate a state of tension that challenges emotional stability and makes it harder to adjust to the sociocultural context in which the migrant is situated.

For many migrants, the migratory stay is reduced to work life, investing most of their time in long working days; the scarce spaces they have for leisure are mainly used to spend time with other Mexicans, leaving little time to rest and take care of their health. During moments of solitude, the feelings of longing for the family and the place of origin flourish and strongly contribute to the generation of a melancholic environment, of mourning, and the idea of only staying for a limited period and going back to the place of origin is constant. These characteristics have begun to be analysed by Joseba Achotegui, a Spanish psychiatrist and specialist in the migratory phenomenon, who states that the presence of a feeling of solitude in the migrant for not being able to bring the family; the internal feeling of failure for not being able to access the labour market immediately and securely; and the constant feeling of fear, many times related to the migratory situation, practically lead the migrant to a survival struggle, generating a high level of stress. He calls this the Odysseus Syndrome. The term comes from the old text of Homer '… and Odysseus passed the days seated on the rocks, at the seashore, consumed by the strength of cries, sight and pity, fixing his eyes in the sterile sea, crying relentlessly…' (The Odyssey, Book IX; cited in Achotegui 2004). This notion is very descriptive of what is experienced by many Mexican migrants in the United States.

Just as these topics have started to be addressed by some researchers, others have gone unnoticed, for example sexuality. What happens with the sexuality of migrants during their migratory stay? Understanding the term in a wide sense, meaning not reducing it to genitality, reproductive health has been explored, especially related to sexually transmitted diseases and the presence of HIV/AIDS, but the emotional part, the part of feelings, the need for tenderness, of receiving and giving affection to loved ones, has not been addressed. Without doubt, traditional gender roles mean that migrant males are more likely than migrant females to get involved genitally with other partners, but what happens with the need to give and receive tenderness? How to manage the absence of the partner and children (for those who have them) who remain in the place of origin, and the absence of maternal affection that is so characteristic of Mexicans? In the case of a woman, the exercise of genitality is socially restricted, as if the fact of migrating implies cancelling the genitalia; what happens with the need to give and receive love when she is the one who is abroad? With the exercise of motherhood when separated from their children? As a matter of fact, it is equally interesting that there are children who initially stayed behind but suddenly decide to migrate in order to be reunited with their father or the mother. They do not mind jeopardising their own life as long as they are d with their parents.

For those who remain, what happens? What does the physical absence of the person who migrates mean for the members of a social group? How do they elaborate the loss—at least partially—of one of their members? Who assumes the responsibilities that were assigned to the person who is not there anymore?

The absence of one or various members of the family leads to a reconfiguration of its functioning in all senses. Even if the symbolic presence can continue to give him or her a place—at least temporarily—through the discourse of those who remain, this is achieved through words; the truth is that, physically, he or she is not there anymore, which inevitably generates a rearrangement of behaviour and affections in the familiar group.

In the case of the migration of the father, the mother tends to exercise the double role of being father-mother in the family, but sometimes the eldest child is the one who is assigned the role of being the 'man of the house' in the father's absence.

This emotional charge can have some negative effects on the individual. For example, being assigned the role of person 'responsible for the house' can lead the eldest son to an emotional imbalance that sparks off situations of anguish, leading to problems in school, with friends or even with the other members of the family, since this is a role he did not ask to exercise and it implies changing from being a son to being the 'carer' of the rest of the members of the family. In the majority of cases where the mother is the one who migrates, the father is not the one who assumes the double parental role of being father-mother. This function is normally assigned to one of the grandmothers. In Mexico the performance of care and upbringing of the children in a family group still strongly crosses a traditional gender perspective.

And finally, what happens emotionally with those who return to their place of origin? How do they feel once they return... how do they integrate again in a familiar dynamic that has been reorganised to 'work' during his or her absence? Depending on the situation in which the return occurs, many of the migrants who return to their place of origin tend to show an attitude of success, of triumph to others through the use of clothes, electronic goods, the purchase of cars, the construction of houses or starting a business. The idea of achieving the so called 'American Dream' in migrants, particularly in males, is still valid; on returning they talk about the great deeds, achievements and abundance that exists in the United States. It is only a few who accept that they went through difficult times, adversities and troubles that they had to deal with to be able to stabilise their stay.

Some Mexicans who stay in the United States for long periods face problems when reintegrating into the community, family and group of friends once they are back in their place of origin; they do not feel they belong either here or there... what is going on? It seems that the interaction between the own culture and the American one contributes to a psychological process in which the feeling of belonging and self-identity are in conflict. There may also be a crisis of loyalty, generating ambivalence when in one context or another. This general analysis of emotional scenarios that involve the migrant and his or family demonstrates the need to widen the study of the migratory phenomenon from a psychological perspective. In the scheme of the health-disease process, for example, in accordance with the previous paragraph, there is still a lot to do; it should not be denied that there can exist a high occurrence of emotional disorders related to the migratory experience; nonetheless, it is important not to categorise the migrant as an agent who will suffer some sort of emotional disorder from the moment the place of origin is left. On the contrary, it can be seen that the majority of our Mexicans adapt to the social context while in the United States and actively contribute to the development of that country. Moreover, those who remain in their place of origin may initially experience diverse situations of crisis, but the fragmentation or division in family functions is not permanent. In the majority of cases, adaptation mechanisms and emotional regularisation allow migrants and family members that remain to maintain an emotional stability that is more or less manageable.

References

Achotegui, J. (2004). Emigrar en situación extrema: el Síndrome del inmigrante con estrés crónico y múltiple (Síndrome de Ulises). *Norte de Salud Mental*, 21, 39–52.

Arellanez-Hernández, J.L. & Sánchez-Huesca, R. (2008). Migración femenina a Estados Unidos y consumo de drogas. *Revista Digital Universitaria*, 9(8), 1–12.

Arellanez-Hernández, J.L. (2010). *Factores psicosociales de aculturación para el consumo de drogas en migrantes mexicanos en Estados Unidos*. Doctoral Thesis, Mexico: UNAM.

Bettini, L. (1999). *La vulnerabilidad femenina frente al VIH/SIDA en un contexto de migración*. Master Thesis. Mexico: FLACSO.

Borges, G. & Cherpitel, C.J. (2001). Selection of screening items for alcohol abuse and alcohol dependence among Mexicans and Mexican Americans in the emergency department. *Journal of Studies on Alcohol and Drugs*, 62, 277–285.

Borges, G., Medina-Mora, M.E., Breslau, J. & Aguilar-Gaxiola, S. (2007). The effect of migration to the United States on substance use disorders among returned Mexican migrants and families of migrants. *American Journal Public Health*, 97(10), 1847–51.

Bronfman, M., & Minello, M. (1995). Hábitos sexuales de los migrantes temporales mexicanos a los Estados Unidos. Prácticas de riesgo para la infección por VIH. In M. Bronfman (Ed.), *SIDA en México. Migración, adolescencia y género. Información Profesional especializada* (pp. 3–89). Mexico: Consejo Nacional para la Prevención y Control del VIH/SIDA.

Bronfman, M., Sejenovich, G. & Uribe, P. (1998). *Migración y SIDA en México y América Central. Serie Ángulos del SIDA*. Mexico: Consejo Nacional para la Prevención y Control del VIH/SIDA.

Chiarotti, S. (2003). *La trata de mujeres: sus conexiones y desconexiones con la migración y los derechos humanos*. Serie Población y Desarrollo, 39. Santiago de Chile: CELADE/CEPAL.

Comisión Económica para América Latina y el Caribe (2006). Migración Internacional, Derechos Humanos y Desarrollo. Síntesis y conclusiones. CEPAL. http://repositorio.cepal.org/bitstream/handle/11362/4206/1/S2006047_es.pdf. Accessed 25 Jun 2014.

Consejo Nacional de Población (2014). *Anuario de migración y remesas. México 2014*. Mexico: CONAPO, Fundación BBV BANCOMER, A.C.

De La Fuente, R. (1979). El ambiente y la salud mental. *Salud Mental*, 2(1), 6–9.

Foladori, G., Moctezuma-Longoria, M., Márquez, H. (2004). La vulnerabilidad epidemiológica en la migración México-Estados Unidos. *Migración y Desarrollo*, 3, 19–44.

Gómez-Tagle López, E. (2004). Migración internacional, explotación laboral y trata de blancas en el siglo XXI. *Revista Venezolana de Ciencias Sociales*, 8(2), 193–212.

Hernández-Rosete, D., Sánchez-Hernández, G., Pelcastre-Villafuerte, B., Juárez-Ramírez, C. (2005). Del riesgo a la vulnerabilidad. Bases metodológicas para comprender la relación entre violencia sexual e infección por VIH/ITS en migrantes clandestinos. *Salud Mental*, 28 (5), 20–26.

Macías, G. (2002). *La recurrencia de los eventos migratorios como factor de riesgo para la manifestación de enfermedades de transmisión sexual*. Master Thesis. Tijuana: El Colegio de la Frontera Norte.

Magis, C., Del Río, A., Valdespino, J.L., García, M. de L. (1995). Casos de SIDA en el Área Rural en México. *Salud Pública de México*, 37(6), 615–623.

Moya, J. & Uribe, M. (2007). Migración y Salud en México: Una aproximación a las perspectivas de investigación: 1996–2006. Organización Panamericana de la Salud. http://www.biblioteca.cij.gob.mx/Archivos/Materiales_de_consulta/Drogas_de_Abuso/Articulos/migracion45.pdf. Accessed 10 Sept 2014.

Organista, K.C. (2004). Culturally and socially competent HIV prevention with Mexican Farm workers. In R.J. Velásquez & L.M. Arellano (Eds.), *The handbook of Chicana/o psychology and mental health* (pp. 353–369). EEUU: McNeill, Brain W.

Pérez-Islas, V., Diaz-Negrete, D.B., Arellanez-Hernández, J.L. (2004). *Impacto de la emigración del padre en la conyugalidad y parentalidad en familias de jóvenes usuarios de drogas (Informe de Investigación 04–06).* Mexico: CIJ.

Salgado de Snyder, V.N. (1990). Estrés psicosocial en la mujer migrante y su relación con malestar psicológico. In Asociación Mexicana de Psicología Social (Ed.), *La psicología social en México III* (pp. 51–55). Mexico: AMEPSO.

Salgado de Snyder, V.N. (1991). Las que se van al Norte y las que se quedan: el estrés y la depresión en las mujeres migrantes y en las no migrantes. *Anales del Instituto Mexicano de Psiquiatría*, 153–159.

Salgado de Snyder, V.N. (1992). El impacto del apoyo social y la autoestima sobre el estrés y la sintomatología depresiva en esposas de migrantes a los Estados Unidos. *Anales del Instituto Nacional de Psiquiatría*, 83–89.

Salgado de Snyder, V.N. (1996). Problemas psicosociales de la migración internacional. México. *Salud Mental*, 19, 53–59.

Salgado de Snyder, V.N. (1998). Migración, sexualidad y SIDA en mujeres de origen rural: sus implicaciones psicosociales. In I. Szasz y S. Lerner (Eds.), *Sexualities in Mexico: Some Approximations From the Social Science Perspectiva* (pp. 155–71). Mexico: El Colegio de México.

Salgado de Snyder, V.N., Cervantes, R.C., Padilla, A.M. (1990). Migración y Estrés postraumático: El caso de los mexicanos y centroamericanos en los Estados Unidos. *Acta Psiquiátrica y Psicológica de América Latina*, 36(3–4), 137–145.

Salgado de Snyder, V.N., Díaz, M.J., Maldonado M. (1996). AIDS: Risk behaviors among rural Mexican women married to migrant workers in United States. *AIDS Education Prev,* 8(2), 134–142.

Sánchez-Huesca, R. & Arellanez-Hernández, J.L. (2009). *Adolescentes migrantes repatriados de Estados Unidos. Análisis de factores psicosociales de la migración y consumo de drogas con perspectiva de género.* Mexico: DIF/CIJ.

Sánchez-Huesca, R. & Arellanez-Hernández, J.L. (2011). Uso de drogas en migrantes mexicanos captados en ciudades de la frontera noroccidental México-Estados Unidos. *Revista de Estudios Fronterizos*, 12(23), 9–26.

Sánchez-Huesca, R., Arellanez-Hernández, J. L., Cielo-Meléndez, D.B. (2007a). *Uso de drogas y factores asociados en migrantes a Estados Unidos captados en la Frontera Nororiental (Informe de investigación: 07–08).* Mexico: CIJ.

Sánchez-Huesca, R., Arellanez-Hernández, J.L., Cielo-Meléndez, D.B. (2007b). *Consumo de drogas en Niños y Adolescentes migrantes a Estados Unidos captados en la Frontera Nororiental (Informe de investigación: 07–09).* Mexico: CIJ.

Sánchez-Huesca, R., Arellanez-Hernández, J.L., Ramón-Trigos, E.M. (2006). *Estudio comparativo del consumo de drogas en menores migrantes y repatriados captados en Tijuana, Mexicali, Nogales y Ciudad Juárez (Informe de Investigación: 06–15).* Mexico: CIJ.

Sánchez-Huesca, R., Arellanez-Hernández, J.L., Cielo-Meléndez, D.B., Ramón-Trigos, E.M. (2008). *Consumo de drogas en adolescentes migrantes a la frontera norte y Estados Unidos captados en la frontera noroccidental.* Mexico: CIJ/DIF.

Sánchez-Huesca, R., Arellanez-Hernández, J.L., Pérez-Islas, V., Rodríguez-Kuri, S.E. (2006a). Estudio de la relación entre consumo de drogas y migración a la frontera norte de México y Estados Unidos. *Salud Mental*, 29(1), 35–43.

Sánchez-Huesca, R., Arellanez-Hernández, J. L., Ramón-Trigos, E. M., Ortiz-Encinas, R. M. (2009). Consumo de drogas en niños y adolescentes migrantes a Estados Unidos. In G.C. Valdéz-Gardea (Ed.), *Achicando futuros, actores y lugares de la migración.* Mexico: COLSON.

Sánchez-Huesca, R., Pérez-Islas, V., Arellanez-Hernández, J.L. (2007). *Relación entre Migración y Consumo de Drogas en Mujeres del Estado de Michoacán (Informe de Investigación: 07–05).* Mexico: CIJ.

Sánchez-Huesca, R., Pérez-Islas, V., Arellanez-Hernández, J.L. (2008). *Mujeres migrantes en retorno del estado de Michoacán y consumo de drogas (Informe de Investigación: 08–11)*. Mexico: CIJ.

Sánchez-Huesca, R., Pérez-Islas, V., Rodríguez-Kuri, S.E., Arellanez-Hernández, J.L., Ortiz-Encinas, R.M. (2006b). El consumo de drogas en migrantes desde una perspectiva de género. Un estudio exploratorio. *Revista Región y Sociedad*, 18(35), 131–164.

Secretaría de Salud (2002). *Programas de acción: Migrantes "Vete Sano Regresa Sano"*. Mexico: SSA.

Secretaría de Salud (2008). Programa de acción específico 2007–2012. Salud del Migrante. Dirección General de Relaciones Internacionales. https://www.ssaver.gob.mx/transparencia/files/2011/11/Programa-de-Acci%C3%B3n-Espec%C3%ADfico-2007-2012-Salud-del-Migrante.pdf. Accessed 9 Jan 2012.

Trigueros, P. & Rodríguez, J. (1988). Migración y vida familiar en Michoacán. In G. López (Ed.), *Migración en el occidente de México* (pp. 201–232), Morelia: El Colegio de Michoacán.

Part II
Environment and Disasters

Chapter 9
Intervention and Security Strategies for Persons with Disabilities in Emergency and Disaster Situations

Iris W. Cátala Torres

Abstract A natural disaster represents a situation of threat or danger that influences the vulnerability of the human being. The effects of natural disasters result in a rise in mortality, morbidity and disability rates that severely impact the affected country. The vulnerability caused by a natural disaster promotes a state of susceptibility to loss, damage or injury, and people with disabilities are considered the most vulnerable population. This is particularly relevant when it comes to protecting those most disadvantaged in emergency situations. Moreover, emergency services are put to the test when it comes to caring for disabled people. In order to minimize risks and prevent damage, it is important to make studies that deeply address the specific needs of this population. Taking care of vulnerable victims in a disaster, such as people with disabilities, is an issue that illustrates our evolution as a society in the provision of services and the consideration of human rights.

Keywords Vulnerability · Natural disasters · Psychosocial intervention
People with disability

9.1 Introduction

A natural disaster can be defined as a sudden ecological disruption that exceeds the capacity adjustment of the affected community and requires immediate assistance (World Health Organization 1980; Lechat 1979). Natural disasters include earthquakes, landslides, tsunamis, floods, hurricanes, volcanic eruptions and droughts (Lechat 1979). A disaster represents a situation of threat or danger that influences the vulnerability of the human being and has consequences in terms of damage, loss and interruption of daily activities. These impacts can be so extensive that it is impossible for affected people to have sufficient mechanisms to recover effectively (Wisner et al. 2012).

Dr. Iris W. Cátala Torres, Professor-Researcher, University of Puerto Rico, Medical Sciences Campus. Email: iris.catala@upr.edu

© Springer International Publishing AG, part of Springer Nature 2018   115
Ma. L. Marván and E. López-Vázquez (eds.), *Preventing Health and Environmental Risks in Latin America*, The Anthropocene: Politik—Economics—Society—Science 23, https://doi.org/10.1007/978-3-319-73799-7_9

The vulnerability caused by a natural disaster promotes a state of susceptibility to loss, damage or injury. Characteristics reflected in a person, in a group and in situations adversely influence their ability to anticipate, adjust, resist and recover from the impact of a disaster (Wisner et al. 2004). The population is vulnerable to a natural phenomenon when it is susceptible to damage and has difficulty recovering from it.

Vulnerability to disasters depends on the type of event (hurricanes, flash floods, landslides and earthquakes), the physical characteristics of the affected area, and the social, cultural, psychological, demographic and economic characteristics of the population (Britton 1987). People with disabilities, children and the elderly are considered the most vulnerable in an emergency or disaster situation. This is particularly relevant when it comes to protecting the most disadvantaged and vulnerable populations in emergency situations (Barile et al. 2006).

The effects of natural disasters result in a rise in mortality, morbidity and disability rates that severely impact the health of the affected country. Similarly, emergency response and disaster recovery and mitigation efforts related to disabilities are put to the test (Reinhardt et al. 2011). Natural disasters have a significant impact on populations with spinal cord related disabilities, severe brain trauma, amputations, fractures, and injury to the nerves of the peripheral nervous system (Eldar 1997). Similarly, people suffering from conditions related to mental health, including anxiety, depression, post-traumatic stress, and others are affected (Fan et al. 2011; Yun et al. 2010). Because of natural disasters, people with acute disabilities may also experience exacerbations of pre-existing chronic conditions (Chan/Griffiths 2009; Chan/Sondorp 2007).

In the International Classification of Functioning, Disability and Health (ICF) of the World Health Organization, disability is defined by the interaction between people who have a disease (e.g. cerebral palsy, Down's syndrome and depression) and personal and environmental factors (WHO 2013). The Convention on the Rights of Persons with Disabilities (CRPD) aims to 'promote, protect and ensure the full and equal enjoyment of all human rights and fundamental freedoms for all persons with disabilities, and promote respect for their inherent dignity' (WHO World Disability Report 2011).

9.2 Empirical Data on the Conditions that Reveal the Most Attention Within the Spectrum of Vulnerability to Disasters

According to the World Health Survey, about 785 million people aged 15 and over (15.6%) live with a disability, while the World Morbidity Burden project estimates a figure close to 975 million (19.4%). The World Health Survey indicates that of the estimated total of persons with disabilities, 110 million (2.2%) have very significant difficulties of functioning, while the World Morbidity Burden amounts to 190 million (3.8%) people with 'severe disability' (the equivalent of disability associated with conditions such as quadriplegia, severe depression or blindness) (WHO 2011).

Some research suggests that more than one-third of adults and children affected by a disaster suffer from post-traumatic stress disorder after the event (Norris et al. 2002; Vernberg et al. 1996). After disasters, there is an increase in the risk of developing problems such as substance abuse, anxiety and depression, as well as issues with adjustment, interpersonal relationships, vocational difficulties and long-term physiological changes that are subsequently related to an increase in health problems and suicide (Norris et al. 2002). Studies have reported an increase in respiratory, gastrointestinal, and cardiovascular symptoms five years after a natural disaster (Logue et al. 1981; Katsouyanni et al. 1986).

Some epidemiological data on physical and emotional damage caused by natural disasters indicate that the incidence of morbidity and mortality due to disasters is highly variable and depends on several human and environmental factors (Armenian et al. 1997; Centre for Research on the Epidemiology of Disaster 2011). Earthquakes in the Republic of China from 2000 to 2009 resulted in 387,829 injured and 87,947 people killed (CRED 2011). Similarly, a high percentage of people affected were injured (4.4%) or died (2.5%) in the earthquake in Armenia in 1988. In turn, data from the Haiti earthquake in 2010 revealed that about 300,000 people (8.8%) were injured and 222,570 (6.5%) died out of a total of 3,400,000 people affected. In terms of data from natural disasters in the United States of America, from 2000 to 2009, 523 people were injured (1.7%) and only three deaths were reported (CRED 2011).

Consequently, disability in developed regions is relatively poorly studied. The major types of disabilities documented in the wake of a natural disaster include traumatic brain injury, limb amputation, fractures and trauma to the nerves of the peripheral nervous system (Li et al. 2009; Redmond/Li 2011). Damage considered minor includes conditions such as foot and leg inflammation, limb pain, tendonitis, ulcer infections and pain in the skeletal muscle area (Bai/Liu 2009; Centres for Disease Control and Prevention 2011).

9.3 Pre-existing Conditions

People with pre-existing physical disabilities and others with chronic physical or mental conditions, whether young or old, are at risk of additional comorbid conditions because of a natural disaster. Even when these persons have been safely evacuated and relocated, the loss of medication, support equipment and support staff can worsen the condition of this vulnerable population, eventually affecting their functioning and life quality (WHO 2005; Priestley/Hemingway 2007; Ke et al. 2010). Moreover, these people often find it difficult to meet their basic needs after a natural disaster, in the absence of adequate transportation and shelter in response plans at their respective locations (Mori et al. 2007; Chan/Sondorp 2007).

Similarly, compromising both basic and specialised health care following a natural disaster increases the risk of medical complications such as infection or organ failure, which can even cause death (Takahashi et al. 1997; Chou et al. 2004). According to studies conducted after the Hanshin-Awaji earthquake in 1995, it was found that the number of deaths of persons suffering from pre-existing physical conditions was doubled (Osaki/Minowa 1999). In turn, another study found that after the Taiwan earthquake in 1999, a significant number of deaths were found in pre-existing moderate physical conditions and the deaths of people with pre-existing mental conditions doubled (Chou et al. 2004).

9.4 General Aspects in the Principles of Action in Psychosocial Intervention in Catastrophes

A psychosocial perspective assists in reducing the duration of symptoms caused by a natural catastrophe (Fernández 2005). Psychosocial care is a community-based programme that provides basic mental health services in affected countries. This programme seeks to restore security, independence and dignity to individuals; promotes resilience in the community; and prevents morbidity for psychiatric reasons and subsequent social problems (Becker 2007). It is a spectrum of comprehensive care that is useful in the short and long term for meeting the needs of survivors of natural disasters, and that simultaneously provides physical and emotional support (Becker 2007).

It relates to the emergency support of a psychologist specialised in crisis intervention, who treats multiple victims. This kind of intervention is characterised by avoiding traumatisation by taking the victim away from the impact scenario. It also offers physical security, psychological support such as crisis control, physical contact when relevant and allows the victim to focus on the present. In addition to normalising symptoms, the psychologist provides information about what has happened and helps sort the traumatic experience and aftermath (Robles/Medina 2002).

According to the WHO (2012), the three basic principles of action in psychosocial intervention are: to observe, to listen and to connect. These principles serve as a guide when contemplating and approaching a crisis situation, to reach out to those affected and to understand their needs and bring them into contact with practical support and information. The following are some examples describing the principles proposed by WHO (2012):

1. Observe:
   Check security.
   Check for people with obvious urgent basic needs.
   Check for people who have severe distress reactions.

2. Listen:
   Address those who may need help.
   Ask what people need and what worries them.
   Listen to the people and help them calm down.
3. Connect:
   Help people solve their basic needs and have access to services.
   Help people cope with problems.
   Give information.

Reunite the affected people with their loved ones and social support networks.

As part of the psychosocial intervention, it is important to distinguish between adequate and inadequate interventions. This provides effective help that promotes high levels of recovery in the disaster-affected person. According to the WHO (2012), the adequate and inadequate aspects in a psychosocial intervention in catastrophes are:

Adequate aspects in a psychosocial intervention:

– Try to find a quiet place to talk, minimising outside distractions.
– Respect privacy and maintain the confidentiality of the story.
– Stand close to the person but keep the appropriate distance according to age, gender and culture.
– Let the person know that you are listening.
– Be patient and stay calm.
– Offer specific information if you have it. Be honest about what you know and do not know.
– Give the information in a way that the person can understand; that is, in a simple way.
– Let them know that you understand how they feel and regret their loss and what happened to them, such as being homeless or losing a loved one. 'I am so sorry. I imagine this is very sad for you.'
– Let them know you recognise their strengths and the way they are helping themselves.
– Leave room for silence.

Inadequate aspects in a psychosocial intervention

– Do not pressure the person to tell their story.
– Do not interrupt or rush the person telling his/her story.
– Do not touch the person if you are not sure that it is appropriate.
– Do not judge what they have done or failed to do or their feelings.
– Do not say things like 'You should not feel this way,' or 'You should feel lucky for being a survivor.'
– Do not make up things you do not know.

- Do not use excessively technical expressions.
- Do not tell the story of another person.
- Do not talk about your own problems.
- Do not make false promises or give false reassuring arguments.
- Do not think or act as if you have to solve all of the person's problems in his/her place.
- Do not take away their strength, their sense of being able to take care of themselves.
- Do not talk about other people in negative or derogatory terms (for example by calling them 'crazy' or 'disorganised').

Disasters such as Hurricane Katrina in 2005 provided the emphasis of urgency on the need for workers to receive specialised training to assist people with a variety of disabilities during emergencies. People with different types of conditions and disabilities require different types of assistance (Rowland et al. 2007). As expressed by the United Nations (2014), 'persons with disabilities are disproportionately affected in situations of disaster, emergency and conflict because evacuation measures, response (including shelters, camps and Food distribution) and recovery are inaccessible' (United Nations 2014).

The negative consequences of natural disasters usually overload society, including the health system (Kelen et al. 2006), especially in the resource-poor regions of the world where most natural disasters have occurred (Smith/Petley 2009). Providing rehabilitation services turns out to be limited in the immediacy of the emergency. However, incorporating concerted rehabilitation strategies for the vulnerable population can help reintegrate affected people, supporting their families and promoting their reintegration into society (Reinhardt et al. 2011).

Organisations in charge of providing rehabilitation services face significant challenges when they reach the disaster area. Even when relocation is effective, staff should be alert to coordination with other governmental and non-governmental organisations that are providing services at the same time in the emergency (Wisner et al. 2012; Rathore et al. 2008; Raissi et al. 2007). As a result, these leading teams in providing assistance to persons with disabilities should assume the responsibility of coordinating directly with rehabilitation specialists, initially to ensure appropriate interventions during the initial emergency, and subsequently to ensure adequate follow-up (Priestley/Hemingway 2007; Raissi et al. 2007).

It is important not to overlook that sometimes cultural beliefs related to rehabilitation and disability limit the impact of services to the population considered to be vulnerable (Priestley/Hemingway 2007; Raissi et al. 2007). Disability is often viewed as a condition that limits a person's ability to be functional (Ingstad 1995). For example, in several regions of the world where resource scarcity prevails, it is believed that people with an invariable disability are permanently dependent on their family or other lifelong supports (Office for the Coordination of Humanitarian Affairs 2010; Rathore et al. 2011).

9.5 Management Strategies in Psychosocial Intervention for People with Disabilities Before Disasters

Literature reflects as a leading subject the matters involved in the preparation before disasters; namely, the location of people at the time of the crisis, the need to develop an effective rescue and evacuation plan, and the availability of accessible resources to enable disaster interveners to do their work effectively (Thompson/Gaviria 2004). Studies show that the inclusion of the needs and views of people with disabilities in the stages of disaster response, and especially in planning and preparation, can significantly reduce the vulnerability of people and increase the effectiveness of response and recovery efforts (United Nations 2014).

When we work in the care of people with disabilities in the face of a natural disaster, it is important to know in which phases we can be most effective and offer help that meets their needs. According to Moeller (1997), there are four phases to face the disaster: (1) Pre-event, (2) Danger, (3) Response and (4) Recovery. However, this author considers the pre-event phase as one of the most important when it comes to the care of persons with disabilities. Accidents and disasters are anticipated during this phase; planning for the response to the emergency by the designated organisations is carried out, as well as the identification of essential resources, assistance equipment and specialised staff. Also, before the event, such planning focuses on identifying the disabled population within the community. In brief, the pre-event phase is a critical phase for the care team to design guidelines for the creation of training aimed at assisting people with disabilities in the event of disasters or catastrophes (Moeller 1997).

Likewise, the WHO (2005) emphasises post-disaster actions. By way of instruction, it recommends the following attention factors in both the acute and reconstruction phases:

I. Acute phase (usually six to twelve weeks after the disaster)

1. Identify injured people and provide the health care needed to protect life and reduce the future development of a disability or impairment.
2. Carry out a registry and monitoring of the people affected by a disability.
3. Provide care to people who already have a disability at the time of the disaster or event.
4. Implement curative therapies and interventions to prevent the development of another condition.
5. Work collaboratively with rehabilitation-orientated psychosocial intervention staff to prevent future complications and provide the community with the necessary resources to reduce risk.
6. Transfer people with disabilities to referral centres (if they exist) that have better rehabilitation facilities.

7. Establish a multidisciplinary team or helm group to develop plans that meet long-term rehabilitation needs, taking into account the resources available within the socio-economic environment of the country affected by the disaster.
8. Provide psychosocial intervention to the disabled person and his/her family.

II. Reconstruction phase (from the beginning of the fourth month onwards)

1. Identify people who are likely to suffer from short- or long-term disabilities caused by or developed after the disaster.
2. Carry out a needs assessment of people with disabilities.
3. Conduct an assessment of existing resources and other community options for working with the issues of people with disabilities.
4. Carry out programmes focused on rehabilitation.
5. Develop the necessary infrastructure for the implementation of health services such as medicine in rehabilitation, specialised therapies and assistance equipment.
6. Initiate community-based programmes to ensure that people with disabilities have equal access to all their basic needs and are treated fairly.
7. Use the Universal Design as a reference for the creation of projects for the population with disabilities.

Additionally, people with disabilities or other special needs often have particular requirements that require more detailed planning in the event of a disaster. The State Emergency Management Agency of the United States of America/the American Red Cross (2009) consider the following measures important when making preparations for people with disabilities:

1. Know what to do in the event of power outages and injuries. Learn how to connect and turn on an emergency generator for medical equipment.
2. Consider getting a medical alert system that will allow the person with disabilities to call for help if they are immobilised in an emergency.
3. If the disabled person uses an electric wheelchair or battery-operated vehicles, have a manual wheelchair as a reserve.
4. Teach the people who may have to help a person with a disability in an emergency how to handle the equipment.
5. Find a suitable place to store the reserve equipment (of mobility, medical, etc.) in the house of a neighbour, the school or place of work.
6. Plan that more than one member of the personal support network will assist the person with a disability in an emergency.
7. If those who care for people with disabilities belong to an agency, sk if the agency offers special services in case of emergencies.
8. If the person is visually impaired, deaf or hard of hearing, plan ahead of time for someone to transmit essential emergency information in the event that radio or television do not work.

9.6 Dignified Treatment and Consideration for People with Disabilities

According to Corporación Gestión Ecuador/USAID United States (2010), disability refers to the limitations, impairments or difficulties that a human being has to perform certain 'normal' daily activities due to the alteration of his physical, auditory, visual, or intellectual functions. Disability can be acquired or congenital, temporary or permanent; it can also occur due to environmental constraints such as barriers and non-equalisation of opportunities. It is important to note that people with some type of disability are still people. Using terms such as 'disabled', 'impaired', 'disabled people', 'blind', 'deaf', among others, is not an appropriate way to refer to this population. Attending to people with disabilities and, even more, their inclusion in social development processes implies adopting a work approach focused on everyone's right to be treated with dignity. In interventions it is important to emphasise the dignified treatment of people with disabilities by taking into account the following considerations (Corporación Gestión Ecuador/USAID United States 2010; OPPI 2009):

1. People with disabilities deserve the same treatment as people without disabilities.
2. It is important to be patient and keep calm when you want to help a person with a disability in an emergency.
3. To refer to people with disabilities you should use their names.
4. In the event of an evacuation and removal, the person with the disability and his or her relatives must be informed about the staff in charge of the shelter.
5. Within a shelter, greater attention should be given, if possible, to people with disabilities who present greater difficulties.
6. As far as possible, adapt the infrastructure of the shelters so that they are accessible to people with disabilities.
7. Speak directly to the person, not his or her companion, a third person or interpreter.
8. Whenever you are introduced to a person with a disability, give him or her your hand.
9. When a visually impaired person is present, identify yourself and identify your companions.
10. If you offer help, wait for it to be accepted. Then listen, wait and act according to the instructions given by the person with a disability.
11. Treat adults as adults. Do not minimise their cognitive skills.
12. Do not lean or lie in the wheelchair or motorised car of a person with a disability. Remember that most of these people consider these instruments as extensions of their body.
13. Listen carefully when talking to people who have speech or communication problems; always wait for them to finish talking.

14. Stand at the same level as the person you are talking to if he/she is in a wheelchair, so that you can make eye contact.
15. When you talk to a person with hearing disabilities, touch them lightly on the shoulder or call their attention by moving their hand.

The following support techniques in the intervention with people with disabilities (Corporación Gestión Ecuador/USAID United States 2010) are presented below:

1. Ask people with disabilities for information to manage the assistance. The information provided is essential to make an appropriate intervention.
2. Do not mistreat, misuse or make jokes about the assistance devices of people with disabilities.
3. Not all technical aids are necessary in case of an emergency. However, technical aids such as wheelchairs, goggles, canes, walkers and crutches are the most used and recommended to have accessible at the time of emergency.
4. It is important that the person with disabilities carries his/her identification documents as well as the medication prescribed for the management of the emergency.
5. Speak clearly and in plain language to the person with a disability. Provide clear instructions.
6. When the person with disabilities and his/her family are in a shelter, it should be explained where they are, emphasising where location, the estimated stay time and the follow-up plan while living together.
7. Provide safety and procedure recommendations. Give the people who are close by an introduction on the proper treatment and handling of the persons with disabilities.
8. When preventive actions such as drills are carried out, instruct the families of people with disabilities or the people with disabilities on the preparation recommendations and the emergency or disaster actions.
9. A registry of homes of persons with disabilities should be available; if possible, carry out a census to be used as an up-to-date resource to support transportation and mobilisation.
10. If you do not know how to move the person with disabilities and he/she is not at risk, it is best not to move him/her immediately.
11. It is important to exercise caution when mobilising a person with a disability, so as not to cause any harm to the person or his or her technical assistance.

9.7 Final Considerations

We have argued that the events brought about by natural disasters have a significant impact on the population with disabilities. At the same time, intervention and mitigation responses to these circumstances are a reason for constant evaluation, given the different needs of different population sectors. This is manifested in the immediacy of the onset of a disaster, affecting at the time populations with disabilities and scientific data that sustain an increase in adverse health symptoms in post-event years. This invites the design of programmes focused on the post-disaster stage, to address the physical and emotional sequels experienced by the population with disabilities due to the natural disaster.

It is important to consider making studies that deeply address the particular aspects of the disability and the specific needs of the population that suffers it. This is with the intention of preventing risk and damage, addressing directly people with disabilities. An intervention lacking these aspects makes it difficult for the affected person to face the disaster and react appropriately before, during and after the disaster. Similarly, it is necessary to train support staff to avoid generic considerations and prepare them for varied and specialised care in the management of the variety of conditions faced by people with disabilities.

We have emphasised that a psychosocial perspective intervention takes the lead in providing assistance in disaster situations. It is focused on the community and considers multicultural aspects at population level. It also seeks a balance between providing security and promoting self-efficacy, independence and incorporation into society, moving away from inappropriate interventions in a time of high vulnerability.

It is vital to take into account the following major care aspects with regard to attending to people with disabilities: (1) Focused coordination of services between organisations that provide direct attention; (2) Weighting of cultural beliefs related to the disability, as these may limit appropriate care; (3) Development of effective security, rescue and evacuation plans; and (4) Accessibility of resources available for an effective intervention.

Caring for vulnerable victims in a disaster, such as people with disabilities, is an issue that illustrates our evolution as a society in the provision of services and the consideration of human rights. Responsible and effective action by aid groups promotes safety and dissipates ambiguities regarding the particular needs of this population. Efficient use of strategies in rehabilitation and psychosocial intervention not only strengthens a health system, it also facilitates optimal post-disaster recovery and reintegration of this population into society, thus preventing victimisation. Delays in providing emergency assistance can reduce the effectiveness of post-disaster treatment, leading to multifactorial problems.

It is imperative to promote a learning culture that includes prevention, intervention and post-intervention aspects during disasters and emergencies. Added to this is the preparation to address the population with disabilities which must provide specialised care corresponding to all stages of the disaster. This makes it possible to consult more people who, because of their roles, have the experience and expertise to develop, strengthen and implement an effective plan, in tune with the needs of the country's different sectors. This plan must include the input of people with disabilities, as they are a valuable source of reference during the design of specific strategies, and because this consideration is nothing more than providing a space for inclusion, promoting universal design.

Significant progress has certainly been made with regard to the dignified treatment of the population with disabilities, but we need to continue working on awareness-raising when providing services to people with disabilities. Being aware, flexible and actively listening facilitates adequate identification of the population's urgent needs. Respect for human dignity, recognition of limits to issues of difficult management and progress towards affordable alternatives which address adverse situations with integrity, security and effectiveness all make sense when providing services during emergencies and disasters.

References

Agencia Federal para el Manejo de Emergencias & Cruz Roja Americana. (2009). Preparación para casos de desastre para personas con discapacidad y otras necesidades especiales. [Guía]. Jessup, Maryland: Author. http://www.fema.gov/pdf/library/spa_pfd_all476.pdf. Accessed 1 juin 2014.

Armenian, H.K., Melkonian, A., Noji, E.K., Hovanesian, A.P. (1997). Deaths and injuries due to the earthquake in Armenia: a cohort approach. *International journal of epidemiology*, 26(4), 806–813.

Bai, X.D., & Liu, X.H. (2009). Retrospective analysis: the earthquake-injured patients in Barakott of Pakistan. *Chinese Journal of Traumatology* (English Edition), 12(2), 122–124.

Barile, M., Fichten, C., Ferraro, V., Judd, D. (2006). Ice storm experiences of persons with disabilities: Knowledge is safety. *Review of Disability Studies*, 2(3), 35–48.

Becker, S.M. (2007). Psychosocial care for adult and child survivors of the tsunami disaster in India. *Journal of child and adolescent psychiatric nursing*, 20(3), 148–155.

Britton, N.R. (1987). Towards a reconceptualization of Disaster for the enhacement of Social preparedness. In R.R Dynes., B, DeMarchi., P, Carlo (Eds.), *Sociology of Disasters: Contributions of Sociology to Disasters Research* (pp-31–55). Milan: Franco Angel.

Centers for Disease Control and Prevention. (2011). Post-earthquake injuries treated at a field hospital – Haiti, 2010. MMWR: *Morbidity and Mortality Weekly Report*, 59(51), 1673.

Center for Research on the Epidemiology of Disasters. (2011). EM-DAT: *Te international disaster database*. Brussels: CRED.

Corporación Gestión Ecuador and USAID Estados Unidos. (2010). *Guía de Atención a personas con Discapacidad en casos de emergencias y desastres para instituciones que trabajan en gestión de riesgo y población en general*. [Guía]. Ecuador: Authors.

Chan, E.Y.Y. & Sondorp, E. (2007). Medical interventions following natural disasters: missing out on chronic medical needs. *Asia Pacific Journal of Public Health*, 19 (1_suppl), 45–51.

Chan, E.Y., & Griffiths, S. (2009). Comparison of health needs of older people between affected rural and urban areas after the 2005 Kashmir, Pakistan earthquake. *Prehospital and Disaster Medicine*, 24(05), 365–371.

Chou, Y.J., Huang, N., Lee, C.H., Tsai, S.L., Chen, L.S., Chang, H.J. (2004). Who is at risk of death in an earthquake? *American Journal of Epidemiology*, 160(7), 688–695.

Eldar, R. (1997). Preparedness for medical rehabilitation of casualties in disaster situations. *Disability and Rehabilitation*, 19(12), 547–551.

Fan, F., Zhang, Y., Yang, Y., Mo, L., Liu, X. (2011). Symptoms of post-traumatic stress disorder, depression, and anxiety among adolescents following the 2008 Wenchuan earthquake in China. *Journal of Traumatic Stress*, 24(1), 44–53.

Fernández, J.M. (2005). *Apoyo psicológico en situaciones de emergencia*. Madrid: Ediciones Pirámide.

Ingstad, B. (1995). Disability and culture. Univ of California Press.

Katsouyanni, K., Kogevinas, M., Trichopoulos, M. (1986). Earthquake-related stress and cardiac mortality. *International Journal of Epidemiology,* 15(3), 326–330.

Ke, X., Liu, C., & Li, N. (2010). Social support and quality of life: a cross-sectional study on survivors eight months after the 2008 Wenchuan earthquake. *BMC Public Health,* 10(1), 573.

Kelen, G.D., Kraus, C.K., McCarthy, M.L., Bass, E., Hsu, E. B., Li, G., Green, G. B. (2006). Inpatient disposition classification for the creation of hospital surge capacity: a multiphase study. *The Lancet,* 368(9551), 1984–1990.

Lechat, M. F. (1979). Disasters and public health. *Bulletin of the World Health Organization,* 57 (1), 11.

Li, W., Qian, J., Liu, X., Zhang, Q., Wang, L., Chen, D., Lin, Z. (2009). Management of severe crush injury in a front-line tent ICU after 2008 Wenchuan earthquake in China: an experience with 32 cases. *Critical Care,* 13(6), R178.

Logue, J. N., Hansen, H., & Struening, E. (1981). Some indications of the long-term health effects of a natural disaster. *Public Health Reports,* 96(1), 67.

Moeller, D. (1997). Disaster response. In D.W.Moeller (Ed.), *Environmental Health,* Revised Edition, 385–409. Cambridge, MA: Harvard University.

Mori, K., Ugai, K., Nonami, Y., Kirimura, T., Kondo, C., Nakamura, T., Kaji, H. (2007). Health needs of patients with chronic diseases who lived through the great Hanshin earthquake. *Disaster Management & Response,* 5(1), 8–13.

Norris, F.H., Friedman, M.J., Watson, P.J., Byrne, C.M., Diaz, E., Kaniasty, K. (2002). 60,000 disaster victims speak: Part I. An empirical review of the empirical literature, 1981–2001. *Psychiatry: Interpersonal and Biological Processes,* 65(3), 207–239.

OCHA (2010). Humanitarian response Haiti. Port au Prince, OCHA: Author.

OPPI de Puerto Rico (2009). Formas Correctas de Expresión hacia las personas con Impedimento. [Manual]. Estado Libre Asociado de Puerto Rico.

WHO (2013). Discapacidad y salud: Nota descriptiva N°352. Septiembre de 2013. http://www. who.int/mediacentre/factsheets/fs352/es/ Accesed 1 juin 2014.

WHO (2011). Informe Mundial sobre discapacidad. Ginebra: OMS.

WHO (2012). War Trauma Foundation y Visión.

Mundial Internacional. *Primera ayuda psicológica: Guía para trabajadores de campo.* Ginebra: OMS.

Osaki, Y., & Minowa, M. (1999). Factors associated with earthquake deaths in the Great Hanshin-Awaji Earthquake, 1995. [Nihon koshu eisei zasshi] *Japanese Journal of Public Health,* 46(3), 175–183.

Priestley, M., & Hemingway, L. (2007). Disability and disaster recovery: a tale of two cities? *Journal of Social Work in Disability & Rehabilitation,* 5(3–4), 23–42.

Raissi, G.R., Mokhtari, A., Mansouri, K. (2007). Reports from spinal cord injury patients: eight months after the 2003 earthquake in Bam, Iran. *American Journal of Physical Medicine & Rehabilitation,* 86(11), 912–917.

Reinhardt, J.D., Li, J., Gosney, J., Rathore, F.A., Haig, A.J., Marx, M., DeLisa, J.A. (2011). Disability and health-related rehabilitation in international disaster relief. *Global Health Action,* 4.

Rathore, F.A., Farooq, F., Muzammil, S., New, P.W., Ahmad, N., Haig, A.J. (2008). Spinal cord injury management and rehabilitation: highlights and shortcomings from the 2005 earthquake in Pakistan. *Archives of Physical Medicine and Rehabilitation,* 89(3), 579–585.

Rathore, F.A., New, P.W., Iftikhar, A. (2011). A report on disability and rehabilitation medicine in Pakistan: past, present, and future directions. *Archives of Physical Medicine and Rehabilitation,* 92(1), 161–166.

Redmond, A.D., & Li, J. (2011). The UK medical response to the Sichuan earthquake. *Emergency Medicine Journal,* 28(6), 516–520.

Robles, J.I., & Medina, J.L. (2002). Intervención psicológica en las catástrofes. Madrid: Síntesis.

Rowland, J.L., White, G.W., Fox, M.H., Rooney, C. (2007). Emergency response training practices for people with disabilities: Analysis of some current practices and recommendations for future training programs. *Journal of Disability Policy Studies,* 17(4), 216–222.

Smith, K., & Petley, D.N. (2009). *Environmental hazards: assessing risk and reducing disaster*. New York: Routledge.

Takahashi, A., Watanabe, K., Oshima, M., Shimada, H., Ozawa, A. (1997). The effect of the disaster caused by the great Hanshin earthquake on people with intellectual disability. *Journal of Intellectual Disability Research*, 41(2), 193–196.

Thompson, M., & Gaviria, I. (2004). *Cuba: Weathering the storm: Lessons in risk reduction from Cuba*. Oxfam America.

United Nations. (2014). Disability, natural disasters and emergency situations. Enable United Nations development and human rights for all (Bulletin). https://www.un.org/development/desa/disabilities/issues/disability-inclusive-disaster-risk-reduction-and-emergency-situations.html.

Vernberg, E.M., La Greca, A.M., Silverman, W.K., Prinstein, M.J. (1996). Prediction of post-traumatic stress symptoms in children after Hurricane Andrew. *Journal of Abnormal Psychology*, 105(2), 237.

World Health Organization. (1980). Emergency care in natural disasters. Views of an international seminar. *WHO Chronicles*, 34, 96–100.

World Health Organization. (2005). Disaster, Disability and Rehabilitation. Department of Injuries and Violence Prevention. [*Bulletin*]. Geneva, Switzerland.

Wisner, B. et. al. (2004). *Natural Hazards, People's Vulnerability and Disasters* (2 ed). London: Routledge.

Wisner, B., Gaillard, J.C., Kelman, I. (2012). *Handbook of Hazards and Disaster Risk Reduction*. Routledge, Abingdon, Oxfordshire, U.K.

Yun, K., Lurie, N., Hyde, P.S. (2010). Moving mental health into the disaster preparedness spotlight. *New England Journal of Medicine*, 363(13), 1193–1195.

Chapter 10
Social-Individual Behaviour Problems Regarding Early Warning Systems Against Disasters

Jesús Manuel Macías

Abstract This chapter addresses the relationship between early warning systems (EWS) and collective behaviour in Mexico, trying to establish a balance in their integrated meanings. It makes observations on how to understand EWS and introduces the subject of collective behaviour as a privileged thematic area in social sciences. It also introduces some elements to help clarify certain terms arising from theories about the social and individual behaviour that has governed the scope of EWS. It concludes that the design, configuration and operation of all existing EWS in Mexico should be reviewed to at least seek to reconcile these systems with the standards recommended by the International Strategy for the Reduction of Disasters.

Keywords Disasters · Collective behaviour · Warning systems
Natural hazards · Social sciences

10.1 Introduction

A clarification to be made at the outset regarding the subject of early warning systems (EWS) is that, taken in its loosest or broader sense, an EWS can be interpreted as an individual or collective action to anticipate and warn of an eventual event, whether it is an unfortunate or negative event or simply a non-negative but anticipatory event. Austin's contribution (2004) reflects the range of topics associated with EWS, for example, in the areas of civil and political crises and conflicts, humanitarian aid areas or possibilities of epidemic outbreak (ICRC 2009). This chapter refers to EWS that deal with disasters, whether they are mediated by natural or anthropogenic phenomena.

The intention of this contribution is to show the importance of a resource such as an EWS and the ways in which it is rendered unusable or underutilised by false conceptions about the system itself, for instance where insufficient understanding of

Dr. Jesús Manuel Macías, Researcher, CIESAS. Email: macserr@att.net.mx.

© Springer International Publishing AG, part of Springer Nature 2018   129
Ma. L. Marván and E. López-Vázquez (eds.), *Preventing Health and Environmental Risks in Latin America*, The Anthropocene: Politik—Economics—Society—Science 23, https://doi.org/10.1007/978-3-319-73799-7_10

related social phenomena prevails, or where practices of operation induced by probable lucrative interests, such as technology, intervene. Currently this is not a minor problem given the increase in communication resources provided by the Internet and social networks.

The failed implementation of EWSs cannot just be explained as the consequence of voluntarily or involuntarily ignoring the social aspects that determine its success or failure, because the bulk of this knowledge has been widely disseminated and even formalised in international organisations relevant to the United Nations, such as the International Strategy for Disaster Reduction (UNISDR) and the so-called Hyogo Framework for Action (HFA). This leads us to observe other current phenomena, such as the fact that some EWSs can incorporate elements like applications ('apps') for cell phones or radios or sirens, electrical and/or electronic devices that may find asignificant market in the consumption of security or false security.

Early warning systems have principally been adopted and developed because of their potential to save lives, particularly since the so-called International Decade for Natural Disaster Reduction (IDNDR) of the UN (Maskrey 1997), subsequently taken over by the ISDR and projected through the Hyogo Framework for Action. As Rogers/Tsirkunov (2011a) argue, even the World Bank has placed alert systems as one of its three pillars in the policies it promotes to reduce disaster risk.[1]

The introduction and progress of EWS in Mexico has been relatively recent. They have been developed since the second half of the Nineties. My colleagues and I suggested the introduction of the EWS against hurricanes in 1999, but before that we participated actively and critically in what was considered a 'volcanic warning system' with regard to the activity status (*semáforo*) of the Popocatépetl volcano. Observing from a distance the way these EWSs have worked allows us to affirm that their dominant design profile in fact excludes the population at risk and greatly limits the possibility of incorporating the knowledge produced by the relevant social sciences to improve its functioning. Therefore, it is not difficult to reprint the prophetic work of Cassandra of Troy, in anticipation of the eventual failure of these EWS.

This chapter addresses the relationship between warning systems and collective behaviour, trying to establish a balance in their integrated meanings. It begins with some considerations on how to understand EWS and introduces the subject of collective behaviour as the privileged thematic area in the social sciences, scrutinising its relevance and retelling the contributions of the affordable literature, basically dominated by North American authors.

It also contributes some elements intended to clarify certain terms arising from theories about the social and individual behaviour that has governed the scope of EWS. We return to the point of dominance of American sociology, in order to highlight both its conceptual weaknesses and the useful and necessary inputs for the adequate functioning of the EWS.

---

[1]The other two pillars are critical infrastructure and environmental buffers.

The end of the chapter offers an overview of the warning systems in Mexico, as a preamble to final conclusions.

10.2 What Is an Alert System?

An early warning system, in a more precise sense, can be understood as an organised set of elements (organisational, technological) and actors (scientists, officials, etc.) interacting from different levels and functions with a common purpose, which is to save the lives of people who are subject to a potentially destructive phenomenon, which requires anticipating its appearance and evolution. In recent years, a consensus proposal has been reached, promoted by UN agencies involved in the issue of disaster risk, with respect to its composition in four elements, namely: (a) knowledge of risk; (b) monitoring and alerting services; (c) dissemination and communication; and (d) responsiveness. The conjunction of interdependent functioning of these elements gives the adjective 'comprehensive' to the alert system and also indicates that these elements are central to people's well-being (UNISDR 2006).

Although social science research on problems related to early warning systems is already over sixty years old, towards the end of the 1990s, specific work began on analysing the different types of existing warning systems. Perhaps the most meticulous work has been done by Tierney (2000) at the University of Delaware's Disaster Research Centre to contribute a sociological analysis to the problem of creating a seismic warning system in the state of California. In the review of related global literature, she found 'hundreds' of different types of systems, technologies and warning devices. A selection of them accounts for forty different types of systems for various threats. Tierney established nine categories that define their differences, and which are synthesised into eight for our purposes, namely: (1) the technologies used to detect threats and their level of reliability; (2) the time needed to have precision in the forecast or prediction; (3) dependence on human mediation; (4) types of devices to emit warning signals (audible, visible, sensitive, etc.); (5) types of communication channels used to broadcast (multichannel or single channel); (6) familiarity of operators, recipients, institutionalisation of its components; (7) siting, buildings, factories, poles, etc.; and (8) goals and objectives regarding protection behaviour: evacuation of a facility, area or vertical evacuation, shelter, self-protection, etc.

Parallel to Tierney's efforts, Sorensen (2000), a recognised specialist in the field, published a twenty-year advance evaluation of warning systems, although this assessment was only applied to the case of the United States. The results of their assessments were critical for their country, because they noticed that the characteristics on the development of the alert systems were very unequal and differentiated, spatially and organisationally. This was reflected in the fact that the

American population was unequally protected and that, in many cases, they continued to resent the onslaught of potentially disastrous natural phenomena as unpleasant 'surprises'. However, Sorensen (2000) acknowledged progress in predictions for floods, hurricanes and volcanic eruptions, particularly highlighting the achievement in public disclosure of hurricane warnings. But he stated that 'there is no 100% reliable warning system for any hazard' (p. 119). He identified some progress on some issues and stagnation in others, but stressed that that country would not have been able to develop any comprehensive national strategy for a comprehensive alert to cover all threats everywhere.

Another effort to recognise the diversity of EWS, its achievements and problems was the EWS global survey developed by UNISDR and published in 2006. When effective, EWS have been considered one of the most cost-effective or profitable disaster prevention measures by the UNISDR (2006) and international financial agencies, therefore they have received much attention in these instances; consequently, the UNISDR itself developed a number of activities to promote the widespread implementation of EWS in member countries, and also sought to establish a 'global early warning system for all natural hazards based on existing national and regional capacities' (p. 5).

This organisation used practically all the structures linked to the UN to carry out the survey, especially its representatives in the different countries. In the conclusions of this survey, which was interested in understanding the national capacities and problems surrounding EWS, it was noted that: 'the weakest elements are related to the dissemination of alerts and the level of readiness to act'.

Likewise, the causes of these problems were identified as the inadequate degree of political commitment, weak coordination among the different actors involved and the lack of public awareness and participation in the development and operation of early warning systems. However, the survey also found that there are already many available capacities on which to build a comprehensive global EWS that is truly effective (UNISDR 2006). The recommendations derived from this survey, assuming that the 'people-centred' EWSs have the four components mentioned above (risk knowledge, monitoring, diffusion and responsiveness), were as follows:

1. Develop a comprehensive EWS at global level, rooted to existing early warning systems and capabilities.
2. Build national population-based EWSs.
3. Address the main shortcomings of global early warning capabilities.
4. Strengthen the scientific basis and information for early warning.
5. Elaborate the institutional bases for a global system of early warning.

There is no doubt that there is fundamental consensus on the enormous importance of EWS. It is clear from the tiny selection of examples mentioned that efforts to understand their characteristics and status, whether nationally or globally, have been significant. However, it is important to pause in the case of the ISDR Global Survey to observe some of the elements of its methodology and results.

The first thing to note about the survey is that the most relevant social scientists in the field of alert systems[2] are not identified as being involved in the process of designing questions, making revisions and proposing recommendations, so the emphasis is on the components of known threats. Attention is drawn, for example, to the involvement of experts in atmospheric phenomena, but the relevant sociologists and social psychologists, who have accumulated significant knowledge on the subject, do not appear to have been consulted; this has resulted in a lack of balance. What is particularly disturbing is that this could have easily been avoided.

Since the publication of the Global Survey, it is not possible to observe much progress in the field of EWS, although there are exceptions. A conception such as that proposed by the aforementioned 'people-centred' EWS survey may be key if it is understood that people are the determining factor in the success or failure of any EWS, but not if, though widely accepted, it diffuses as a political slogan only.

10.3   Social Sciences and EWSs

There are several resources which look at both the developments of the social sciences in the field of EWS and its main achievements. Quarantelli (1980) made a 'box cut' to clarify what he called a codification of research related to collective behaviour in evacuations and the relationship with alert mechanisms. He identified 103 relevant papers and discovered that, curiously, they could be divided into two groups, one 'people-centred' and the other 'organisation-centred.' That division, according to Quarantelli, reflected two separate currents that would have gone practically parallel, 'they relatively overlap a little,' he said. He was actually referring to cutting-edge research results from social psychology versus research-related problems that address organisation sociology.

The same separation of 'currents' had already been noted by Perry (1979) in his studies on decision-making in evacuation processes. He argued that important processes in responses to alerts proceed simultaneously at two levels of abstraction that refer both to the aspects of the individual and to the organisations involved. Quarantelli found that organisational aspects of evacuation had received more research attention, at a ratio of about 7 to 4, than aspects of individual behaviour.

It is important to note that, in the Fifties, the impetus for American sociologists to study the subject of disasters was the US army's need to know more about social behaviour in emergencies during the Cold War period, when the military had the imperative to prepare for the threat of a nuclear attack (Macías 2012). Quarantelli (1984: 4) pointed out: 'It is not surprising, therefore, that when social and behavioural scientists began conducting disaster research some 30 years ago, many of

---

[2]In the Global Survey document (UNISDR 2006) the names of those involved in the undertaking can be consulted.

the studies focused on the issue of disaster warnings and the reactions of potential victims. This interest has been maintained over the years.'

Later some aspects of the theoretical components related to that dispute will be considered. The sociology that took the lead in disaster research was identified as belonging to the 'Chicago school'; but, according to Ritzer (1997), it experienced a leadership crisis when its leader then (and creator of the term symbolic interactionism), Herbert Blumer, moved to the University of California at Berkeley.

According to Quarantelli/Dynes (1977), sociology had performed much more psychological-social than social-organisational research, yet a much greater social-organisational emphasis was being imposed after the 1950s. Disaster research, according to the authors mentioned, closely reflected this general picture in the discipline. Initially, research was conducted primarily at the National Opinion Research Centre (NORC) at the University of Chicago, where Quarantelli began his research to later form the Disaster Research Centre at the State University of Ohio, which moved to the University of Delaware in the early 1980s. In the mid-2000s, in 2006, Dennis Mileti and colleagues at the University of Colorado conducted another 'coding' of sociological research related to early warning systems (Mileti et al. 2006a, b), by means of an important effort of bibliographical collection that covered up to that year. It is a bibliography with a very extensive and comprehensive title: *Bibliography of public communication of risks on alerts for the public response of protection actions and public education*. The following chart (Fig. 10.1) shows the quantitative aspects of scientific social production, which point to a clear take-off since the 1950s, reaching their highest levels in the decade 1990–2000, which coincides with the International Decade for Natural Disaster Reduction (IDNDR), driven by the UN, but showing a downward trend after that decade.

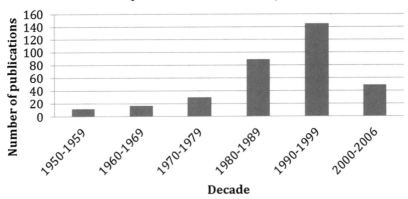

Fig. 10.1 Numbers of publications on disaster warning 1950–2006. *Source* Own elaboration with data obtained from Mileti et al. (2006a)

The bibliography mentioned then covered almost sixty years of knowledge production that included works not only by American authors, although it takes into consideration only literature in English, but also of other countries, with case studies developed in different places of the world. Theorisation on social and individual behaviour in the field of EWS was included.

Any definition of the EWS's reason for existence places the population at risk at the centre: saving the lives of people and reducing the loss of assets due to the materialisation of a destructive phenomenon. This implies the necessary existence of relationships between people to achieve this goal and, from the sociological point of view, it is about individual relationships within a socially definable collective. For this reason, the theoretical apprehension of these social relationships is at once fascinating as well as complex, because it is embedded in theoretical topics that have been the cornerstone of different interpretations and gnoseological positions on the matter. The components related to technology (monitoring, communications) and the forecast of destructive phenomena are very important as parts of the whole but do not determine the success of the overall operation of EWSs.

Social scientific development related to the subject of EWSs maintains a pre-eminence in sociology, and, as we know, there is not one 'sociology' but several, that have substantive differences in terms of conceptions and traditions. It is necessary, then, to locate the thematic axes (in this case of both sociology and psychology) in what is called 'collective behaviour', and a conceptual delineation. The designation focus on the notion of social or collective behaviour[3] (terms used here interchangeably) is that both may suggest an ascription to behaviourist currents that were (in the case of symbolic interactionism) very influential in both sociology and social psychology. My perspective on these issues coincides with the approaches that Marx made to relate the capacities of man and the world, and which Ollman (1975) expresses well as the work of three interconnected processes: perception, orientation (or behaviour) and appropriation. In this interconnection, perception necessarily leads to the orientation itself, which gives meaning and structure to one's perception (from the very moment in which man perceives something, perceived objects possess a certain order and relative value), at the same time pointing to one or more routes to where it should direct its future actions. In short, following Ollman, perception is the immediate contact of the individual with nature; the orientation, the way in which it comprises that contact; and appropriation, in general, the use made of it. But in almost all cases, appropriation requires some 'mediation' for man's abilities to join nature. Therefore, the use of the term 'behaviour' (even 'perception') here is suggested to be devoid of that burden related to some form of 'behaviourism'. The development of this notion in sociology or social psychology is not univocal, but has several conceptual proposals, as we will see.

According to Marshall (1996), the study of collective behaviour in sociology is almost its own disciplinary definition, although a certain parallelism with

---

[3]The current uses of these terms are distinguished more as 'collective behaviour' in sociology and 'social behaviour' in behavioural biology.

psychology is recognised. My approach to the analysis of these disciplines on the subject of collective behaviour and EWS, as has been said, goes more with an assumption of the social and collective in terms of Marx described by Ollman, who assumes that he is an individual in society. This brief phrase implies many things and resolves the apparent opposition between the individual and the collective concerning the effects of behaviour. On the other hand, it is important to note that this resolution is substantive in the case of EWSs. The relationship between the individual and the collective must necessarily be taken into account in the decision-making.

The old argument that divided philosophy—the relationship between consciousness and being—is still observed in social disciplines and in various theoretical variations. In the study of disasters, this issue offers guidance on understanding several of the social phenomena associated with warning systems and is also central to interpretating the empirical data between Marxist currents and symbolic interactionism. For example, it was this current that initiated the investigations related to alert systems with regard to collective behaviour in the United States. The most outstanding analysis of American sociologists on the problem of what they call the 'response' to alerts[4] is, in general, close to the theory of the emergent norm, one of those that are inscribed in the sphere of collective behaviour. According to Perry (Perry 1979; Perry et al. 1981), this theory was proposed by Turner/Killian (1972) as an alternative to the other two recognised in this sphere (theories of 'convergence' and 'contagion'), because it focuses on the development of norms defined by a condition that depends on certain circumstances, and because the expectations it generates as a function of some crisis or change in the social or psychological environment make the traditional norms inappropriate. In other words, the emerging norm theory emphasises the parameters and processes by which a new norm becomes established as the basis of collective behaviour; social conditions where no rules or instructions have been established promote the emergence of a new standard.

Perry and other authors (Drabek 1968; Gillespie/Perry 1976; Aguirre et al. 1998; Dash/Galdwin 2007) make such assumptions that otherwise are very generalised in other academic texts but also among other empiricist observers of disasters. Drabek (1968), quoted by Perry (1979: 27), sums it up very well: 'In disasters, those actions … are largely governed by emerging norms rather than by established norms, but norms after all.' It goes without saying that this theory is not without discussion. Some critical and alternative approaches are now suggested. First:

(a) Most of the researchers mentioned—American sociologists who have made the most relevant contributions to the subject of alert systems and collective behaviour—understand 'disaster' as the moment of the destructive impact, that is, the disaster is identified with the moment of destruction, the conversion of a

---

[4]Generally, the literature mentioned, which is basically in English, refers to the term 'warning', which in Spanish can be translated as 'alerta' or 'advertencia'. The sense in which it is used here is that of an alert.

'normal' state into a state of damage, etc. In this sense, disaster is an 'event' and therefore the possibility of the emergence of an emergent norm is the condition of the damage. Diverse contributions have offered an alternative conception of disaster as a social process and not just a condition that refers to the moment of destruction. A label sometimes used to describe this process is social disruption (Macías 1999).

(b) The theory of the emergent norm works by opposing the established 'traditional norm' to an 'emergent norm', as if the normative aspects of society were interchangeable and did not depend on social conditions. However, taking the arguments of another non-functionalist sociologist, (Poulantzas 1977), 'emerging norms,' new or transitional, are not independent of the conditions that originated the established norms. An alternative argument, which I do not agree with, is that of the social constructors Berger/Luckmann (2001), who attribute order, direction and stability to human existence, which is the product of society. Any temporary rupture of it could not be corrected without the same social productions.

(c) In the field of the study of collective behaviour, researchers have put 'disaster' in the same bag as riots, armed revolutions, etc.; that is, collective actions that not only respond to different social and contextual-environmental circumstances, but have different purposes. Gillespie/Perry (1976) stated: 'The most intense and complex situations that facilitate an emerging norm are those like riots, strikes or outbreaks of violence where the existing social order seeks to be annulled and opposition is shown in order to implement the intended action of change.'

In the analysis of several post-impact conditions in disasters of all kinds of mediating agents, such as hurricanes (Macías/Fernández 1999), floods (Macías 2009), and even anthropogenic mediators such as the gasoline explosion in the drainage system of Guadalajara in 1992 (Macías/Calderón 1994), what we have observed empirically is that social or collective behaviour, both of the people directly affected by the destructive impacts and by those who were indirectly affected by the destructions and by the human suffering that is witnessed, follows the demands of solving immediate needs,[5] and does not seek to change the social order or to oppose obstacles to the 'governability' of established authority, as is mistakenly believed. Rather, it has been observed that behaviours follow the same pattern of reproducing the same social relationships that those affected have maintained during their existence. Many specific examples can be pointed out in activities such as the direct rescue of relatives, the search for shelter, life in a temporary shelter, and recovery activities.

There is another aspect that forces us to exercise caution in the interpretation and transference of concepts in the field of collective behaviour. This is due to the

---

[5]Immediate needs include not only those of a material nature, but also those related to food, housing, health, and other moral needs, such as the identification of those responsible for the death of relatives or loss of belongings.

mistrust of the spatial sense in the theorisation of sociologists: one thing is mass behaviour (manifestations, marches, massive face-to-face protests), where something similar to the de-individualisation or soul of the mass may work, the depersonalisation of Le Bon (2007); and another is the behaviour of social organisations or organised social responses. Disasters, in their moments of destruction, manifest themselves at different spatial and social scales. I am drawing attention to this point of the scale of collective behaviour because it is relevant to the topic of EWS, which I will return to later. An alert population may have tens or thousands or even hundreds of thousands of inhabitants, and the physical contexts are also variable. Despite its limitations, the social science literature discussed in this section is useful with regard to EWS because it shows that people's behaviour in situations related to EWS is not linear or 'automatic'; it is not a stimulus-response relationship, etc., as assumed by many of the governmental and non-governmental agents that take part in it. The scholars mentioned have identified the assumptions of these agents as both generalised and misguided beliefs, and have therefore treated them under the condescending term of 'myths.'

10.4 Useful Results of the Knowledge About the Social Behaviour Before Alert Systems

In 2012, the US National Research Council (NRC 2013) organised a meeting with experts on early warning systems to highlight the main achievements in that country. The purpose of this meeting was extended to the search for answers to certain questions raised by the new social conditions influenced by current technological developments. These new technological incorporations refer to so-called 'social networks' or 'social media', and the widespread use of computer simulations of scenarios at risk and subject to massive alertness.

The substantive conclusions are virtually the same as those reported by the authors mentioned above. Many of them have been incorporated at international level in order to establish certain official standards, as can be seen in the UN International Strategy for Disaster Reduction (ISDR 2006) and in Rogers/Tsirkunov (2011b), namely:

1. Research using hypothetical scenarios does not faithfully represent the public's response.
2. It is necessary to carry out educational activities on the alarm system before an event.
3. Alerts (messages, artefacts) need to attract the attention of the population.
4. Warned people seek social confirmation before taking protective measures.
5. Messages should contain information that is important to the population.
6. Emergency planning and care bodies should consider the demographic data of affected populations during the preparation of warning messages.

7. The development of alert or warning systems should be adequately accessible to people with disabilities.
8. Alerting and warning is a process, not a single act.

The foregoing is placed at the level of efforts to formalise various aspects of alert systems. It is important to underline this because it becomes a reference to establish some guidelines for the developments in progress in Mexico. It is worth noting the conclusions that consider the changes introduced by the incorporation of techno-logical facilities in three areas: transmitting artefacts and alert emitters; applications; and simulations. The common reaction to these new conditions is that they should be used with great caution against the complex social process of response to alerts. In addition, it is essential to synthesise the main research findings that are key to any design or implementation of EWS. The relationship between people at risk and the contact that warns them of imminent danger forms a process that is, like any process, sequential. It begins with the reception of the alert (signal, message) and ends with the response that may be the adoption of protective measures (evacuation, search for shelter or other form of protection) or the dismissal of the alert, which is the decision not to act on the warning Between these two extremes of the sequence (the reception of the alert and the response), there are several stages: belief in the alert message, confirmation of the alert and an assessment of personal or group danger, and then making a decision. It should be noted that only in the decision-making part of this process, which Rohrmann (2000) mentions as the 'risk communication', have seven external variables and seven other internal variables that people use to process an alert message and make decisions been found. The authors mentioned in the following Table 10.1 have identified 32 general and specific factors that affect the process of warning response, including sender and receiver factors, situational and social contact factors (family, religion, ethnicity, socio-economic and educational level, age, sex, relationship with the community, distance and time relationships with respect to the threat, media and alert message).

The sequential parts indicated in the table relate to the individual-collective level that establishes the warning relationships involved in sending and responding to EWS.

The design and operation of the EWS are crucial, in order for it to fulfil its purpose of saving lives and reducing the loss of goods.

To increase the effectiveness of EWS, Sorensen (2000) has established some guidelines that are generally accepted as essential for good organisation, efficient coordination and high chances of success.

General principles that facilitate effective organisational coordination and responses are well defined: (1) all parts of the organisation need to know not only what they do or should do, but also what the other parties do and must do in case of emergency; (2) each party needs to know who is responsible for implementing each procedure. These two principles require organisational and functional integration. For this, the system itself must (3) have and recognise relationships and

Table 10.1 Sequence of the response to alerts

Quarantelli (1983)	Perry (1985)	Mileti/Sorensen (1990)	Lindell/Perry (1991)	Rohrman (2000)
Receiving alert message	Receiving alert	Listen to the alert	Identification of risk	Reception
Definition of the situation	Interpreting the message	Understand the content of the alert message	Risk evaluation	Attention
Perception or belief in the message	Personal risk evaluation	Believe that the warning is true and accurate	Risk reduction	Understanding
Social confirmation of message	Evaluation of threat and protection activity	Customise alert to oneself	Protective response	Confirmation
Other observations of the situation	Response	Confirm that the alert is true and that others are aware of it		Acceptance
Protection response		Respond by taking a protective action		Retention realisation

Source Author's own elaboration

mechanisms of communication; these organisational arrangements should not be confined to the system, but must incorporate other systems and external agents useful in the provision of information. Although the aforementioned requires the best organisation, experts point out that it is very important (4) to maintain flexibility.

10.5 Warning Systems in Mexico

In Mexico, there are several formally organised early warning systems, although with different characteristics in their conceptualisation, components and organisation. It is not possible, for reasons of space, to go into depth about each and every one of these systems (the cases of early warning systems for droughts and forest fires will be omitted), so only some relevant aspects of their main characteristics are pointed out at a general level.

The earliest warning system of massive potential impact is the Earthquake Warning System (SAS, in Spanish) that began operating formally in 1991 and covered only the Mexico City area. The coverage of this system has now been extended, or replicated, to cover the cities of Oaxaca, in Oaxaca; Acapulco and Chilpancingo in Guerrero; and Morelia in Michoacán (CIRES 2014). It is an automated system that links seismic sensors installed on the Pacific Ocean floor with stations that repeat the signals they emit when they detect an earthquake. These signals reach the cities mentioned and are received by devices that can retransmit them in the form of audible signals by various channels, such as electronic mass media (radio, television) or special radio receivers.

There are several scientific evaluations of this system, which has operated with a lot of disorganisation and amid a regulatory vacuum that results in a lack of clear responsibilities regarding the decisions for broadcasting an alert. In fact, the mechanism that triggers the alert signal is the algorithm that regulates the input and output signals of the technological system. The political authorities of the Federal District keep the SAS in an area of uncertainty regarding responsibility for making the decision to alert. The operation of the SAS has been approved for 23 years and has had dubious results, mainly because it has not operated in conditions of a really strong earthquake.

Some evaluations have been made of its functioning, most of them by its authors and promoters (Lee/Espinosa-Aranda 1998; Espinosa-Aranda et al. 1996, 2009). I know of only two independent evaluations (Goltz/Flores 1997; Suárez et al. 2009) that have pointed out several problems that suggest a state of unreliability. Except for these two independent appraisals, evaluations basically refer to the technological functioning of the system. The independent evaluations have observed the operation of the SAS from the perspective of the population that is supposed to be alerted by this system. They note, in particular, that the SAS arrangements only consider the population peripherally, seeing it purely as an alert object. So far, no social knowledge base has been recognised to support such a system under the conditions of a disastrous earthquake.

The warning system of the Popocatépetl Volcano, known as the 'Semáforo Volcánico' (activity status of the volcano), began to function formally in 1996 as a result of the eruptive crisis of December 1994 and the expanded monitoring of different volcanic activity (gases, deformation, earthquakes). The alert system was designed with two facets, one of an internal character, which links monitoring scientists with Civil Protection officials. The other facet is the proper 'semáforo' (literally a traffic-light), which is the public face alert. The link between these facets is derived from the volcano symptoms which are codified to correspond with the Civil Protection organisations and are expressed to the public as colours. The 'public' is assumed, in the organisation of the system, as a recipient for evacuation purposes.

Since the implementation of this system, so called inadequacies have been identified (Macías 1999) in its organisation, and these shortcomings were corroborated by organised evaluations (Macías 2005) in the results of various intervention actions of the federal and state governments in Puebla, Morelos and the State of Mexico, to evacuate communities that are increasingly reluctant to do so. There is, however, a view that defends its design, observes positive results in forced evacuations and, furthermore, recommends it as a replicable alert mechanism worldwide

(De la Cruz/Tilling 2008). The general design of the Popocatépetl warning system has been, with greater or less modification, replicated for the case of the Fire Volcano in Colima (Gavilanes-Ruiz et al. 2009; Gavilanes-Ruiz 2013b), even though, particularly in Colima, Gavilanes-Ruiz et al. (2009, 2010) identified residents' resistance to evacuations, which they interpreted as having no direct relation to the perception of the threat, but rather to their perception of what they considered 'risk management.' The same authors report 'ignorance of the functioning of processes that trigger alerts among members of civil protection systems'; they also identified and corroborated the belief of officials and scientific advisors in the typical EWS myths already mentioned.

The Early Warning System for Tropical Cyclones (SIAT-CT, in Spanish) began operating in 2000. Its construction was proposed in 1999 by the Scientific Advisory Committee of SINAPROC in Social Sciences to the person in charge of the Civil Protection coordination, who required a proposal to carry out an important action in disaster prevention during his brief time in office. The disastrous hurricanes Ismael/Roxana (1995), as well as Paulina (1997), demanded an elementary effort to reduce the adverse effects of these phenomena (Macías/Fernández 1999). Several working meetings were organised for the construction of the hurricane EWS (SIAT-CT).

A design centred on the evolution of meteorological systems, supported by information from the National Hurricane Centre of the United States, was created, in order to organise the activities of civil protection officials at all levels of government, but the incorporation phase of the population into the system was not achieved. A sound-logo was also devised for the identification of the alert through electronic means of mass communication.

The SIAT-CT was actually tested in the case of Hurricane Isidore in 2002, and had the anticpated effect of significantly reducing the loss of human life, largely because forced evacuation was used; but these lifesaving effects ceased to be replicated because of the failure to make adequate provision for the evacuated population once the alert had been issued. People are less inclined to act on alerts and if poorly supported shelters are the only accommodation available when they have evacuated the risk area.

In 2003, the SIAT-CT was reviewed and some of its technical characteristics were modified: 'The result was a new version of the SIAT-CT which includes as main parameters the Saffir-Simpson hurricane scale, the ability of the system to produce remote rains (measured by the circulation), tropical cyclone arrival time and, in the event that it is moving away from the national coasts, the distance to which it is located.' (Prieto/Oropeza 2007). In the year 2013, with the appearance of tropical cyclones (Ingrid and Manuel), the failures of the SIAT-CT became evident, both in its technical and technological arrangements and in the substantive, which is the communication with and between human beings[6] (see Fig. 10.2).

---

[6]See: 'Mayors agree: they received alert but did not know the size of the storms' In: http://mexico.cnn.com/nacional/2013/09/26/las-alertas-por-manuel-si-se-emitieron-pero-no-se-dimensionaron.

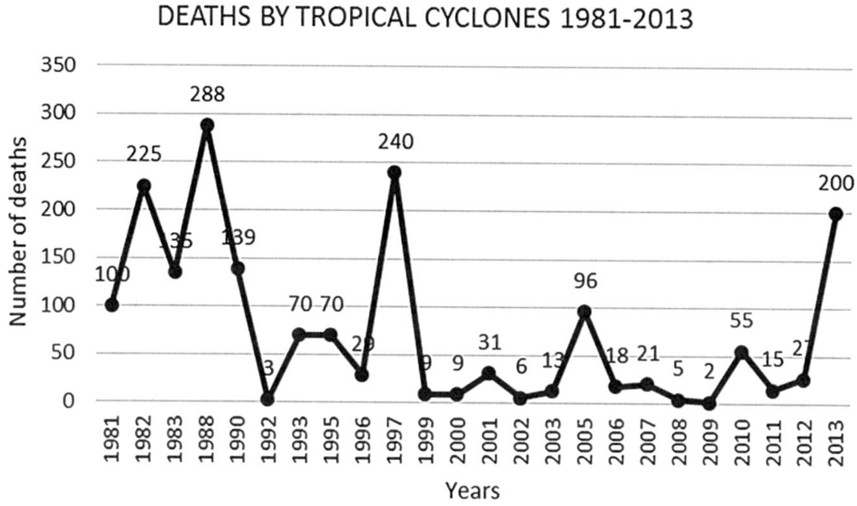

DEATHS BY TROPICAL CYCLONES 1981-2013

Fig. 10.2 Deaths from tropical cyclones from 1981 to 2013. *Source* Author's own elaboration with data obtained from CENAPRED (2010)

Local flood warning systems have been designed by the National Centre for Disaster Prevention (CENAPRED 2001) in response to the catastrophic floods of 1999 that affected several states of the country, such as Tabasco, Veracruz, Hidalgo and Puebla. The design of these systems is based on the same logic of most of the cases treated before, that is, on technological components (precipitation sensors and current expenditure) that are installed in rivers or streams and send signals that are processed by computer to calculate the time and distance of the arrival of a flood. The same system warns Civil Protection officials of the imminent flood. In this system, it is agreed that local levels of Civil Protection assume the tasks of alerting the population by conventional means (loudspeakers, home visits, sirens, etc.). These systems were installed in several regions such as northern Veracruz, Acapulco, Coahuila, Guerrero and Piedras Negras, among others. However, they have two major problems. The first is that the population is not considered as a determining part of the system; the second, that technological devices have lost their effectiveness due to lack of maintenance.

The National Tsunami Warning System (SINAT in Spanish) was formally created in 2012, under the Secretariat of the Navy's 'operational coordination'. This system includes several federal government agencies (SEGOB, Secretariat of Communications and Transportation and the Secretariat of the Navy, in addition to UNAM, CICESE and the National Seismological Service). A Tsunami Central Alert is an integral part of the this system. Before its formal creation, CIESAS had the opportunity to interact with the main managers of this system[7] to try to contribute to a better design, incorporating the social component missing from the previously discussed alert systems. SINAT is 'in development'. To date, the construction of the Alert Centre in the Secretariat of the Navy has not been completed, and the most recent exhibitions of its progress by its immediate supervisors (Montoya et al. 2013) reflect the exclusive interest in technological arrangements for seismic monitoring, sea level and mathematical forecasting models. The case of the earthquake and tsunami of 11 March 2011 that affected Japan, and the eventual arrival in Mexico of tsunami waves, was a very relevant opportunity to look at the operation of this alert system. It was accurately analysed by Gavilanes-Ruiz et al. (2011) in his response in Colima and Southern Jalisco; they observed 'an efficient communication flow' of the danger-detecting part that included a rapid reaction of federal, state and municipal civil protection systems, but socialisation of the information still requires a lot of work, and a curious and unusual process of 'alert verification' by unofficial channels was detected by state and municipal authorities. They also identified that community social organisation and its knowledge are under-dimensioned by decision-makers.

10.6 Conclusions

I must offer some criticism of the idea that collective behaviour is opposed to individual behaviour, especially when observed in adherence to social norms (Muñoz/Vázquez 2002). Collective behaviour sociology offers useful insights when it discusses the typologies of the masses and behaviours which more or less demarcate the spatial and social conditions of a community. It is therefore necessary to revise outdated notions and establish fairer connections with the social phenomena that can be observed in the disaster process.

Many statements found in the sociologists' literature discussed here are based on what they call the 'disaster,' which refers only to the moment in the disaster process that corresponds to the disastrous impact of some natural or anthropogenic phenomenon. Ambiguities in the conception of collective behaviour and its transfer to 'disaster', when viewed at its moment of destructive impact, generate the impression that the set of advances is not necessarily applicable to the case of alert systems.

---

[7]For this purpose, a formal session of the Permanent Seminar on Social Vulnerability to Disasters was organized by CIESAS in 2011 to warn of the importance of understanding the characteristics of the social groups that this EWS must address. Critical reflections on the SINAT design highlighted the autonomous function of component parts and the lack of an integrated approach which ensured, at the outset, a malfunction extremely recognized by experts in warning systems (Vid. Mileti/Sorensen 1990). Details of the seminar at: http://ciesas.wordpress.com/2011/06/10/los-sistemas-de-alerta-contra-desastres-en-mexico-y-su-vinculacion-con-la-poblacion-en-riesgo/.

On the other hand, other researchers' observations turn out to be extremely valuable, for example, the assertions made by Lindell/Perry (1992) that 'emergency managers' or senior managers and operatives of emergency services, as well as other social and lay actors (authorities), assume that the population at risk (the masses) is 'dumb', without organisational capacity and dependent, (cf. Dynes 1999 and the Model of the three 'Cs'; Gavilanes-Ruiz et al. 2011), and therefore, in accordance with this conception,[8] treat people as mere objects in the design and implementation of public policies. The EWSs, in this sense, are deficient since such disregard for human psychology results in failure to treat people like people.

On the other hand, when EWSs initially communicate with the public, in general it is in relatively calm circumstances, rather than as a 'reaction to emergencies'. This is a fundamental fact that must be kept in mind to distinguish between individual and collective action. Certain conditions, such as an imminent volcanic eruption, will probably generate conditions that, without being understood as 'calm', will neither be considered an 'emergency'. This must be very well identified by the agents involved in designing, operating and monitoring the alert systems.

The problem posed by the action of an efficient EWS with respect to an appropriate response of the population can be defined in the following terms:

(1) In general, there will be no communication exchange in crisis conditions (although there may be circumstances of action in times of crisis or emergency).
(2) It may be that there is an obvious threat, but there may be uncertainty about whether or when to activate a warning. That is the essential, defining core problem of the relations of and between EWSs.
(3) EWSs can deal with different sizes of social units, from a family to entire communities, and also with areas of differing size. This means that a warning can be an action directed at small communities, or a mass action. It may even be from or for several individual cases not spatially connected. The differential of the operation is a subject that demands the flexibility of the organisation in charge of the EWS.
(4) What is uniform (in organisational and coordination terms) is external action (whatever the level of intervention of the EWS organisation) to populations at risk.

Finally, the compelling conclusion is that the design, configuration and operation histories of all existing EWSs in Mexico should be reviewed to at least seek to reconcile these systems with the standards recommended by the International Strategy for the Reduction of Disasters of the UN, and that they incorporate the results of research carried out in Mexico and other countries.

---

[8]'Cognitive prejudice', Orbach would say elegantly (2014).

References

Aguirre, B., Wenger, D. & Vigo, G. (1998). A test of the emergent norm theory of collective behavior. *Sociological Forum*, 13(2): 301–320.

Austin, A. (2004). Early Warning and The Field: A Cargo Cult Science? Berghof Research Centre for Constructive Conflict Management. http://www.berghof-foundation.org/fileadmin/redaktion/Publications/Handbook/Articles/austin_handbook.pdf. Accessed 13 February 2006.

Berger, P. & Luckmann, T. (2001). *La construcción social de la realidad*. Buenos Aires, Argentina: Amorrortu Editores.

CENAPRED, (2010). *Elementos para sistemas de alerta efectivos/Experiencia de México*. Mexico City. Centro Nacional de Prevención de Desastres (National Center for Disaster Prevention). Secretaría de Gobernación [SEGOB] (Ministry of Government).

CIRES, (2014). *Boletín del sistema de alerta sísmica mexicano*. http://www.cires.org.mx/sasmex_reporte_alerta_es.php. Accessed 20 June 2015.

Dash, N. & Gladwin, H. (2007). Evacuation Decision Making and Behavioral Responses: Individual and Household. *Natural Hazards Review*. 8 (3), 69–77.

De la Cruz, S. & Tilling, R.I. (2008). Scientific and public responses to the ongoing volcanic crisis at Popocatepetl Volcano. Mexico; importance of an effective hazards-warning system. *Journal of Volcanology and Geothermal Research*, 170 (1–2): 121–134.

Drabek, T. (1968). *Disaster in Isle 13*. Columbus Ohio. Disaster Research Center. Ohio State University.

Dynes, R. (1999). Planificación de emergencias en comunidades. *Cuadernos de Extensión* (Extension Journals). No. 2. CUPREDER-BUAP.

EIRD. (2006). *Desarrollo de Sistemas de Alerta Temprana: Lista de comprobación*. Third International Conference on Early Warning (EWC III). Geneva. UN Secretariat of the International Strategy for Disaster Reduction (UN/ISDR). www.unisdr-earlywarning.org. Accessed 22 April 2009.

Espinosa-Aranda, J.M., Jimenez, A., Ibarrola, G., Alcantar, F., Aguilar, A., Inostroza, M., Maldonado, S. (1996). Results of the Mexico City early warning system. *Proc. 11th World Conf. Earthq. En.*, Paper No. 2132.

Espinosa-Aranda, J.M., Cuellar, A., Garcia, A., Ibarrola, G., Islas, R., Maldonado, S. (2009). Evolution of the Mexican Seismic Alert System (SASMEX). *Seismological Research Letters* 80, (5), 694–706.

Gavilanes-Ruiz, J.C. (2013a). Indispensable fortalecer coordinación y respuesta social ante amenazas naturales. Informativo. University of Guadalajara. Mexico. Recovered from http://148.202.105.20/prensa/boletines/2013/septiembre/1192edu.pdf. Accessed 12 December 2013.

Gavilanes-Ruiz, J.C. (2013b). Roles y conceptualizaciones de los actores sociales en la eficiencia de los sistemas de alerta temprana por tsunamis y lahares en colima y el sur de Jalisco. X Coloquio del Doctorado en Geografía (Tenth Colloquium of the Doctorate in Geography): From 11th to 15th of November 2013. http://posgrado.aplikart.com/coloquio/assets/juan_carlos_gavilanes_ruiz.pdf. Accessed 19 March 2015.

Gavilanes-Ruiz, J.C., Cuevas-Muñiz, A., Varley, K., Gwynne, G., Stevenson, S., Saucedo-Girón, R., Pérez-Pérez, A., Aboukhalil, M., Cortés-Cortés, A. (2009). Exploring the factors that influence the perception of risk: The case of Volcán de Colima, Mexico. *Journal of Volcanology and Geothermal Research*, 186(4), 238–252.

Gavilanes-Ruiz, J.C., Cuevas-Muñiz A., Varley N., Gwynne G., Stevenson J., Pérez-Pérez A., Aboukhalil M., Cortés-Cortés A. (2010). Exploring risk perception and warning effectiveness: The case of Volcán de Colima, México. *Cities on Volcanoes 6 Conference* p. 166, Tenerife, Spain.

Gavilanes-Ruiz, J.C., Cuevas, A., Calderón, G., Mayoral, W., Larios, Z., Merlo. Fernández, C.J., Urzúa, S. (2011). Hacia una propuesta urgente para mejorar la prevención y la reducción de desastres en México desde la comunidad científica. *Geos*, 31(1), p. 167.

Gillespie, D. & Perry, R. (1976). *An Integrated System and Emergent Norm Approach to Mass Emergencies*. Amsterdam: Elsevier.

Goltz, J.D. & Flores, P.J. (1997). Real-time earthquake early warning and public policy: A report on Mexico City's Sistema de Alerta Sismica. *Seismological Research Letters* 68: 727–733.

ICRC. (2009). *World Disasters Report 2009. Focus on early warning, early action*. International Committee of the Red Cross. http://www.ifrc.org/Global/WDR2009-full.pdf. Accessed Aug 29, 2013.

Le Bon, Gustave. (2007). Psicología de las masas. https://seryactuar.files.wordpress.com/2012/12/psicologc3ada-de-las-masas-gustave-le-bon-1895-pdf.pdf. Accessed 13 February 2018.

Lee, W.H. & Espinosa-Aranda, JM. (1998). *Earthquake Early Warning Systems: Current Status and Perspectives*. International IDNDR-Conference on Early Warning Systems for the Reduction of Natural Disasters. Potsdam, Federal Republic of Germany September, 7–11.

Lindell, M.K. & Perry, R.W. (1992), *Behavioural Foundations of Community Emergency Planning*. Washington, DC, Hemisphere Publishing.

Macías, J.M. (1999). *Desastres y Protección Civil. Aspectos sociales, políticos y organizacionales*. Mexico. CIESAS.

Macías, J.M. (2005). *La Disputa por el Riesgo en el Volcán Popocatépetl* (The Dispute for the Risk in the Volcano Popocatepetl). Mexico: CIESAS.

Macías, J.M. (2009). ¿Quién y porqué decidieron reubicar poblaciones por las inundaciones de 1999 y el huracán Isidoro? In: G. Vera (editor) *Devastación y éxodo. Memoria de seminarios sobre reubicaciones por desastre en México*. CIESAS, Mexico. Papeles de la Casa Chata. 275–292.

Macías, J.M. (2012). Estado y desastres. Deterioro, retos y tendencias en la reducción de Desastres en México 2011. Cambio climático y políticas de desarrollo sustentable (Climate change and sustainable development policies). José Luís Calva (Coordinator). México. Juan Pablos Editor/ Consejo Nacional de Universitarios. 14. 368–394.

Macías, J.M. & Calderón, G. (1994). *Desastre en Guadalajara. Notas Preliminares y Testimonios*. Mexico: CIESAS.

Macías, J.M., & Fernández, A. (1999). Las Enseñanzas del Huracán Paulina. CUPREDER/ BUAP. Puebla, Mexico. *Cuadernos de Extensión* (Extension Journals).

Marshall, G. (1996). Sociology. Oxford. Oxford University Press.

Maskrey, A. (1997). *IDNDR Early Warning Programme Report on National and Local Capabilities for Early Warning*. Geneva: IDNDR Secretariat.

Mileti, D., Bandy S.R., Bourque L.B., Johnson, A., Kano, M., Peek, L., Sutton, J., Wood, M. (2006a). *Annotated Bibliography for Public Risk Communication on Warnings for Public Protective Actions Response and Public Education*. (Revision 4). NHAIC. University of Colorado. http://www.colorado.edu/hazards/publications/informer/infrmr2/pubhazbibann.pdf. Accessed Mar 20, 2009.

Mileti, D., Bandy S.R., Bourque, L.B., Johnson A, Kano, M., Peek, L., Sutton, J., Wood, M. (2006b). Bibliography for Public Risk Communication on Warnings for Public Protective Actions Response and Public Education. (Revision 4). NHAIC. University of Colorado. Recovered from http://www.academia.edu/7678602/bibliography_for_public_risk_communication_on_warnings_for_public_protective_actions_response_and_public_education_revision_4. Accessed 20 March 2009.

Mileti, D. & Sorensen, J. (1990). *Communication of emergency public warnings*. ORNL-6609, Oak Ridge National Laboratory, Oak Ridge, Tenn.

Montoya, J.M., Ortiz M., Martínez de Pinillos, L. (2013). *Diseño de un Sistema de Alerta de Tsunamis para la Costa Occidental de México. Memorias del Encuentro internacional de manejo de inundaciones* (Report of the International Meeting on Flood Management). UNAM. http://www.iingen.unam.mx/es-mx/BancoDeInformacion/MemoriasdeEventos/RiesgoporInundaciones/03_Miercoles/SISTEMAALERTASUNAMIS.pdf. Accessed 20 June 2014.

Muñoz, J. & Vázquez, F. (2002). Processos col-lectius y acció social. In: Félix Vázquez *Pscologia del comportament collectiu*. Barcelona. Universitat Obertura de Catalunya. 1–63.

NRC. (2013). 1. Fundamentals of Alerts, Warnings, and Social Media. in: *Public Response to Alerts and Warnings Using Social Media: Report of a Workshop on Current Knowledge and Research Gaps*. Washington, DC. National Research Council. The National Academies Press. https://www.nap.edu/read/15853/chapter/1. Accessed 13 February 2018.

Ollman, B. (1975). *Alienación. Marx y su concepción del hombre en la sociedad capitalista*. Buenos Aires: Amorrortu Editores.

UNISDR. (2006). *Global Survey of Early Warning Systems. An assessment of capacities, gaps and opportunities toward building a comprehensive global early warning system for all natural hazards*. Report prepared at the request of the Secretary-General of the United Nations. Ginebra. http://www.unisdr.org/2006/ppew/info-resources/ewc3/Global-Survey-of-Early-Warning-Systems.pdf. Accessed May 15, 2008.

Orbach, B. (2014). Accountability for Predictable Disasters. http://www.orbach.org/publications/ (Accessed 6 November 2013).

Perry, R. (1979). Evacuation decision-making in natural disasters. *Mass Emergencies* 4:25–38. The Netherlands. Elsevier.

Perry, R. (1985). *Comprehensive Emergency Management: Evacuating Threatened Populations*. Greenwich, JAI Press Inc. p. 75.

Perry, R., Lindell, M., Greene, M. (1981). *Evacuation planning in emergency management, Lexington Press*, Lexington, Mass.

Poulantzas, N. (1977). *Clases sociales en el capitalismo actual*. Madrid. Siglo XXI.

Prieto, R. & Oropeza, F. (2007). Estudio para el desarrollo de nuevas herramientas para el Sistema de Alerta Temprana de Ciclones Tropicales. Proyecto: TH0767.3. Informe final Estudio solicitado por la Secretaría de Gobernación a través del CENAPRED. Subcoordinación de Hidrometeorología. Coordinación de Hidrología (Project: TH0767.3. Final report Study requested by the Ministry of Government through CENAPRED. Subcoordination of Hydrometeorology. Coordination of Hydrology). Mexico. www.imta.gob.mx/historico/instituto/. Accessed 3 June 2008.

Quarantelli, E. (1980). *Evacuation Behavior and Problems: Findings and Implications from the Research Literature*. Disaster Research Center. Ohio State University. Miscellaneous Report No. 27.

Quarantelli, E. (1983). *People's Reactions to Emergency Warnings*. University of Delaware. Disaster Research Center. Article #170.

Quarantelli, E. (1984). *Evacuation, Behavior and Problems: Findings and Implications from the Research Literature*. Disaster Research Center. Miscellaneus Report #27.

Quarantelli. E. & Dynes, R. (1977). Response to Social Crisis and Disaster. *Annual Review of Sociology*, (3), 23–49.

Ritzer, G. (1997). *Teoría Sociológica Contemporánea*. Mexico. McGraw-Hill.

Rogers, D. & Tsirkunov V. (2011a) Cost and Benefits of Early Warning Systems. Global Assessment Report (GAR 2011). ISDR/ World Bank. http://www.preventionweb.net/english/hyogo/gar/2011/en/bgdocs/Rogers_&_Tsirkunov_2011.pdf. Accessed 23 April 2012.

Rogers, D. & Tsirkunov V. (2011b). Implementing Hazard Early Warning Systems. Global Facility for Disaster Reduction and Recovery (GFDRR). https://www.preventionweb.net/files/24259_implementingearlywarningsystems1108.pdf. Accessed 13 February 2018.

Rohrmann, B. (2000). A socio-psychological model for analyzing risk communication processes. *The Australasian Journal of Disaster and Trauma Studies*. Vol. 2. Recovered from http://www.massey.ac.nz/~trauma/issues/2000-2/rohrmann.htm. Accessed 23 June 2003.

Sorensen, J. (2000). Hazard Warning Systems: Review of 20 Years of Progress. *Natural Hazards Review*. 1, (2), 119–125. http://www.colorado.edu/geography/class_homepages/geog_4173_f11/Sorensen_warning_systems.pdf. Accessed 12 August 2007.

Suárez, G. Novelo, D., Mansilla, E. (2009). Performance Evaluation of the Seismic Alert System (SAS) in Mexico City: A Seismological and a Social Perspective. *Seismological Research Letters* 80, (5), 707–716.

Tierney, K. (2000). Implementing a Seismic Computerized Alert Network (SCAN) for Southern California: Lessons and Guidance from the Literature on Warning Response and Warning Systems. Disaster Research Center. University of Delaware. Report prepared for task 2, TRINET Studies and Planning Activities in Real-Time Earthquake Early Warnings. http://dspace.udel.edu/bitstream/handle/19716/1155/FPR45.pdf?sequence=1. Accessed 20 July 2005.
Turner, R. & Killian, L. (1972). *Collective Behavior*. Englewood Cliffs, NJ: Prentice-Hall.

Chapter 11
Social Networks and Disaster Risk Perception in Mexico and Ecuador

Eric C. Jones, A. J. Faas, Arthur Murphy, Graham A. Tobin, Linda M. Whiteford and Christopher McCarty

Abstract We examine social aspects of risk perception in seven sites among communities affected by a flood in Mexico (one site), as well by volcanic eruptions in Mexico (one site) and Ecuador (five sites). We conducted over 450 interviews with questions about the danger people feel at the time (after the disaster) about what happened in the past, their current concerns, and their expectations about the future. We explored how aspects of the context in which people live have an effect on the relationship between risk perception and social network factors. Levels of risk perception for past, present, and future aspects of a specific hazard were similar across these two countries and seven sites. However, specific network factors varied from site to site across the countries, thus there was little overlap between sites in the variables that predicted the past, present, or future aspects of risk perception in each site.

Keywords Comparative research · Disaster · Resettlement · Latin America
Social support · Recovery · Wellbeing

Dr. Eric C. Jones, Assistant Professor, University of Texas Health Science Centre at Houston. Email: eric.c.jones@uth.tmc.edu

Dr. A. J. Faas, Assistant Professor, San Jose State University. Email: aj.faas@sjsu.edu

Dr. Arthur Murphy, Professor, University of North Carolina at Greensboro. Email: admurphy@uncg.edu

Dr. Graham A. Tobin, Professor, University of South Florida. Email: gtobin@usf.edu

Dr. Linda M. Whiteford, Professor, University of South Florida. Email: lwhiteford@usf.edu

Dr. Christopher McCarty, Department Chair, Associate Professor, Center Director, University of Florida. Email: ufchris@ufl.edu

This previously published paper was abridged and edited to focus on social network factors, and is reproduced here with permission of Springer Science+Business Media and the editors of the journal *Human Nature*. The citation for the previously published paper is as follows: Jones, E.C., A.J. Faas, A.D. Murphy, G.A. Tobin, L.M. Whiteford, C. McCarty. 2013. Cross-Cultural and Site-Based Influences on which Demographic, Individual Well-being, and Social Network Factors Predict Risk Perception in Hazard and Disaster Settings. *Human Nature* 24(1):5–32.

151

Ma. L. Marván and E. López-Vázquez (eds.), *Preventing Health and Environmental Risks in Latin America*, The Anthropocene: Politik—Economics—Society—Science 23, https://doi.org/10.1007/978-3-319-73799-7_11

11.1   Introduction

In this study of hazards experienced by people in Ecuador and Mexico, we examine
how the structure and composition of social networks are associated with risk
perception in different affected sites.[1] Although insights from social network
analyses are relatively new to disaster and hazard studies—and still unexplored in
the study of risk perception—our effort can be seen to build upon research on the
'culture of response' (e.g. Dyer/McGoodwin 1999) that investigates how people in
different places respond differently to similar hazards or disasters. Because there is a
lack of research on social networks and risk perception—especially comparative
research on social networks and risk perception—our study is necessarily
exploratory.

In between inter-societal and individual differences in risk perception, there are
communities that can produce different pressures for individuals responding to
hazards and disasters. We think that there are site differences in the experience of
the same hazard that encourage people to begin to form different senses of risk after
engaging with the hazard, especially when actually faced with a disaster, since
disasters often result in relocation of individuals and/or communities. For this
study, we interviewed relocated and non-relocated populations faced with volcanic
hazards and landslides.

11.2   Social Factors in Risk Perception

People's relationships have been found to play important roles in individual and
community recovery from disasters (e.g., Hobfoll 2002), which in turn could rea-
sonably be expected to influence risk perception after the experience of disaster. In
a study of many factors of risk perception in a disaster setting, Tobin et al. (2011)
called for further research on social aspects of risk perception. A study in Malawi
on perception of health risks (HIV/AIDS) found that network effects are mediated
by gender, marriage, and geographic region, but generally can be characterised on
the one hand as people seeking information from their networks and, on the other
hand, that having many people in your network concerned about a risk can increase
your own concern with the risk (Helleringer/Kohler 2005). In a short pioneering

[1]Data collection and management for this project were supported by US National Science
Foundation grants BCS-CMMI 0751264/0751265 and BCS 0620213/0620264. Special thanks to
Brittany Burke and Olivia Pettigrew for editorial support in preparation of this manuscript; to
Fabiola Juárez Guevara, Isabel Pérez Vargas, Brittany Burke, and Audrey Schuyler for their
considerable efforts in the field helping collect data; to Jason Simms for feedback on analytical
procedures; and to research partners at the University of Puebla's disaster centre
(BUAP-CUPREDER) in Puebla, Mexico and at the National Polytechnic University's
Geophysical Institute (EPN-IG) in Quito, Ecuador. Preparation of this manuscript was supported in
part by a School for Advanced Research Team Seminar in 2012.

piece on social networks and risk perception, Scherer/Cho (2003) studied perceived risks from a hazardous waste clean-up site and found that the strength of ties between actors predicted similar risk perceptions but did not predict similar attitudes towards a control question about belief in science. Kitchovitch/Liò (2010) sought to include social network impacts in an existing model of risk perception in order to study possible reduction in risky behaviours once at least some members of a social network are made aware of the risks. Regarding network structure, such studies generally consider only network density and size and the strength of ties between actors in the network. We have chosen in this chapter to test a number of theoretically relevant network measures.

Community recovery from disaster depends in part on individuals feeling that they are part of a strong network and can thus overcome adversity (Hobfoll 2002; Tobin/Whiteford 2002; Hall et al. 2003; Reissman et al. 2004). However, dense networks of strong ties might create redundant feedback loops not conducive to the introduction of new information regarding evolving risk conditions. Relatedly, when an individual's network does not include different subgroups, the potential exists for restrictive norms to limit a person's choice of how and from whom to seek help (Avenarius 2003; Unger/Powell 1980; cf. Avenarius/Johnson 2004 as an example of a disaster study). However, the presence of subgroups in personal networks might present a vulnerability to opinion leaders in the development of risk perception. Density, because it is associated with trust within the network (Buskens 1998), but not between individuals from different networks, could be expected to have a negative association with perceived risk. Our goal in this chapter is to better understand predictors of risk perception that vary cross-culturally, particularly social network structure and content.

11.3 Methodology

In the state of Puebla, Mexico, we collected data from April to August 2007 in San Pedro Benito Juárez (n = 62), and from April 2008 to March 2009 in Ayotzingo (n = 139). In Ecuador, we interviewed in five sites on western and southwestern flanks of Mt. Tungurahua between April and December 2009 (with between 30 and 99 interviews in each site; see Table 1). First, we administered a half-hour preliminary socio-demographic survey to a random sample of households in all sites. The data from this questionnaire were used to establish the distribution and basic attributes of each reference group and provided the basis for the randomly sampled individual in each household to which we administered a 90-min well-being and personal network survey.

Well-being included scales covering economic status and employment status, mental health, health, and household conditions. Personal networks involved the interviewee naming several individuals and the interviewee reporting on the relationships between those people named by the interviewee. We asked participants to 'Please list the people you know by sight or by name with whom you have had contact or could have had contact if you needed to, in the past two years (we would like you to list 45 names)' (after approaches recommended by Bernard et al. 1990, McCarty 2002, McCarty et al. 2000). We then asked the interviewee for basic demographic information about a randomly chosen pre-selected sequence of 25 individuals named (same 25 numbers on each list of 45), since a random subsample of ~20–30 individuals from the larger list of individuals named by a respondent (~40–60) provides accurate structural representations or structural measures of a personal network (McCarty/Killworth 2007; Chris McCarty, personal communication). Interviewees were also asked to indicate the presence and strength of interactions between individuals in the random subsample of people they named.

11.3.1   Measures of Dependent Variable of Risk Perception

Data on risk perception were collected by asking respondents if: they were concerned about living where another disaster event could happen (Currently Concerned); they believe that their life or the lives of their family were in danger because of a specific disaster event (Perceives Past Threat to Life); another disaster event could happen during their lifetime (Expects Future Event); and plans for evacuating if another event occurs (Plans to Evacuate). We also created a five-point overall risk perception variable that combined these three measures plus desire for future assistance from an institution in evacuating.

11.3.2   Independent Measures

Site-Based Characteristics: In order to measure the effects of different types of site in our sample, we developed four different variables to account for site-based variation in risk perception: (a) urban versus rural; (b) resettled versus non-resettled; (c) low versus high impact sites; (d) Mexico versus Ecuador.

Non-specified relationships: Our survey employed perceived support measures in the Provisions of Social Relations Scale—believing that people are really there for you—including subscales for perceived support from friends (7 items), family (7 items), and spouse (8 items) (Turner/Marino 1994). We also asked about the number of close kin living abroad, whether their work/institution or boss helped them, whether they had a spouse/partner, and whether they had worked outside the city/region in which they currently live.

Network content: We collected demographic variables for 25 randomly selected network members (referred to as 'alters' in social network analysis) named by each interviewee after each had first named 45 alters. The interviewee—or focal individual—is known as ego in social network research. We calculated the average age, as well as the percentage of each network constituted by each the following: females in network, higher/same/lower socio-economic status compared to interviewee, bilingual (as a measure of ethnicity), religion, very/somewhat/not close to ego emotionally, having given and/or received material support, informational support, emotional support, and work/labour with ego.

Network structure: In addition to the demographic and socio-economic composition of each personal network and the incidence of support exchanges in each personal network, we created ratio measures of network structure. To create networks for each of the egos, we asked ego whether each of individuals in the ego's network interacted with one another a lot, some or little/none. Delphi-based EgoNet 2.0 (www.mdlogix.com) was used in Mexico and the Java-based EgoNet (http://sourceforge.net/projects/egonet/) was used in Ecuador to collect and analyse the data to produce the following measures:

- Components, or the number of sets of alters in which each alter is tied to every other alter directly or indirectly (where each set is totally disconnected from the others), is a measure of disconnected subgroups;
- Normalised average degree (i.e., Density), or the mean for all alters of the direct ties between them and other others, implicates the roles of homogeneity (everyone knows everyone) but also varied potential paths for transmission of information and opinion about risk (lots of ways to get from A to B);
- Average betweenness, or the mean for all alters of the proportion of times each alter lies on the shortest path between all sets of two alters in the network, can show the importance of bridging or unique paths through a personal network for influencing aspects of risk perception;
- Degree centralisation, or the extent to which the network has only one or a few people who know most people, can be important for information/opinion gatekeeping and influence on the respondent's risk perception;
- Betweenness centralisation, or the extent to which the network is dominated by a few alters that lie on the paths to all other alters, can highlight the role of networks with one or very few unique bridging people that tie together the respondent's personal network;
- Isolates, or the number of isolated alters with no ties to any alters in the network, shows us how fragmentation or disconnectedness in a network is associated with risk perception;
- Dyads, or the number of times two alters are connected but neither is connected to any third alter, is another measure of fragmentation.

11.3.3 Analysis

We approached the analysis with an interest in describing differences between sites in terms of: (1) demographic and contextual factors (Table 11.1); (2) level of perceived risk (Table 11.2); (3) testing the relationship between contextual variables and risk perception in each site; (4) testing the relationship between non-specified relationship variables and risk perception in each site; and (5) testing the relationship between network variables and risk perception in each site.

Table 11.1 Characteristics of study sites

Site (Country)	Sample size	Disaster impact	Settlement pattern	Hazard	Pop	Percent male/female	Time since last evacuation
Penipe Viejo (EC)	46	Low, not evacuated	Urban village	Eruption	710[iii]	50/50	Not applicable
San Pedro Benito Juárez (MX)	62	Low, evacuated	Rural village	Eruption	3512[i]	44/56	7 yrs
Pillate (EC)	48	High, evacuated	Rural village	Eruption	193	49/51	3 yrs
San Juan (EC)	30	High, evacuated	Rural village	Eruption	172	53/47	3 yrs
Pusuca (EC)	40	High, resettled	Rural village	Eruption	161	48/52	3 yrs
Penipe Nuevo (EC)	99	High, resettled	Urban village	Eruption	1405	50/50	3 yrs
Ayotzingo (MX)	139	High, resettled; dozens of deaths	Urban neighbourhood	Flood	1609[ii]	45/55	9 yrs

Source All data based on results from study interviews, unless otherwise noted
[i]Local Health Centre Census (2005);
[ii]Local Health Centre Census (2008);
[iii]INEC (2001)

11.4 Sites

We chose study sites in the State of Puebla, Mexico, and the provinces of Chimborazo and Tungurahua, Ecuador, as two Spanish-speaking Latin American contexts because of important similarities and differences in their hazard experiences. Besides a general cultural similarity in language and colonial history, a major difference in context is that Ecuadorians have a much lower expectation of the capabilities and desirability of intervention by National Guard-type forces, while Mexicans have seen the intervention by these forces as more normal and acceptable (at least prior to subsequent government warfare with drug mafias in distant northern Mexico after our study). However, while principles associated with disasters are common, it is appreciated that no two disaster experiences are completely alike, plus local hazardous conditions are usually dissimilar. Thus local context remains an important component of this research.

In Tungurahua and Chimborazo Provinces in Ecuador, we chose to vary the sites by degree of impact from the eruptions of the strato volcano, Mt. Tungurahua: low impact involved occasional volcanic ash fall; high impact/evacuated involved chronic heavy ash fall, occasional rock fall from eruptions, evacuation in 1999, and

Table 11.2 Comparison of study sites in levels of perceived risk, percentage answering yes

	Perceives past threat to life (%)	Currently concerned (%)	Expects future event (%)	Plans to evacuate (%)	Overall perceived risk (0–5)***
Penipe Viejo (EC) *Urban* *Low impact, not evacuated*	56	22	93	71	3.8
San Pedro Benito Juárez (MX) *Rural* *Low impact, evacuated*	21	40	73	71	3.4
Pillate (EC) *Rural* *High impact, evacuated*	77	55	91	84	4.2
San Juan (EC) *Rural* *High impact, evacuated*	70	77	90	79	4.4
Pusuca (EC) *Rural to rural resettlement* *High impact, resettled*	92	45	100	83	4.2
Penipe Nuevo (EC) *Rural to urban resettlement* *High impact, resettled*	87	62	88	72	4.2
Ayotzingo (MX) *Urban* *High impact, resettled*	84	80	76	85	4.3

Source The authors

***(Kruskall Wallis; $X^2 = 37.52$, df = 6, $p = .000$)

two evacuations in 2006; and high impact with resettlement involved houses destroyed by lahars, evacuation in 1999, two evacuations in 2006, some temporary housing for years for some people, and eventually resettlement in 2008. In Mexico, we chose a minor disaster that involved chronic volcanic eruptions and evacuations in 1994 and 2000, and a major disaster involving landslides in 1999 that were too sudden for evacuation and that resulted in resettlement of around 300 households in 2000 (see Table 1). Most of the people from the study sites in both Mexico and Ecuador are primarily agriculturalists.

Penipe Viejo is the county seat of Penipe County in Chimborazo Province, Ecuador. Penipe sustained moderate ash fall during major eruptions in 1999 and 2006, and occasional light ash fall in the interim and ensuing periods that caused minor damage to buildings, crops, roads, and utility infrastructures, as well as presenting some public health risks. Penipe has served as a base of emergency response operations, and several local buildings were repurposed as shelters. The 2008 resettlement added 285 houses to the township's previously existing 190 households. Most Penipeños make a living from small businesses in town (e.g., restaurants, stores, trades, etc.) or wage employment in the nearby city of Riobamba.

Penipe Nuevo is a resettlement community of 285 houses built in 2007–2008 by the Ministry of Housing and Urban Development along with Samaritan's Purse (a US-based religious disaster relief organisation) as an extension of the urban centre of Penipe Viejo. Houses were granted to villagers displaced from more than a dozen villages after the major eruptions of Mt. Tungurahua in 1999 and 2006. Though some have sought limited employment in nearby Riobamba and even fewer have created small businesses in the resettlement (usually small convenience stores), the majority of the residents of Penipe Nuevo still travel daily to their lands in the high risk zone.

Pusuca is a resettlement community of 45 households built by the non-governmental organisation Esquel Foundation approximately 5 km south of Penipe Nuevo. Resettlers in Pusuca hail from the same villages as those in Penipe Nuevo. Unlike Penipe Nuevo, each household in Pusuca was provided with a little over one half hectare of land for agricultural production and/or livestock, and there are additional communal plots of land for cooperative agricultural production. Although some resettlers in Pusuca, like those in Penipe Nuevo, have sought wage employment in nearby cities, agricultural production is the primary economic activity of nearly two-thirds of households in Pusuca. Some of them farm in Pusuca, and many continue to farm in the high risk area from which they were relocated.

Pillate and San Juan are two adjacent villages in Tungurahua Province, of approximately 40 and 30 households respectively, to the north of Penipe and directly across the River Chambo from the western flanks of Mt. Tungurahua, well within the high risk zone. They were evacuated for eruptions in 1999 and 2006 and suffered immense damage as a result of heavy ash fall, incandescent material, and tremor-induced landslides. In spite of this damage, approximately 70% of the former residents of these communities returned to live in and rebuild the villages after each eruption. Like their neighbours in the northern parishes of Penipe

County, the productive capacities of their soil and fruit trees have been greatly reduced by continued chronic ash fall.

San Pedro Benito Juárez is an agricultural village of approximately 850 Nahuatl and mestizo households that lies on a fracture zone on the south-eastern flanks of Mt. Popocatépetl, directly west of the city of Puebla, Mexico. The village is also known for high rates of migration to urban centres in Mexico and the United States. The village is the closest of its neighbours to the crater of the volcano, although ash fall and prevailing winds more commonly flow to the east and north-east of the volcano. An eruption in December 1994 deposited ash over a wide area and led to the evacuation of San Pedro Benito Juárez and neighbouring villages. In December 2000, Popocatépetl erupted again, more powerfully than before, resulting in a second evacuation, though many villagers chose to remain (Tobin et al. 2007).

Ayotzingo is a resettlement of just over 300 houses. Families relocated from various neighbourhoods in the mountain city of Teziutlán (pop. ∼50,000) after flooding and landslides destroyed significant parts of the city in 1999. Teziutlán is located on the eastern slopes of the Sierra Madre approximately 250 km northeast of Puebla, Mexico. Over 400 people died and over 200,000 lost their residences (Garcia 2000) along the Mexican Gulf Coast. In Teziutlán, entire sections of the city were washed away, causing tens of millions of pesos of damage. As part of the recovery, approximately 350 families were given plots and building materials in Ayotzingo, a state-funded resettlement community several kilometres away from town, between four and twelve months after the disaster. (e.g., Norris et al. 2005).

11.5 Results and Discussion

11.5.1 Levels of Risk Perception

Table 11.2 shows the percentages of people answering yes to the four risk perception questions in each of the study sites, and suggests there is some uniformity to risk perception regardless of rural-urban setting, relocation/non-relocation, and the country or cultural context, as long as the hazard has had a large impact. Nonetheless, this does not mean risk perception works in the same way in each place. Table 11.2 has been organised to put the lowest exposure to risk (from our perspective) at the top of the table and the highest exposure to risk at the bottom, with evacuated sites separate from resettled sites. The significant difference between sites in overall risk in the far right column is likely attributed to the lower risk perceived in the two least affected sites of Penipe Viejo and San Pedro Benito Juárez. The three relatively urban sites are the first one in the list and the last two in the list.

We next evaluated the difference in overall perceived risk (scale 0–5) between different site types in our sample: (a) urban versus rural; (b) resettled versus non-resettled; (c) low versus high impact sites; (d) Mexico versus Ecuador (see

Table 3 in Jones et al. 2013). Urban sites had higher rates of past and present risk perception than rural sites, though there was no difference for future perspective. Similarly, resettled sites had significantly higher rates of past and present risk perception than non-resettled sites, but there was no difference for rates of perception that a future event is likely. High impact sites have higher rates of past and current risk perception, as well as plans to evacuate, but no difference for expecting it to happen again. Ecuadorian sites have a higher rate of past and future risk perception, while Mexican sites have a higher rate of current risk perception. For the overall perceived risk scale, differences are noted between resettled and non-resettled sites, low impact and high impact sites, but not between rural and urban sites nor between Mexican and Ecuadorian sites.

Urban and resettlement sites have higher rates of current risk perception than their rural and non-resettlement counterparts despite the residents of the urban and resettlement sites having been spatially removed from the risks they faced in the past. Both resettlement sites in Ecuador continue to rely heavily on agricultural production and animal husbandry in the hinterland near or within the volcanic high risk zone, which could contribute to their comparatively heightened perception of current risk. Past impact being higher for Ecuador than Mexico could be explained by the fact that all but one of the five Ecuadorian sites were recently affected and continue to experience ash fall, but Mexico having a higher rate of current risk perception may owe to the loss of life that was involved and the high proportion of the sample that was relocated.

11.5.2 Non-network Measures of Social Factors and Risk Perception

In general, non-network measures of social support are relatively minor in their prediction of risk perception (Table 11.3). Being married seems to have mixed impacts. In the urban resettlement of Penipe Nuevo, marriage is negatively associated with perception of past risk. On the other hand, marriage may produce higher concern in the present due to family responsibilities, especially in rural areas. In the rural non-resettled site of San Pedro Benito Juárez, married people were equally likely to be currently concerned as not, but only 20% of single people expressed current concern about living where this might happen again.

Having worked outside the area did predict lower concern about current risk in Ayotzingo, such that people with wider geographic networks and experiences in other places feel they have options in the case of another extreme event. The second measure of geographic reach of one's life—having any closely related kin (siblings, parents, children) living abroad—predicted higher values in the current concern about the risk (urban relocated Ayotzingo), and expectations that it will happen again (rural evacuated, low impact San Pedro Benito Juárez). Both are Mexican sites with more frequent international emigration than Ecuador. This association of

Table 11.3 Significant relationships between non-specified relationships and risk perception (yes/no), by site (Mann-Whitney U for ordinal variables; Chi-square for dichotomous variables; $p < 0.05$; underlined variables show negative correlations)

	Perceives past threat to life	Currently concerned	Expects future event	Plans to evacuate
Penipe Viejo (EC) *Urban Low impact, not evacuated*	–	–	–*	–
San Pedro Benito Juárez (MX) *Rural Low impact, evacuated*	Institution or boss helped them	Married	Any close kin living abroad	–
Pillate (EC) *Rural High impact, evacuated*	–	–	–*	–
San Juan (EC) *Rural High impact, evacuated*	–	Perceived support from family	–*	–
Pusuca (EC) *Rural to rural resettlement High impact, resettled*	–	–	–*	–
Penipe Nuevo (EC) *Rural to urban resettlement High impact, resettled*	Married	–	Perceived support from family	Perceived support from partner
Ayotzingo (MX) *Urban High impact, resettled*	–	# close kin living abroad worked outside the area	–	–

Source The authors

*Indicates insufficient variation in risk perception variable family abroad with current and future concerns suggests that having family members living elsewhere increases one's risk perception, perhaps due to reduced insularity of information, or due to concern resulting from lack of physical proximity of emotionally close family members. Overall risk perception was higher in San Pedro Benito Juárez when the number of family living abroad was higher, and the same trend seemed to occur in Pusuca and Ayotzingo, again suggesting that Mexico has a different experience in this regard. In the least impacted evacuated site of San Pedro Benito Juárez, people who received help from an institution were less likely to perceive past threat, suggesting that the receipt of help may reduce the strength of a perceived past threat. Perceived support appears only relevant in Ecuador when considering a linear relationship and not controlling for other factors. In terms of perceived support, family or spouse/partner support was associated with increased people's perception that a similar eruption will happen again and that they plan to evacuate in Penipe Nuevo (where friend support also predicted perception it will happen again), plus family support was associate with increased current concern in the evacuated site of San Juan.

11.5.3 Social Networks and Risk Perception

In general, personal network composition or content played a bigger role than did network structure, although we did use a few more network composition measures than structural measures (Table 11.4). The importance of support exchanges suggests, as mentioned earlier, that there may be shared cultural/cognitive models of past disaster events emerging in the denser networks and where support is exchanged more frequently, contributing to the reinforcement of collective memories of past events, current concerns, and future scenarios.

Having a higher percentage of females in one's personal network was negatively associated with perception of past impacts in the rural Ecuadorian resettlement site of Pusuca and positively associated with perception of current risk in the high-impact site of Pillate. When disasters occur, it is a confirmed generalisation that women's networks and access to resources are more adversely impacted than those of men (e.g., Norris et al. 2005). However, because the percentage of females in personal networks is negative in a high impact rural resettlement site and positive in a high impact rural evacuation site—and the two sites are otherwise very similar culturally and socio-demographically—it is possible that being resettled is the mitigating factor (Faas 2012).

We see little direct and independent influence of social factors on future-oriented risk perception. However, in Ecuador, the expectation that it will happen again is associated with degree centralisation for urban-resettled Penipe Nuevo. This positive association of a centralised network with the perception of future risk suggests the presence of opinion leaders in many of the personal networks. Degree centralisation has the opposite relationship with perception of the past, decreasing it in San Pedro Benito Juárez, nonetheless suggesting another,

Table 11.4 Significant relationships between network content/structure variables with risk perception by site, columns supplying yes/no values (Mann-Whitney U, $p < 0.05$; underlined variables show negative associations; % designates per cent of a network with that attribute)

	Perceives past threat to life	Currently concerned	Expects future event	Plans to evacuate
Penipe Viejo (EC) *Urban* *Low impact, not evacuated*	–	% alters received material support % very close ties to alters	–*	–
San Pedro Benito Juárez (MX)** *Rural* *Low impact, not evacuated*	% not close ties to alters % very close ties to alters % alters gave ego material support % alters gave ego emotional support degree centralisation # dyads	% alters invited ego to work % alters invited by ego to work	% not close ties to alters % somewhat close ties to alters density	% alters invited ego to work % alters invited by ego to work
Pillate (EC) *Rural* *High impact, evacuated*	–	% ties with women alters density Average betweenness centrality	–*	–
San Juan (EC) *Rural* *High impact, evacuated*	–	% alters invited ego to work % alters invited by ego to work	–*	–
Pusuca (EC) *Rural to rural resettlement* *High impact, resettled*	% ties with women % alters gave ego emotional support % alters received emotional support	% alters received material support %alters received emotional support % alters gave emotional support	–*	
Penipe Nuevo (EC) *Rural to urban resettlement* *High impact, resettled*	% alters invited ego to work	% not close ties % ties with alters of higher economic status	Degree centralisation	% Evangelical alters

(continued)

Table 11.4 (continued)

	Perceives past threat to life	Currently concerned	Expects future event	Plans to evacuate
Ayotzingo (MX)** Urban High impact, resettled	% somewhat close ties to alters	% ties with alters of higher economic status	–	–

Source The authors
*Indicates insufficient variation in risk perception variable
**We did not ask about ego giving support to alters in the two Mexico sites

different role for opinion leaders in one's network. However, coupled with the negative relationship with number of dyads, it appears that decentralised and fragmented networks may be more of an effect than a cause of high perceived past threat.

Work exchange—asking neighbours to work in your fields for you and vice versa—is always associated with increases in risk perception in this dataset. In the urban relocated site of Penipe Nuevo, perception of past threat was higher with work exchange. Also higher were current concern and plans to evacuate in San Pedro Benito Juárez, with a similar increase in current concern in another rural evacuated site, that of San Juan in Ecuador. Interestingly, work exchange tended to be more reciprocal (i.e., higher percentage of ego and alters both offering work to one another in a personal network) in the cases where it was associated with risk perception. It is possible that, in the rural sites, working closely with someone exacerbates existing perceptions of risk; alternatively or thinking of causality in the other direction, reciprocal relationships might be part of a suite of social support practices and collective approaches to disaster recovery and coping with chronic hazard.

In the Mexican site of Ayotzingo, having a wealthier personal network was associated with decreased current concern, while it was associated with increased current concern in the Ecuadorian site. Reduced wealth means reduced options for dealing with the hazards or with disaster recovery, and thus would increased concern. Why this is not the case in Penipe Nuevo may have something to do with the fact that people are still in the throes of deciding how to proceed—whether to continue farming via a daily commute, whether to move back into the high risk zone, whether to invest in the small town of Penipe, or whether to move elsewhere for work or farming—such that those with more might feel they have more to lose while things are still a bit unsettled.

Received social support (emotional, material, and informational) plays a major role in perception of past threat and current concern. It appears to some extent that for past threat, ego receiving support is more relevant, while for current threat ego giving support is more relevant. It is really in Pusuca and San Pedro Benito Juárez that these relationships exist, along with material support for Penipe Viejo's current concern. In addition to receiving support from one's network, it is clear that the degree to which someone feels close to the people in their network is implicated in risk perception and expectations. Despite low variation in religious affiliation of network members in all the sites except San Pedro Benito Juárez, having more evangelical Christians in one's network was associated with plans to evacuate in the relocated urban site of Penipe Nuevo.

11.6 Conclusion

Relatively little variation exists between the two countries. There is some interesting variation across some risk perception questions about the past, present, and future in terms of the variables associated them. Site characteristics, such as urban versus rural, resettlement versus non-resettlement, high impact versus low impact, and country, do appear to be associated with variation in risk perception and to be related to which kinds of variables predict risk perception in each site. We must also remember that the variation is fairly low in risk perception measures in Ecuador, suggesting that being exposed to a disaster or major hazard is a totalising experience. Our findings suggest that, where support exchanges are more frequent and there is a common experience of past disasters, risk perception tends to be higher— at least for the past. As we noted, this suggests the emergence of shared cognitive or cultural models (a collectivisation of memory or a redundant feedback loop of information in a dense network) of past events that may be contributing to the perception of current and future risk in ways that are not obvious in our current data.

References

Avenarius, C. (2003). The structure of constraints: Social networks of immigrants from Taiwan. Dissertation, Institut für Völkerkunde Universität Köln.

Avenarius, C., & Johnson, J.C. (2004). Recovery from natural disasters and the 'lack of weak ties.' Paper presented at the International Sunbelt Social Network Conference, Portoroz, Slovenia.

Bernard, H.R., Johnsen, E.S., Killworth, P.D., McCarty, C., Shelley, G.A., Robinson, S. (1990). Comparing four different methods for measuring personal social networks. *Social Networks*, 12(3), 179–215.

Buskens, V. (1998). The social structure of trust. *Social Networks*, 20(3), 265–289.

Dyer, C.L., & McGoodwin, J.R. (1999). The culture of response: The political ecology of disaster assistance and its impact on the fishing communities of Florida and Louisiana after Hurricane Andrew. In A. Oliver-Smith & S. Hoffman (Eds.), *The angry earth: Disaster in anthropological perspective* (pp. 211–231). New York: Routledge.

Faas, A.J. (2012). Reciprocity and political power in disaster-induced resettlements in Andean Ecuador. Paper presented at the Annual Meeting of the Society for Anthropological Sciences. Las Vegas, NV.

Garcia, J.L.O. (2000). *Teziutlan: Historia y tragedia*. Puebla, Mexico: Benemérita Universidad Autónoma de Puebla.

Hall, M., Norwood, A., Ursano, R., Fullerton, C. (2003). The psychological impacts of Bioterrorism. *Biosecurity & Bioterrorism*, 1, 139–44.

Helleringer, S., & Kohler, H.P. (2005). Social networks, perceptions of risk, and changing attitudes towards HIV/AIDS: New evidence from a longitudinal study using fixed-effects analysis. *Population Studies*, 59(3), 265–282.

Hobfoll, S.E. (2002). Social and psychological resources and adaptation. *Review of General Psychology*, 6(4), 307–324.

Kitchovitch, S., & Liò, P. (2010). Risk perception and disease spread on social networks. *Procedia Computer Science*, 1(1), 2345–2354.

McCarty, C. (2002). Measuring structure in personal networks. *Journal of Social Structure*, 3(1).

McCarty, C., & Killworth, P. (2007). Impact of methods for reducing respondent burden on personal network structural measures social networks. *Social Networks*, 29(2), 300–315.

McCarty, C., Killworth, P.D., Bernard, H.R., Johnsen E.C., Shelley, G. A. (2000). Comparing two methods for estimating network size. *Human Organization*, 60(1), 28–39.

Norris, F.H., Baker, C., Murphy, A.D., Kaniasty, K. (2005). Social support mobilization and deterioration after Mexico's 1999 flood: Effects of context, gender, and time. *American Journal of Community Psychology*, 36(1–2), 15–28.

Reissman, D., Spencer, S., Tanielian, T., Stein, B. (2004). Integrating behavioral aspects into community preparedness and response systems. In Y. Danieli and D. Brom (Eds.), *The trauma of terror: Sharing knowledge and shared care* (pp. 707–720). Binghamton, NY: Haworth.

Scherer, C.W., & Cho, H. (2003). A social network contagion theory of risk perception. *Risk Analysis*, 23(2), 261–27.

Tobin, G.A., & Whiteford, L.M. (2002). Community resilience and volcano hazard: The eruption of Tungurahua and evacuation of the Faldas in Ecuador. *Disasters: The Journal of Disaster Studies, Policy & Management*, 26(1), 28–48.

Tobin, G.A., Whiteford, L.M., Jones, E.C., Murphy, A.D. (2011). The role of individual well-being in risk perception and evacuation for chronic vs. acute natural hazards in Mexico. *Journal of Applied Geography*, 31(2), 700–711.

Turner, R.J., & Marino, F. (1994). Social support and social structure: A descriptive epidemiology. *Journal of Health & Social Behavior*, 35(3), 193–212.

Unger, D.G., & Powell, D.R. (1980). Supporting families under stress: the role of social networks. *Family Relations*, 29(4), 566–574.

Chapter 12
Human, Gender and Environmental Security at Risk from Climate Change

Úrsula Oswald Spring

Abstract This chapter analyses the risks of extreme hydrometeorological events with the concept of dual vulnerability: environmental and social vulnerability, which focuses on people affected by global environmental change and climate change. The understanding of dual vulnerability orientates the policy to promote resilience that may mitigate impacts of extreme events, since it is only recently that the factors that create, increase or limit risks have been analysed. Improving adaptation and mitigation may reduce the impact of disasters and the loss of life and livelihood. This chapter explores an integrated human, gender and environmental—a HUGE—security approach. The gender perspective allows the differential susceptibility between men and women during an extreme event to be understood, which reflects gender relations consolidated during thousands of years by the patriarchal system characterized by violence, authoritarianism, exclusion and discrimination. This integrated security opens analytical perspectives for policy reflections that could enhance resilience and facilitate the empowerment of men and women before, during, and after a disaster. Governments will achieve greater success in disaster management when they promote participatory governance where authoritarian arenas, agendas, activities and actors are replaced, the dual vulnerability addressed, and adaptation and resilience embraced.

Keywords HUGE-security · Dual vulnerability: social and environmental
Disaster risk reduction · Gender security · Environmental security
Resilience · Participative governance

Prof. Dr. Úrsula Oswald Spring, Research Professor, Regional Centre for Multidisciplinary Research at the National Autonomous University of Mexico. Email: uoswald@gmail.com.

© Springer International Publishing AG, part of Springer Nature 2018   167
Ma. L. Marván and E. López-Vázquez (eds.), *Preventing Health and Environmental Risks in Latin America*, The Anthropocene: Politik—Economics—Society—Science 23, https://doi.org/10.1007/978-3-319-73799-7_12

12.1 Introduction

The present chapter analyses the risks of extreme hydrometeorological events starting with the concept of dual vulnerability (Bohle 2002; Oswald Spring 2013a): the environmental and social one. The text focuses on people affected by global environmental and climate change, which affects human, gender and environmental security (HUGE security; Oswald Spring 2009, 2013b). The chapter investigates the creation of resilience that may reduce the impacts of extreme events, since it is only recently that the factors that create, increase or reduce disaster risks have been scrutinised (Cardona 2007; Beck 2001, 2007, 2011). The impacts of global environmental change[1] (Brauch et al. 2008, 2009) and climate change (IPCC 2012, 2013, 2014a, b; McBean/Ajibade 2009) indicate that extreme events are becoming increasingly unpredictable, non-linear, abrupt and chaotic, especially when adverse natural factors are triggered by social and political problems.[2] On the environmental side, there are advances in understanding the risks that can lead to irreversible changes or tipping points (Lenton et al. 2008), such as the collapse of the Gulf Stream, the destruction of the Amazon, the disappearance of monsoons in India, Africa and Latin America, thaw in the Arctic, Antarctica and Greenland, etc.

In socio-political aspects, there is less progress on potential tipping points in the global, regional and local social system. There is a lack of rigorous analysis of the negative interactions between environmental migration (Sánchez et al. 2013; Oswald Spring et al. 2014) and desertification (Ahmed et al. 2009), flooding (Paul 2005), crop failure and famine (Messer/Cohen 2011; Rowhani et al. 2011), as well as complex emergencies IMDM (2014). Other studies have shown that extreme water scarcity has aggravated upheavals (Scheffran et al. 2012) and can trigger regional conflicts and civil wars (Sunga 2011; Verhoeven 2011; IPCC 2014a).

In the meantime, few studies have been published on the vulnerability to disasters with a gender perspective (Ariyabandu/Fonseka 2009; Oswald Spring 2008; Anttila-Hughes/Hsiang 2013; Jung et al. 2014), which indicates the differential susceptibility between men and women or children, adults and elderlies during a disaster. This behaviour reflects gender relations consolidated over thousands of years within the patriarchal system, characterized by violence, authoritarianism, exclusion and discrimination. However, the origin of inequality is rarely explained, and it is a key factor that increases the risks of women and marginal people to extreme events.

Given this socio-environmental complexity, the existing concepts are insufficiently defined and only partially analysed. Thus, this paper asks: How can a

---

[1]Global environmental change is more than climate change, since it interrelates physical aspects (atmosphere, hydrosphere, biosphere and pedosphere) with anthropogenic factors (urbanisation, population growth, production processes, transportation and consumption).

[2]I would like to acknowledge the support received from the research project funded by PAPIIT-UNAM IN 300213 'Integral management of a basin affected by climate change: risks, adaptation and resilience', as well as careful anonymous review by academic peers.

disaster risk reduction (DRR)[3] policy with a gender perspective reduce the impacts of hazards in the face of increasing and more frequent extreme events? Is it possible to promote prevention, recovery and adaptation activities for the affected population before, during and after a disaster, with increasingly complex socio-economic conditions, in order to reduce their vulnerability?

The chapter begins with an exploration of the concepts of human, gender and environmental security, what I call a HUGE security. It also analyses the concept of dual vulnerability and investigates why women, girls and others vulnerable are more highly exposed to risks during an extreme event. Through various examples, the text illustrates not only the vulnerability of women, but also the potential of women to adapt efficiently when authorities promote a gender perspective. The chapter also reviews the new risks related to changing climate conditions. The text investigates potential policies for DRR that would avoid that an event being converted into a disaster. It explores examples of preventive actions, adequate responses during the event and participatory actions in the reconstruction phase that improve DRR actions in the hands of authorities and directly help the affected population. A more comprehensive adaptation could potentiate responses from below and support the efficiency of government actions in DRR. By refining both processes, the resilience of exposed communities may be improved. In addition, adaptation actions identified by the Intergovernmental Panel on Climate Change (IPCC 2014a, b) could raise awareness of the new, unfamiliar and related hazards of global environmental change in each region and produce co-benefits. By locally understanding socio-environmental vulnerability and gender discrimination, policies may be developed to address behaviours that trigger more dangerous conditions. The chapter ends with a proposal that synthesizes the results of the reflection and explores how arenas, agendas, actors and activities of vulnerability could be converted into arenas, agendas, actors and activities of welfare, in order to improve human, gender and environmental security.

12.2   Conceptualization of Human, Gender and Environmental Security (HUGE Security) in the Face of Global Environmental Change

The term security has undergone changes throughout its conceptual history. Traditionally, it has been linked to military and political control over the territory and the legitimate defence of the State to protect national sovereignty in the face of threats from other states or subversive groups. The end of the Cold War, global risk

---

[3]UNISDR (2009: 10–11) defines DRR 'The concept and practice of reducing disaster risks through systematic efforts to analyse and manage the causal factors of disasters, including through reduced exposure to hazards, lessened vulnerability of people and property, wise management of land and the environment, and improved preparedness for adverse events.'

theory (Beck 2007, 2011), reflexive postmodernity (Giddens 1994) and the review
of the neoliberal globalization (Stiglitz 2003) have led scientists to reconceptualise
security. Wolfers (1962) has distinguished first between an objective and subjective
approach to security. Wendt (1992) has introduced intersubjective security, but it
was the Copenhagen School that has reviewed the various actors who are intervening in the security process and who have transformed the arena in which these
changes occur. This School (Buzan et al. 1998) defined security as a 'speech act'.
Wæver (1997) proposed a securitisation process, where an intersubjective understanding with the political community is created to declare something as an existential threat (e.g. George W. Bush declaring the 'war on terrorism' after the 11th of
September 2001). In this securitisation process a politician usually assigns crucial
importance to a reference object (an event or a basic value) in order to justify the
use of extraordinary measures (war). In a successful securitisation process, the
audience accepts these measures, sometimes at the cost of limiting some of their
own basic rights.

12.2.1 Human Security

United Nations has intervened in the discussion of the reconceptualization of
security by promoting human security. UN has changed the reference object of
military security, understood only as sovereignty and defence of the territory, and
has focused on human beings (UNDP 1994). Its frame of reference refers to people,
facts and changing historical conditions, where the basic value at risk is individual
and collective well-being. Humanity, therefore, becomes the articulating axis of
politics, expressed in experiences and perceptions that are culturally diverse in each
region. In human security, the value at risk is survival, quality of life and cultural
integrity, where the threats come from the neoliberal State, the multinational
enterprises, the speculative globalization processes (Amin 1974), the global environmental change (Brauch et al. 2009), climate change (IPCC 2013), poverty
(Coneval 2015), inequality (Oxfam 2017) and fundamentalism (Sousa-Santos 2010).

   Step by step, the concept of human security has been clarified, first as 'freedom
from fear', when the Nobel prize was awarded in 1999 to social groups that fought
for the abolition of personal mines and small arms in Canada and also insisted on
promoting sustainable socio-economic conditions that limit migration. Ogata and
Sen, in Human Security Now (CHS 2003), created the second pillar, called 'freedom from want', where these authors promoted the reduction of social vulnerability
through combating poverty and respecting human rights. Policies of equality and
equity would improve and consolidate human security. Bogardi/Brauch (2005)
proposed a third pillar as 'freedom from hazard impacts', by reducing environmental and social vulnerability, which could prevent extreme events from turning
into disasters. In the same year, Annan (2005) proposed as a fourth pillar 'freedom
to live in dignity'. In the Report 'In Larger Freedom', the Secretary General of UN

insisted on strengthening the rule of law, respect for human rights, equity and empowerment to consolidate human security. Human security represents a deepened process of security that impacts from the individual and its home to the world. The four pillars on human security have left intact the roots of violence, discrimination and exclusion, which are embedded in the patriarchal system (Mies 1985; Lagarde 1990; Lamas 1996) and have been consolidated all over the world for more than five thousand years.

12.2.2 Gender Security

Given some analytical deficiencies of the concept of security, Oswald Spring (2013b) reviewed different currents of feminism and developed the concept of 'gender security'. 'Epistemological feminism' has first deconstructed androcentric concepts within the paradigm of military and political security, underpinning the deep, manifest and subtle mechanisms of power exercise. These mechanisms are based on discrimination, where gender violence is central to the reproduction of the whole system and the maintenance of the patriarchal *status quo*. Bordo (1990) has shown how metaphors and ideals of masculinity are at the root of positivism by glorifying the ideals of objectivity and rationality. Harding (1986, 1988) argued that the dualism of nature-culture; subject-object; female-male is at the root of all discrimination and justifies violence and exploitation. Haraway (1988) demonstrates the false dichotomy between object and subject that justifies the present systems of control, but where subjectivity always defends positions of domination and control.

From the 'empirical feminism', women have generated new questions, theories and methods of research. They have played a crucial role in the scientific transformation and built up a defence against a socially unfair, exclusive, hierarchical, and violent system. Their life experience has provided them with more comprehensive knowledge that is not pigeonholed in separate disciplines, universal methods, ex ante assumptions, and globalizing currents. Empirical feminism has shown that gender is constructed from sex, or from what each culture historically recognizes as sexual difference. This approach justifies the characteristics that classify sexed beings in various genres. The number of sexual characteristics may vary interculturally, although the generic classification is manifested in all known societies, thus it is a universal classification. This favoured classification axis is the genital difference (sexual dimorphism: female-male), therefore gender has been transformed into a bio-sociocultural construction of being (Lagarde 1990).

'Post-structuralism and postmodernist theories' have questioned the universality of laws, theory of needs, objectivity, rationality, essence, unity, totality, fundamentalism or the ultimate truth and the unique reality that has justified neoliberalism as the only economic option, which has led to an exclusionary globalization (Stiglitz 2003). To maintain the *status quo* and fight against resistance, the oligarchy in power generates violence, repression and military interventions. All these essentialist theories of late capitalism have prevented us from seeing the underlying mechanisms that produce inequality and structurally impede greater equity.

The fourth current, the 'Feminism of the point of view', has highlighted the position of women as oppressed, which has allowed criticism of the dominant system and practices of exploitation. It has promoted criticism and processes of liberation from the household to the global world. Feminists have also been associated with other social struggles of the oppressed (class, environmental and indigenous struggles). The location of women and other oppressed groups generates actions based on their daily life that is located in a certain space and time. By this introspective vision, different positions and theories about the transformation of reality are analysed and more effective mechanisms to overcome violence and discrimination are explored. Feminist knowledge is also expressed in the first versus the third person, since expressing knowledge in the third person is intended to give an 'objective positivist' vision. Feminists also interpret external symptoms to contextualize the process of 'I am here and now' (Hartsock 1983).

These four currents of feminism have inspired my concept of gender security, which analyses the structural changes, the underlying structures and the limited access of women and other social groups to public and economic power. These discriminatory practices have been developed over thousands of years by patriarchy and have generated differential regional expressions, but always related to exploitation, discrimination, subordination and violence against other human beings, increasingly also against nature. Thus, the concept of gender security explores the change of the reference object, especially when the nation-state has given its exclusive control of power to transnational corporations and the oligarchs have privatised the sphere of power (military, jails, strategic research). Gender security has guided the analysis of socially constructed gender relations over thousands of years. It investigates the values at risk, where gender security no longer focuses on sovereignty and territorial control, but on the values of equality and equity within global and local society. Gender security explores Northern domination of low-paid workers in the South, but also allows the analysis of the appropriation of income by economically and politically dominant sectors within each country and region. The sources of threats are no longer other States or only partially. The threats come directly from patriarchal society, characterized by totalitarian institutions such as oligarchies, authoritarian governments, and hierarchically organized churches. Its mechanisms of control are exclusion, exploitation, discrimination, authoritarianism, violence, wars and financial crises, which have allowed power concentration in an increasingly smaller elite (Oxfam 2017).

Acceptance of the concept has faced obstacles, since world society has been organized for thousands of years by vertical power relations, where the male gender has appropriated the public space and dominated over the female gender, labelling it the weaker sex. The symbolic space has assigned to the man the external sphere of production and power, that is, man has traditionally been responsible for the *res publica* and is considered in most societies as *homo sapiens*. At the same time, women have been confined to the private sphere of reproduction, at home and as *homo domesticus* and their work is turned invisible and without economic value.

Thus, gender insecurity is socially constructed and systemic within the present society (Serrano Oswald 2009). The reference object refers to the social relations built from gender socialization (masculinity and femininity), which have been weakly questioned. The multiple interactions of power have been combined, confronted, aligned and sometimes summed or subtracted, but finally an oligarchic power has been built that exerts a global hegemony over capital, politics, labour, violence, wars, the environment and the people. This power subordinates other men with less power and all women, since this vertical power is anchored in the home, but extends to the global sphere. It also explains the high number of feminicides, where women are not only killed, but tortured, raped and then brutally slaughtered. It also indicates the violence that exists for young men, who suffer from unemployment, insecurity and mass murders related to organized crime and drug wars.

12.2.3 Environmental Security

Feminist analysis also indicates that violence is not only against humans, but also against nature. Ecofeminists (D'Eaubonne 1974; Mies 1985) established a parallelism between gender violence and violence against environment. Just as human beings with less power suffer from violence, there are the same mechanisms of exploitation against natural resources and its ecosystem services, which affects the environmental security of the planet. There is some consensus among researchers that environmental safety was consolidated through three phases and that a fourth phase is being developed (Dalby et al. 2009; Oswald Spring et al. 2009). The first phase of environmental security has analysed the impacts of wars, weapons and military movements on ecosystems. Long-term damage from the use of the orange agent in the Vietnam War (Westing 2013) indicated a causal relationship to its mutagenic effects among American soldiers and the Vietnamese population. Environmental destruction of flora, fauna, soils, water and seas was also demonstrated by the long-term effects of atomic bombs and bullets and bombs with depleted uranium in various conflicts (e.g. UN Secretary General 2008).

In the second phase, the research group of Homer-Dixon (1991, 1994, 1999) analysed the increase of disputes over scarce resources. Bächler (1999) encountered local and regional wars and conflicts also when an abundance of natural resources occurs, such as new irrigation projects, where peasants were displaced by landlords or transnational corporations. Recently, massive purchases of land and water rights by foreign nations and companies have led to multiple regional and local conflicts (IPCC 2014a). In the third phase of environmental security, a number of disciplinary and multidisciplinary investigations were developed that empirically and theoretically documented the interrelation between access and use of resources or their lack, producing protests, guerrilla warfare, civil wars and regional conflicts. In this third phase of environmental security emerged global projects such as the Nile project of UNESCO, ECONILE and ECOMAN, where multilateral institutions promoted policy initiatives to raise awareness among the international community and specific regions of the potential of conflicts and the urgency to share scarce natural resources. One example is transboundary water agreements. Agreements is in 1944 the Treaty between Mexico and the United States for the Utilisation of Waters of the Colorado and Tijuna Rivers and of the Rio Grande. However, conflicts can be aggravated by problems of social vulnerability (Bohle 2002), where disasters and conflicts may cause 'complex emergen-cies' in several parts of the world (Sudan, Liberia, Nigeria and others).

Specific policies, human actions and insufficient measures to reverse pollution and environmental degradation have also increased environmental conflicts. In addition, population growth has generated a greater demand for natural goods and ecosystem services (MA 2005), which has sharpened the scarcity of multiple natural resources, often exacerbated by pollution. The increase in greenhouse gases after the industrial revolution and especially the massive use of fossil fuels over the last six decades has produced global warming. Anthropogenic activities on forests, land, air and water have increased global environmental change and climate change, due an increase in the emissions of greenhouse gases (GHG). Intensive burning of fossil energy has changed the physical-chemical composition of the air, increased the global temperature and altered the traditional precipitation patterns, but is also producing increases in extreme hydrometeorological events (IPCC 2012, 2013, 2014a, b).

The actors involved in this environmental deterioration are the State, entrepreneurs who extract raw materials and pollute natural goods, but also all citizens, who consume, pollute and produce waste and GHG. The values at risk in environmental security are sustainability and biodiversity, and the sources of threat are nature itself (extreme events), but also human beings, with their unsustainable productive system, their consumerism, GHG emissions and waste production. For the first, time human beings—we—are both the cause and the victims of global environmental change. Due to GHG emissions we reinforce the conditions for extreme events and then we suffer the consequences in the form of disasters. Against environmental insecurity, especially GHG emissions, waste and consumerism, military or weapons are unable to help to reduce environmental threats. Faced with a progressive deterioration of environmental security, mitigation and adaptation measures are necessary. Restoration of ecosystem services, fairer access to natural resources, reduction of overexploitation and pollution of aquifers, safe freshwater, recovery of ecosystem services, and reduction of GHG together with resilience may reduce the risks to extreme events for exposed people.

From 2005 onwards, the term of 'environmental security' has been linked to human security in the projects of GECHS (1999), UNU-EHS and IHDP, although it is now clear that the notion of vulnerable human beings is directly linked to the lack of governmental support due to the shrinking of the State related to the neoliberal model and the privatisation of public services. An alternative global model of sustainability has not been developed, nor have existing binding commitments to recover the atmosphere, despite multiple meetings and proposals to control GHG emissions and some success at COP 21 in Paris.

By rethinking the causal chains between ecology, human security, gender security and controversies, the term governance appears as a crucial element for peaceful negotiations of conflicts, where transdisciplinary teams are proposing a fourth phase of environmental security, which investigates the complex interrelationships among physical, social, cultural, political and governance factors to reduce the dual vulnerability (Oswald Spring 2013a) and the environmental-induced migration (Oswald Spring et al. 2014; IPCC 2014a) and to resolve differences peacefully through negotiation within a frame of win-win.

In short, it was clear that threats to environmental safety are related to our way of producing and consuming. We are our own enemies by causing the deterioration and, at the same time, we are the victims through disasters and loss of livelihood. To reduce the risks associated with our anthropogenic behaviours, weapons or army do not serve, thus the narrow military security approach is unable to resolve these new threats. It is necessary to radically transform our way of consumption and production by reducing GHG emissions, deforestation, and cleaning polluted water, soil and air, recover biodiversity, and reduce the destruction of environmental services. This requires a new revolution, a fourth one[4]: a 'revolution of sustainability' (Oswald Spring/Brauch 2011; Brauch et al. 2016).

12.2.4 Human, Gender and Environmental: A HUGE Security

Confronted with this deplorable physical, environmental, human and gender insecurity, several researchers have asked: which security do we need? Security for whom, for what, of what and security from whom and from what? (Wæver 1997). Oswald Spring (2008) proposed a comprehensive security that integrates human, gender and environmental security. This security analyses violent and exclusive structures, from the family to the global system existing in human, gender and environmental relations. At micro level, human gender and environmental security (HUGE) offers criteria to explore the conditions of job creation, salaries and welfare. At meso-level, the HUGE concept understands the conditions of well-being, to clean air and the elimination of water pollution, so as to overcome soil depletion. At

---

[4]The first revolution, the agricultural one, began 7000 years ago, when humans settled in villages and small towns and developed irrigation systems to produce their food. Subsequently, with the surplus of several harvests, began an internal division of labour, a social stratification, and struggles for power with violence, slavery and conquests. The industrial revolution followed around 1750, when the majority of the peasantry in northern countries moved to cities and became salaried workers. In the South these changes have occurred from 1950 onwards and in Africa there still exist a majority of rural people, however most of the world population is now living in cities. Fifty years ago, a third revolution began, called IT: technological and communicative, which facilitated the globalization of communication systems, trade of goods and financial flows, but did not include the free flow of people. The acute environmental and social deterioration, as well as the threats caused by global environmental change and climate change, require a new revolution, a sustainable one, as soon as possible, in order to guarantee the survival of humanity and Planet Earth.

macro level, the HUGE concept suggests variables to analyse economic stability, the transformation of roles and social representations of gender, welfare, quality of life, images of success, beauty and desirable futures, but also the limits of the present neoliberal model. Within countries highly exposed to extreme events, such as the Philippines, Mexico, Bangladesh, Madagascar, China and others, the HUGE approach reinforces the interdisciplinary exploration of environmental and social vulnerability factors that are intertwined with specific local, regional and global environmental risks. From empirical researches, comparative studies are able to evaluate the deterioration of the quality of life for the majority, as well as the loss of the purchasing power of the working class and the concentration of wealth in the hands of a thin oligarchy.

Thus, the concept of human, gender and environmental security, which I call HUGE security, offers guidelines for the analysis of environmental and gender security through a widened and deepened understanding of human, gender and environmental security, its impacts on human beings, on environment and its challenges for peace. Going back to the patriarchal roots of violence (patriarchy), the unequal appropriation of natural resources (environmental security), the lack of respect for human rights, and the loss of the rule of law (human security), the structures of vertical and violent power (gender security) with its underlying structures are searched. The HUGE-security approach also opens the analysis for alternatives that examine the processes of citizen participation and the potential of a participatory democracy, where postmodern governance (In't Veld 2011) could facilitate conflict prevention and resolution. HUGE security finally allows solidarity during disasters to be examined and the sustainable practices that would consolidate the transition towards a local and global sustainability to be understood (Grin et al. 2010).

12.2.5 Dual Vulnerability: Environmental and Social Vulnerability that Increase Risks

Dual vulnerability refers to a predisposition to be affected by adverse natural conditions (abrupt slopes, flood areas, drylands, etc.) and social situations that increase these risks, such as discrimination, inequality, abandonment, lack of support and violence. Frequently, dual vulnerability is combined with weak resilience that limits the ability to recover or to adapt to socio-environmental deterioration and to extreme hydrometeorological events. It relates to:

(a) fragility in communities, social groups and environments exposed to dangerous extreme events, either because of their location or their lack of physical or social resistance;
(b) socio-economic and environmental instability (high levels of marginality, dangerous physical conditions and/or severe environmental deterioration).

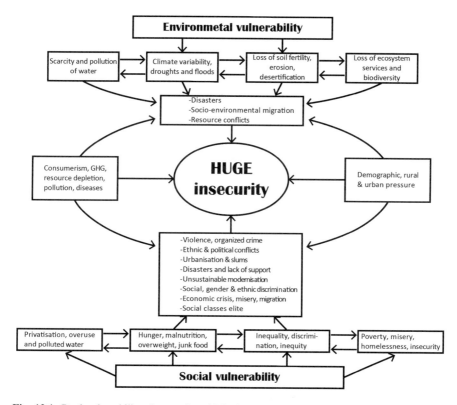

Fig. 12.1 Dual vulnerability. *Source* Oswald Spring (2013a: 21); Inspired by Bohle (2002)

Figure 12.1 shows, in the area of environmental vulnerability, the negative interrelations of pollution and water scarcity with climatic variability, loss of soil fertility and deterioration of ecosystem services and biota. All these factors set negative scenarios to extreme events that turn them into disasters. The consequences can induce migration and conflicts over scarce resources. On the side of social vulnerability, the neoliberal model is privatizing basic public services and there are no compensation mechanisms for the poorest to pay for these services. Often young people are left without job opportunities and poverty and inequality have increased, which can lead to violence, public insecurity and economic stagnation, sometimes aggravated by organised crime. In the absence of a participative governance, high crime and inefficient weak judicial systems, there is the danger of failed states. Potential socio-environmental disasters negatively increase this dual vulnerability (Haiti), and both vulnerabilities are aggravated by the present consumer model, the generation of waste and the pollution of water, soil and air. Demographic pressure and chaotic urbanization further aggravate both vulnerabilities.

12.3   Global Risks with a Gender Perspective to Extreme Impacts

Risk refers to the closeness, imminence or contiguity of possible damage caused by an adverse event, which may be climatic. It can be aggravated by the socio-environmental conditions in which it occurs. Risk also depends on the severity of the event, its danger and complexity, the socio-environmental vulnerability in which it occurs and the population's ability to prevent, adapt and recover quickly. Beck (2011) insists that the risks are complex, unpredictable, non-linear, chaotic, and can produce cascading effects (see the thermonuclear disaster in Fukuyima, Japan after an earthquake and a tsunami). Measuring the risk of an extreme event refers not only to the possibility of a misfortune occurring, but also predicts its impact—extreme or not. However, conditions of environmental and social vulnerability can turn a minor event into a disaster, because the community or the individual do not have the ability to avoid harm to its life and property (Haiti). In most extreme cases, the Government has failed to promote preventive measures, processes of adaptation and resilience, thus the people are unable to reduce the expected impact. Cardona (2007) insists that dimensioning a risk also includes organizational factors and social and institutional networking. Therefore, risk analysis is a multidisciplinary tool and requires a holistic framework for analysis.

In human perception certain risks tend to be underestimated, but there are also factors of socialization that influence this perceptive process of denial. For example, cyclones with female names throw more deaths than those with male names, although the first ones have a smaller impact on the Saffir-Simpson scale. It seems that the female name reduces the perception of risk and, therefore, increases the vulnerability of being affected. In addition, potential severe risks are sometimes denied by social groups, in the fear of losing welfare, status or leadership. In other cases they are exaggerated to control social groups through fear (drug war promoted by Felipe Calderón). But not all risks are related to perception, and past events have left practical learning among the impacted population. In 2005, with 28 tropical storms and 14 hurricanes in the Atlantic, the first events of some severity prepared the Mexican population exposed to more extreme ones later. When Wilma appeared, with a barometric pressure of 882 hPa at its centre and winds of 295 km/h, the population had learned to evacuate pre-emptively to safe refuges. Emily and Rita, previous hurricanes, had taught the people that they could not resist Wilma within their homes. Moreover, in the fear of losing their lives, they overcame the anguish that their belongings might be looted. Despite the destruction of Hurricane Wilma in the Yucatan Peninsula, the Cancun resort and the Riviera Maya, there was almost no loss of human life, although the material damage was substantial.

The contrary happened with hurricane Stan, a week before, in the same country and during the same month. The lack of early warning, no preventive evacuation and the lack of shelters still four days after Stan impacted, caused numerous deaths and material damage in the indigenous zone of Chiapas. A decade later, most of the damaged houses had not been rebuilt, and the existing social vulnerability pushed the indigenous population into indigence and migration. In the case of Wilma, the international tourist resort Cancún was rebuilt in months, thanks to Government support and insurance that allowed some risk transfer (Oswald Spring 2012). These objective and subjective factors, related to Government support and knowledge of potential risks, the experience of past events, the possibility of transferring risks, the level of schooling among those exposed, early warning, Government intervention and resilience by past experiences, influenced a correct or false assessment of risks. These factors facilitated, in the case of Wilma, a preventive evacuation, considerable prudence during the extreme event and intense cooperation in the reconstruction. Finally, risk transfer with insurance companies is a common practice among real estate investors, but is almost non-existent among the low-income indigenous population. Government support facilitated a rapid reconstruction of the tourist centre, while the lack of prevention and corruption within the Government increased the dual vulnerability of the most marginal people in Chiapas.

12.4 World Disasters

During 2013, 308 events were reported globally, almost all of them turned into disasters due to lack of prevention and institutional support. They caused 25,000 deaths, and the estimated costs amounted to 140 billion dollars (BD; SwissRe 2014). Typhoon Haiyan in the Philippines alone produced more than 7,500 dead, and flooding in the Himalayan region in India generated another 6,000 deaths. Both disasters occurred in emerging countries. During 2016, MunichRe (2017) reports that the total damage caused by natural extreme events amounts to 175 BD, of which only 30% was insured. Worldwide, during 2016 8,700 people were killed, less than the 25,400 in 2015. Two earthquakes in Japan produced 31 BD in losses, and floods in China caused another 20 BD in damage. After an earthquake in North Korea's northernmost province, Typhoon Lionrock caused widespread flooding and killed more than 130 people. Hurricane Matthew produced 500–1,000 deaths in Haiti, due to lack of preventive evacuation and produced 10 BD in loss. North America was hit during 2016 by 160 disasters, most of them related to climate change prone extreme events.

Figure 12.2 illustrates the hydrometeorological events that occurred during the last 30 years. While the number of geological events has remained stable, hydrometeorological events have increased. In the last six decades, Asia was the region with the highest number of extreme events, deaths, people affected and

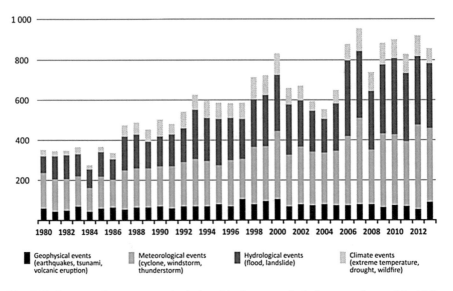

Fig. 12.2 Impacts of extreme geophysical and hydrometeorological events. *Source* MunichRe (2012)

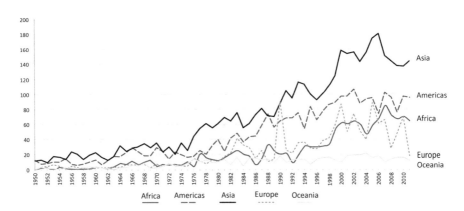

Fig. 12.3 Natural extreme events by continent. *Source* Guha-Sapir et al. (2013)

economic damage (Figs. 12.3 and 12.4). It is the most populated region on the planet, and also contains countries highly exposed to extreme hydrometeorological and geological events (China, Philippines, Japan and Indonesia).

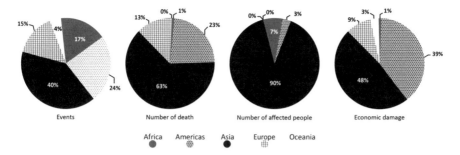

Fig. 12.4   Events, deaths, affected y damage during 2002–2011. *Source* Guha-Sapir et al. (2013)

12.5   Disasters and Gender

These global data overlook another vulnerability, the one of women and girls. During the Asian tsunami, 63–68% of deaths were female; in the earthquake in Pakistan, women accounted for 80% of deaths (Aruyabandu/Fonseka 2009), and during hurricane Stan in Chiapas for 72% (Oswald Spring 2012). Generally, these deaths are poor women, often heads of households. The highest number of deaths is related to poverty and social mechanisms of discrimination, which include lack of education and training. But it is also about social relations that were developed within each society and family, where dress, long hair, role at home, education, training (not knowing how to swim), lack of access to early warning and lack of knowledge of the official language (indigenous communities) increase the social vulnerability of these women (Oswald Spring 2008). Throughout the world, gender identity is crucial to understanding the differences of gender in the number of deaths. Women were and are socialized and assumed the role of 'caring for others', even at the cost of their own lives (Serrano Oswald 2010, 2013). This identity is important in any society, although it is not always recognised. In several countries, women receive little or highly inadequate help in times of emergency, and assistance is usually given to men, heads of households.

Anttila-Hughes/Hsiang (2013) shocked public opinion with a study carried out three years after several cyclones in the Philippines. With the aid of death certificates, they showed that a year after thirteen cyclones of different intensities, fifteen times more female babies have died, compared with male babies. Deaths a year after a major typhoon far exceeded deaths which directly occurred during the cyclone. In addition, during the extreme event no differences were found in numbers of deaths among male and female children (Anttila-Hughes/Hsiang 2013: 40). The two authors state that 'economic factors are key, because almost half of the babies who died later were not even born when the cyclones occurred' (see Fig. 12.5).

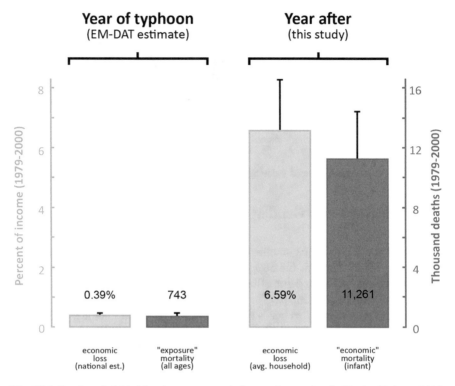

Fig. 12.5 Deaths of girl babies due to economic losses. *Source* Anttila-Hughes/Hsiang (2013: 71). Permission was granted by the authors

The number of deaths of female infants is even higher when they had older male siblings, as the family prioritizes feeding and care of them. The authors found a 6.6% reduction in income after a disaster, which reduced food expenditure by 16.9%. Through this difference they explain the cause of 15.1 times more deaths among female babies. When households cannot adjust their food expenses with subsidies or other income, the widespread practice is to reduce the consumption of nutrient-rich foods, although tobacco, alcohol or recreation expenditures have not decreased. Education and coping strategies could help sensitize families to dramatically reduce these deaths and thus change the survival rate of female babies. All these poor families would have to incorporate the impact of cyclones in their spending schedules, which would enable them to build greater resilience, especially when governmental policies are inadequate or insufficient. Additional alternatives could be micro-insurance, direct governmental support for nutritious foods, poverty reduction programmes, microcredit for food consumption, education on nutrition, adapted technologies for reconstruction, and information in the mass media about the risks to female babies.

12.6 Adaptation Within the Concept of a HUGE Security

The IPCC (2014a) has deepened the concept of adaptation and defines it as gradual or transformative adjustment processes, depending on the changing conditions of the climate system, global environmental change and its regional impacts. In order to achieve a more effective adaptation, various factors are involved: citizen and governmental practices, training, availability of resources, short- or long-term benefits, as well as costs, effectiveness and viability of actions, where socio-cultural factors play a crucial role. Local constraints limits and collaboration between different levels of Government can limit or improve adaptation conditions. But it is always the participation of the directly affected that guarantees greater success. Involving those exposed in management plans and programmes would improve the chances of achieving protection. In addition, the affected people provide valuable local knowledge, which allows plans to be adapted to suit their own idiosyncrasies. Generally, they help to choose the most feasible and most effective options, often also the cheapest. This adaptation plan embraces broad actions involving the social, institutional and structural background. However, mitigation measures are often required in addition to adaptation processes, which may include the relocation of populations exposed to excessive risk.

A crucial issue is the training of public officials in DRR programmes and adaptation plans. But at the same time, it is also necessary to train those affected, where the gender perspective is decisive in reducing impacts, and also in optimising local resources, since women are more enthusiastic participants. A successful adaptation is part of participatory planning, local experimentation, reduction of local dual vulnerability, adjustment of the strategy to changing conditions and, above all, high flexibility in the face of increasingly less predictable, stronger and greater events with higher uncertainty. Adaptive management involves, therefore, gradual approaches that respond to changing situations, which also resolve the structural factors of environmental and social vulnerability. Despite all these practices, it is important to know that any adaptation has limits. When risks become intolerable, or get so unpredictable and dangerous, it is necessary for the population to emigrate by their own will, or through a forced relocation by the Government. The latter will be more successful when the affected population is directly involved, because without collective efforts to systematize the risks and dangers to which the population is exposed, it is difficult to carry out a successful relocation. Generally, the population refuses to leave its community, its culture and its natural surroundings to be relocated in a new place where the uncertainty, the local conditions and the new threats are unknown.

In order to consolidate adaptation, resilience is crucial. This refers to the ability of a socio-political and natural system to absorb disturbances, while consolidating basic operating structures, improving their activities and strengthening self-organization and the capacity to overcome new stresses, threats and changes (IPCC 2014a). Since extreme events cannot be avoided, it is important to promote practical learning from past difficulties and mistakes in order to avoid them in the future, although concrete conditions may change. This approach increases flexibility in social organization and DRR, associated with the changing conditions of global environmental change. In addition to social organization, it is important to restore ecosystem services (MA 2005), which not only provide services (water, air, pollination), support production, disintegrate and clean up wastes, mitigate winds and infiltrate water to aquifers, but also generate cultural immaterial services that consolidate peaceful coexistence, the beauty of the environment and the creativity of those affected.

Faced with a growing complexity of extreme events and additional impacts by global environmental change, the IPCC (2012; Fig. 12.6) proposed the integration of various epistemic communities to unite efforts: DRR (McBean/Ajibade 2009), that of climate studies and prediction of atmospheric phenomena and that of development. Multiple studies have shown that DRR-orientated development processes and adaptation make it possible to reduce environmental vulnerability and social vulnerability, especially for women and girls. Although events are becoming increasingly extreme, DRR and disaster prevention management can reduce negative outcomes and avoid disasters, especially when the climate scientific community makes accurate forecasts and appropriate models that prevent changes in the medium and long term. In food production affected by drought, early warning could prevent the decrease of well-being and future famines. Trained societies and those exposed to previous experiences can acquire learning processes that enable them to consolidate their resilience and face fewer human and material losses during new extreme events. Therefore, transdisciplinary partnerships also consolidate prevention, timely care, and restore normal conditions as soon as possible.

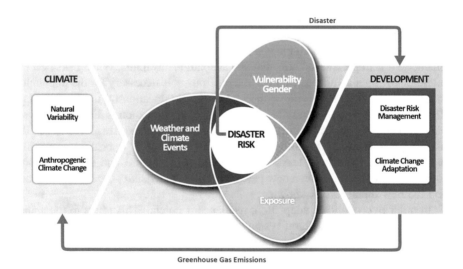

Fig. 12.6 Collaboration among epistemic communities. *Source* Adapted by the author based on IPCC (2012: 4)

12.7   Conclusions

By relating the examples presented with the conceptual elements and the research question, we can conclude that an integrated HUGE-security concept allows processes of mitigation, adaptation and resilience to be explored. It may help to reduce present and future risks related to global environmental change and climate change. Figure 12.7 systematizes this complex task by locating the necessary actions in a political sustainable arena that depends on local and global geopolitical conditions. The political and social agendas, the public, private and social actors and the activities that are producing dual vulnerability must be overcome. Through collective efforts, an adverse arena of vulnerability can be transformed into an arena of well-being with citizen participation. In developing and emerging countries, the vulnerable arena lacks or has generally weak governmental support. Institutions are

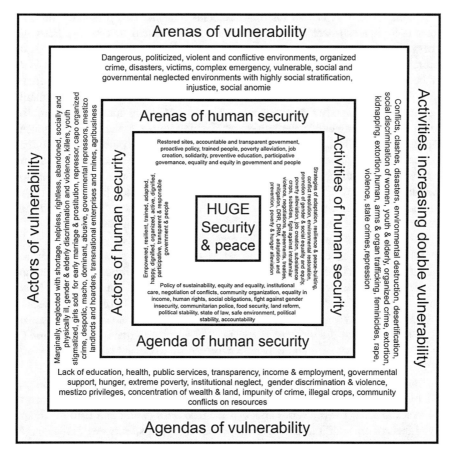

Fig. 12.7 Transformation of the vulnerable towards well-being arenas, actors, agendas and activities. *Source* The author

uncoordinated, and the Government has been reduced due to the privatization of basic public services. In addition, there is a systematic loss of purchasing power among people and an increase in the cost of living and food, thus people live in more and more precarious conditions. Support for science and technology is limited in the Southern Governments due to the high payment of domestic and foreign debt services, which are also generating a lack of innovation. In this unfavourable arena, the transnational and national oligarchy appropriates the national rent and internal inequality increases rapidly. Latin America, despite major efforts undertaken by several progressive Governments in South America, remains the most unequal region in the world.

In this arena of vulnerability there also exists serious environmental deterioration related to 'extractivism' in the hands of transnational and national companies, which received Government contracts to extract oil, gas and minerals from the subsoil. Their activities left rampant environmental destruction and pollution (open pit mines), which have increased environmental vulnerability. At the same time, these companies have caused multiple accidents (arsenic, lead poisoning, drinking water pollution and accidents). They have also generated diverse conflicts over access to land and water resources (Willaarts et al. 2014). An extreme case was the deforestation of 92% of the tropical rainforest in the 1970s and 1980s, related to the oil boom in Tabasco, Mexico, which devastated an invaluable natural heritage (Oswald Spring/Flores 1985). The World Bank had financed this environmental destruction for promoting extensive livestock in the Mexican humid tropic. Another example of environmental deterioration is related to tourist and urban developments in coastal areas, which have destroyed mangroves that have protected people from hurricanes and high waves. Especially dangerous is the construction of urban and tourist infrastructure on barriers and coastal lagoons (Tacloban in the Philippines or Cancun in Mexico), river beds (Atitlán in Guatemala, affected by Stan) and lakes (Mexico City). These predatory activities have increased the vulnerability of millions of people and are associated with short-term profits for a tiny elite.

During the last three decades, the governmental 'agendas' in Latin America started from a dependent economic model, characterized by a massive sale of land to transnational corporations and often failed privatization processes, which have delegitimized governments and companies (Suez, Agua de Barcelona and Banco Galicia in Argentina during the Government of Menem). Corruption and lack of governmental and social prevention have created insecure environments. Organized crime has increased the violence in most of the Latin American countries, and the so-called 'ninis', who are young people deprived by work and study, are easy prey by these criminal groups.

The 'activities of vulnerability', imposed by global, national and criminal capital, have sharpened local and regional conflicts over the access to strategic natural resources (Sudan, Libya, Syria) and to basic resources (water and land in Chiapas and Liberia). In addition, the sale of public companies, services and banks to transnational oligarchs has displaced millions of workers and limited public funds required for the development of these emerging countries.

Faced with these socio-economic difficulties and the unknown threats of global environmental change, it is necessary in most countries to strengthen the arena of HUGE security from below. At local level, affected people and governments can start with the restoration of soils, water and forests. Through negotiated agreements for providing ecosystem services, women and young people should be incorporated into the environmental restoration with dignified salaries. Other activities of HUGE security may promote resilience among the affected population, improve the capacity of early warning, and create community policies with cooperatives for healthy food supply. Through productive linkages, microenterprises and economy of scale, an entire region can be developed with the supply of basic products. In turn, the Government can improve its governance with the participation of people and control better the abuses of economic agents. As mediator, the Government should negotiate during conflicts in order to open spaces for democratization processes in national and local agendas. Together public and social actors may combine prevention policies, especially in regions often affected by climate change, thus these activities would improve the 'arena' of a HUGE security. Through education, principles of peaceful coexistence, preventive health care and DRR can be taught. The 'agenda' of HUGE security is based on a new social agreement between Government, exposed people and responsible companies, where the ethical behaviour of the latter must limit the abuses and exploitation of humans and nature. Empowerment from bottom-up among the vulnerable enables them to demand the accomplishment of their human rights. When they are trained in resilience-building and adaptation practices, these vulnerable groups are able to face more difficult climatic and social conditions.

These practices related to HUGE security are more easily developed at local level. As pilot projects, they may later expand to regional, national and global spaces. Threats and risks from extreme events do not respect national boundaries, as evidenced by the tsunami in 2004, where the victims came from many Asian, African and European countries, and a large part of the local population and tourists were affected. The reaction of global solidarity to this global catastrophe indicates that basic human values exist among citizens that allowed local communities to recover and achieve a situation of normality as soon as possible. In many countries, this solidarity helped to improve living conditions. In addition, cutting-edge international technology (tsunami warning) was installed and combined with local efforts, which created an efficient early warning system that may improve human, gender and environmental security.

Not only the post-tsunami recovery, but multiple other examples have shown that in times of crisis, women maintain social networks and establish new ones to support communities and people affected or in mourning. They also collaborate with the Government to overcome the existing crisis as soon as possible. They are sensitive to new risks that arise when family ties, living and working conditions are broken and sexual violence is more visible. As shown in Fig. 12.8, the interaction

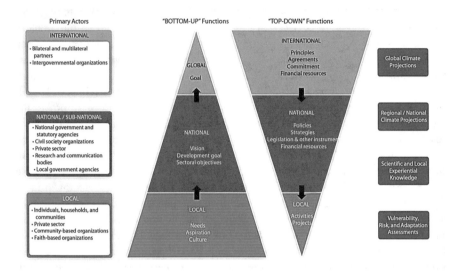

Fig. 12.8 Interaction of global and local actors with diverse functions. *Source* IPCC (2012: 346.12)

between governmental efforts and international organizations from global to local, combined with actions and agendas from below, may allow the creation of arenas, agendas, actors and activities of HUGE security that are aware of the present threats and that would help to prepare for future risks, increasingly more unpredictable.

In short, a HUGE security approach opens up analytical perspectives which may influence policy decisions that, when correctly applied, can enhance resilience and facilitate the empowerment of men and women before, during and after a disaster. This approach represents the analytical basis for creating political tools that can empower the vulnerable and convert them from passive and deprived objects to active agents of change. These trained people—women and men—have the capacity to build and improve their resilience and adapt to diverse cultural contexts.

The understanding of this extended and deepened security approach from the household to the global scenario offers some conceptual reflections that may guide the policies towards a complex understanding of risks and DRR. These policies may open the field for a comprehensive management of extreme events, where authorities and citizens are together able to protect the survival of those exposed and guarantee their human rights. Undoubtedly, it is the responsibility of every citizen to collaborate in this noble task for the benefit of themselves, humanity and the future of the Earth. There is no doubt that there exists only one planet and that together we must all help to recover the conditions for a sustainable life that continues to offer conditions of dignity to the present and the future generations.

References

Ahmed, S.A., Diffenbaugh, N.S., Hertel, T.W. (2009). 'Climate volatility deepens poverty vulnerability in developing countries'. *Environmental Research Letters* 4(3), 1–8.

Annan, K. (2005). *In Larger Freedom, New York: Report form the Secretary General*, UN.

Anttila-Huge, J. K., Hsiang, S. M. (2013). Destruction, Disinvestment, and Death: Economic and Human Losses Following Environmental Disaster http://papers.ssrn.com/sol3/papers.cfm?abstract_id=2220501.

Ariyabandu, M., Fonseka, D. (2009). 'Do Disasters Discriminate? A Human Security Analysis of the Tsunami Impacts in India, Sri Lanka and Kashmir Earthquake', in Hans Günter Brauch, Úrsula Oswald Spring, John Grin et al. (eds.), *Facing Global Environmental Change. Environmental, Human, Energy, Food, Health and Water Security Concepts*, 1223–1236. Berlin: Springer.

Bächler, G. (1999). Environmental Degradation and Violent Conflict: Hypotheses, Research Agendas and Theory-Building, in Mohamed Suliman (ed.), *Ecology, Politics and Violent Conflict*. London: Zed Books.

Beck, U. (2001). *Políticas ecológicas en la edad del riesgo*. Barcelona: El Roure.

Beck, U. (2007). *Weltrisikogesellschaft auf dem Weg in eine andere Moderne*, Frankfurt a. M: Surkamp.

Beck, Ulrich (2011). 'Living in and Coping with the World Risk Society', in Hans Günter Brauch, Úrsula Oswald Spring, Czeslaw Mesjasz et al. (ed.), *Coping with Global Environmental Change, Disasters and Security – Threats, Challenges, Vulnerabilities and Risks*. Berlin: Springer-Verlag.

Bogardi, J., Brauch, H. G. (2005). 'Global Environmental Change: A Challenge for Human Security – Defining and Conceptualising the Environmental Dimension of Human Security', in Rechkemmer, A. (ed.), *UNEO-Towards and International Environmental Organization-Approaches to a sustainable reform of global environmental governance*. Baden-Baden: Nomos.

Bohle, H. G. (2002). 'Land Degradation and Human Security', Paper presented to the UNU/RTC Workshop on 'Environment and Human Security', Bonn, 23–25 October 2002.

Bordo, S. (1990). 'Feminism, Postmodernism and Gender-Scepticism', in L. Nicholson (ed.), *Feminism/Postmodernism*. New York: Routledge.

Brauch, H. G. (2005a). *Threats, challenges, vulnerabilities and risks of environmental and human security*, UNU-EHS, Source 1. Bonn: UNU-EHS.

Brauch, H. G. (2005b). "Environment and Human Security. Towards Freedom from Hazard Impacts", *Intersection*, Bonn, UNU-EHS.

Brauch, H. G., Oswald Spring, Ú., Grin, J., et al. (2009). *Facing Global Environmental Change: Environmental, Human, Energy, Food, Health and Water Security Concepts*, Hexagon Series on Human and Environmental Security and Peace, vol. 4, Berlin Heidelberg: Springer.

Brauch, H. G., Oswald Spring, Ú., Mesjasz, C. et al. (2008). *Globalization and Environmental Challenges: Reconceptualizing Security in the 21st Century*, Hexagon Series on Human and Environmental Security and Peace, vol. 3, Berlin-Heidelberg: Springer.

Brauch, H. G., Oswald Spring, Ú., Grin, J., Scheffran, J. (Eds.) (2016). *Transition to Sustainability and Sustainable Peace Handbook*, Cham, Springer International Publishing.

Buzan, B., Wæver, O., de Wilde, J. (1998). *On Security. A Framework of Analysis*, Boulder: Lynne Rienner.

Cardona, O. D. (2007). *Indicadores de Riesgo de Desastre y Gestión de Riesgos. Programa para América Latina y El Caribe, Informe Resumido*, Washington: BID.

CHS [Commission on Human Security] (2003). *Human Security Now: Protecting and Empowering People*, Human Security Unit, New York: United Nations Office for the Coordination of Humanitarian Affairs (OCHA).

Coneval [National Council of Evaluation of the Social Development Policy] (2015). *Medición de la pobreza*, Mexico, Coneval.

Dalby, S., Brauch, H.G., Oswald Spring, Ú. (2009). 'Towards a Fourth Phase of Environmental Security', in H. G. Brauch, Ú. Oswald Spring, J. Grin, et al., *Facing Global Environmental Change. Environmental, Human, Energy, Food, Health and Water Security Concept.* Berlin: Springer: 781–790.

D'Eaubonne, F. (1974). *Le Féminisme ou la Mort*, Paris, Pierre Horay.

GECHS [Global Environmental Change and Human Security] (1999). *GECHS Science Plan.* Bonn: IHDP.

Giddens, A. (1994). *Beyond Left and Right – the Future of Radical Politics.* Cambridge: Polity.

Grin, J., Rotmans, J., Schot, J. (2010). *Transitions to Sustainable Development. New Directions in the Study of Long-term Transformative Change.* New York: Routledge.

Guha-Sapir, D., Santos, I., Borde A. (2013). *Reported Natural Disasters 1950–2011.* Leuven: EMDAT CRED.

Haraway, D. (1988). 'Situated knowledge: The science question in feminism and the privileged of partial perspective', *Feminist Studies*, No. 14, Fall: 575–599.

Harding, S. (1986). *The Science Question on Feminism*, Ithaca: Cornell University Press.

Harding, S. (1988). *Is Science Multicultural? Postcolonialism, Feminism, and Epistemologies*, Indiana: Indiana University Press.

Hartsock, N. (1983). 'The feminist standpoint: Developing the ground for an especially feminist historical materialism', in S. Harding and M. B. Hintikka (eds.), *Discovering Reality: Feminist Perspectives on Epistemology, Metaphysics, Methodology, and Philosophy of Science.* Dordrecht: D. Reidel Pub.: 283–310.

Homer-Dixon, T. F. (1991). 'On the Threshold. Environmental Changes as Causes of Acute Conflict', *International Security*, 16(2), Fall: 76–116.

Homer-Dixon, T. F. (1994). 'Environmental Scarcities and Violent Conflict. Evidence from Cases', *International Security*, 19 (1), Summer: 5–40.

Homer-Dixon, T. F. (1999). *Environment, Scarcity, and Violence.* Princeton: Princeton University Press.

IMDM (2014). "2nd International Workshop on In-Memory Data Management and Analytic", http://imdm.ws/2014/.

In't Veld, R. J. (2011). *Transgovernance. The Quest for Governance of Sustainable Development*, Potsdam: IASS Institute for Advanced Sustainability Studies.

IPCC [Intergovernmental Panel on Climate Change] (2012). *Report on Extreme Events.* Cambridge: Cambridge UP.

IPCC [Intergovernmental Panel on Climate Change] (2013). *The Fifth Assessment Report (AR5)- The Physical Science Basis*, Cambridge: Cambridge UP.

IPCC [Intergovernmental Panel on Climate Change] (2014a). *Climate Change 2014: Impacts, Adaptation, and Vulnerability. Working Group II Contribution to the IPCC Fifth Assessment Report*, Cambridge, Cambridge University Press.

IPCC [Intergovernmental Panel on Climate Change] (2014b). *Climate Change 2014. Mitigation of Climate Change. Working Group III Contribution to the Fifth Assessment Report of the Intergovernmental Panel on Climate Change*, Cambridge, Cambridge University Press.

Jung, K., Sharon, S., Viswanathan, M., Hilbe, J. M. (2014). "Female hurricanes are deadlier than male hurricanes", *PNS*, Vol. 111, No. 24, June 17, pp. 8782–8787.

Lagarde y de los Ríos, M. (1990). *Los cautiverios de las mujeres. Madresposas, monjas, putas, presas y locas*, México, D.F., PUEG/UNAM.

Lamas, M. (1996) (Ed.). *El género. La construcción cultural de la diferencia sexual*, Mexico, D.F., PUEG-Porrúa.

Lenton, T., Held, H., Kriegler, E., Hall, J.W., Lucht, W., Ramstorf, S., Schellnhuber, H. J. (2008). "Tipping elements in the Earth's climate system", in *Proceedings of the National Academy of Science*, PNAS, Vol. 105, No. 6, 12 February, pp. 1786–1793.

MA [Millennium Ecosystem Assessment] (2005). *Ecosystems and Human Wellbeing: Desertification Synthesis*, Washington, D.C., Island Press.

McBean, G., Ajibade, I. (2009). "Climate change, related hazards and human settlements", *Current Opinion in Environmental Sustainability*, Vol. 1, No. 2, December, pp. 179–186.

Bostenaru, M., D. Aldea Mendes, Thomas Panagopoulos (2013). "Assessing the costs of hazards mitigation", *JBU*, Vol. 3, No.122, pp. 51–68.

Mies, M. (1985). *Patriarchy & Accumulation on a World Scale. Women in the International Division of Labour*, London, Zed Books.

MunichRe (2017). Quarterly Statement 3, Munich, MunichRe.

MunichRe (2012). *MunichRe Annual Report 2012*, Munich, MunichRe.

Oswald Spring, Ú. (2008). *Gender and Disasters. Human, Gender and Environmental Security: A HUGE Challenge*, Source, No. 8, Bonn, UNU-EHS.

Oswald Spring, Ú. (2009). "A HUGE Gender Security Approach). Towards Human, Gender and Environmental Security", in Hans Günter Brauch et al. (Eds.), *Facing Global Environmental Change). Environmental, Human, Energy, Food, Health and Water Security Concepts*, Berlin-Heidelberg, Springer, pp. 1165–1190.

Oswald Spring, Ú. (2012). "Environmentally-Forced Migration in Rural Areas. Security Risks and Threats in Mexico", in Jürgen Scheffran et al. (Eds.), *Climate Change, Human Security and Violent Conflict. Challenges for Societal Stability*, Berlin-Heidelberg, Springer, pp. 315–350.

Oswald Spring, Ú. (2013a). "Dual vulnerability among female household heads", *Acta Colombiana de Psicología,* Vol. 16, No. 2, pp. 19–30.

Oswald Spring, Ú. (2013b). "Seguridad de género", in Fátima Flores (Ed.), Representaciones Social y contexto de investigación con perspectiva de género, Cuernavaca, CRIM-UNAM, pp. 225–256.

Oswald Spring, Ú., Brauch, H. G., S. Dably (2009). "Linking Anthropocene, HUGE and HESP: Fourth Phase of Environmental Security Research", in Hans Günter Brauch et al. (Eds.), *Facing Global Environmental Change: Environmental, Human, Energy, Food, Health and Water Security Concepts*, Berlin, Springer-Verlag, pp. 1277–1294.

Oswald Spring, Ú., A. Flores (1985). *Gran Visión y Avance de Investigación del Proyecto Integrado del Golfo*, México, D.F., UAM-X, UNRISD, CONACYT, CINVESTAV, IFIAS, COPLADET, PEMEX.

Oswald Spring, Ú., Brauch, H. G. (2011). 'Coping with Global Environmental Change – Sustainability Revolution and Sustainable Peace' in: Brauch, H.G., Oswald Spring, U. et al., *Coping with Global Environmental Change, Disasters and Security – Threats, Challenges, Vulnerabilities and Risks*. Hexagon Series on Human and Environmental Security and Peace, vol. 5, Berlin Heidelberg: Springer: 1487–1504.

Oswald Spring, Ú., Serrano Oswald, S.E., Estrada-Álvarez, A., Flores-Palacios, F., Ríos M., Brauch, H. G., Ruíz, T., Lemus, et al. (2014). *Vulnerabilidad Social y Género entre MigrantesAmbientales*. Cuernavaca: CRIM-DGAPA-UNAM.

Oxfam (2017). 'Una economía para el 99% Es hora de construir una economía más humana y justa al servicio de las personas' https://www.oxfam.org/sites/www.oxfam.org/files/file_attachments/bp-economy-for-99-percent-160117-es.pdf.

Paul, B. K. (2005). Evidence against disaster-induced migration: the 2004 tornado in north-central Bangladesh, *Disasters*, 29(4), 370–385.

Rowhani, P., Degomme, O., Guha-Sapir, D., Lambin, E. (2011). 'Malnutrition and conflict in East Africa: the impacts of resource variability on human security', *Climatic Change*, 105, 207–222.

Sánchez-Cohen, I., Oswald-Spring, Ú., Díaz-Padilla, G., Cerano-Paredes, J., Inzunza-Ibarra, M., López, R., Villanueva-Díaz, J. (2013). 'Forced migration, climate change, mitigation and adaptive policies in Mexico: some functional relationships', *International Migration*, 51, 53–72.

Scheffran, J., Brzoska, M. Brauch, H. G., Link, P., Schilling, J. (2012). *Climate Change, Human Security and Violent Conflict: Challenges for Societal Stability*. Berlin: Springer.

Serrano Oswald, S. E. (2009). 'The impossibility of Securitizing Gender vis a vis Engendering Security', in Hans Günter Brauch, Úrsula Oswald Spring, John Grin et al. (eds.), *Facing Global Environmental Change. Environmental, Human, Energy, Food, Health and Water Security Concepts*. Berlin: Springer: 1151–1164.

Serrano Oswald, S. E. (2010). *La Construcción Social y Cultural de la Maternidad en San Martín Tilcajete, Oaxaca*, Tesis Doctoral, México D.F.: Instituto de Antropología de la UNAM.

Serrano Oswald, S. E. (2013). 'Migration, woodcarving and engendered identities in San Martín Tilcajete, Oaxaca', in Thanh-Dam Truong, Des Gasper, Jeff Handmaker, Sylvia Bergh (eds.), *Migration, Gender and Social Justice. Perspectives on Human Insecurity*. Berlin: Springer: 173–192.

Sousa Santos, B. (2010). *Decolonizar el saber; reinventar el poder*. Montevideo: Trilce.

Stiglitz, J., E. (2003). *Globalization and Its Discontents*. New York: W. W. Norton & Company.

Sunga, L.S. (2011). 'Does climate change kill people in Darfur?', *Journal of Human Rights and the Environment,* 2(1), 64–85.

SwissRe (2014). Global Risks Report 2014: finding a path for 'Generation Lost' http://www.swissre. com/media/news_releases/Insured_losses_from_disasters_below_average_in_2014.html.

UNDP [United Nations Development Programme] (1994). *Human Development Report 1994: New Dimensions of Human Security*. New York: Oxford University Press.

UNISDR (2009). *UNISDR Terminology on Disaster Risk Reduction*, Geneva, UNISDR.

United Nations Secretary General (2008). *The Republic of Serbia: Positions on the Effects of the Use of Armaments and Ammunitions Containing Depleted Uranium*. http://www. bandepleteduranium.org/en/united-nations.

Verhoeven, H. (2011). 'Climate change, conflict and development in Sudan: global Neo-Malthusian narratives and local power struggles', *Development and Change*, 42(3): 679–707.

Wæver, O. (1997). *Concepts of Security*, Copenhagen, Department of Political Science.

Wendt, A. (1992). 'Anarchy is what states make of it: the social construction of power politics', *International Politics*, 42(2), 391–425.

Westing, A. H. (2013). *Arthur H. Westing: Pioneer on the Environmental Impact of War*. Berlin: Springer.

Wolfers, A. (1962). 'National Security as an Ambiguous Symbol', in: Arnold Wolfers (ed.) *Discord and Collaboration. Essays on International Politics*. Baltimore: John Hopkins University Press.

Chapter 13
Adolescents' Perceptions and Behaviours Regarding Volcanic Risk

Dayra Elizabeth Ojeda-Rosero, Melissa Cepeda-Ricaurte
and Esperanza López-Vázquez

Abstract The Popocatepetl volcano is the most dangerous one in Mexico, given the fact that it makes more than 20 million people vulnerable to a major eruption. We conducted a qualitative research in an educational institution located on the southern slope of Popocatepetl, in which there are five student risk brigades. Here we describe the most important topics of risk perception that the members of the brigades talked about. The educational community is an area of special interest in relation to volcanic risk that concerns many aspects of the community life and different perceptions and representations in their way of thinking. School brigades, beyond being groups formed by students trained to respond to a possible emergency, can be scenarios of collective-value-building to guide new ways of understanding human relations and risk prevention.

Keywords Volcanic risk · Risk perception · Students brigades
Educational community

13.1 Introduction

According to the United Nations International Children's Emergency Fund (UNICEF) and the International Strategy for Disaster Reduction (ISDR 2009), children and adolescents who attend school at the time of a disaster represent one of

Dayra Elizabeth Ojeda Rosero, Ph.D., Research Professor, Centre of Transdisciplinary Research in Psychology, Autonomous University of the State of Morelos and University of Nariño (Colombia). Email: deliza75@yahoo.com.

Melissa Ricaurte Cepeda, M.Sc., Centre of Transdisciplinary Research in Psychology, Autonomous University of the State of Morelos. Email: meli1254@hotmail.com.

Dr. Esperanza López-Vázquez, Research Professor, Centre of Transdisciplinary Research in Psychology, Autonomous University of the State of Morelos. Email: esperanzal@uaem.mx.

© Springer International Publishing AG, part of Springer Nature 2018   193
Ma. L. Marván and E. López-Vázquez (eds.), *Preventing Health and Environmental Risks in Latin America*, The Anthropocene: Politik—Economics—Society—Science 23, https://doi.org/10.1007/978-3-319-73799-7_13

the most vulnerable groups. Therefore, drawn from a student brigade research, this chapter seeks to present significant aspects to be taken into account in proposing guidelines for volcanic risk prevention.[1]

Risk perception can be understood as a process of integrating information from the environment based on the individual's understanding of the world, and behaviours are the actions taken in relation to the risk. It is a complex and changing psychosocial process that cannot be reduced to the interpretation of the intensity of a danger. The factors involved are multiple and are closely related to the context, the level of perceived vulnerability, the ability to control the situation, the familiarity, the novelty of the hazard, the subject's and the social group's previous experience, the perception of the cost-benefit, as well as other psychosocial, institutional and cultural factors at stake (Slovic 2000; Sjöberg 2000; Fischhoff et al. 2000). Many of these factors are explained in this book's first chapter.

Considering that our interest is focused on the processes of community perception, it seems pertinent to clarify that the concept of *community* can be understood as the confluence of interests and affections, especially the sense of belonging, from which individuals and groups organise and build their identities (Krichesky 2006). This allows the social group to be in a space of common exchanges that have a known organisation and are full of group and individual meanings that give a sense of belonging to each person. Given that the context in which we developed the present study was in an educational community, it is important to highlight the concept of an *educational community* as a community where interrelationships are generated around the educational process among different actors (teachers, students, administrators, parents, relatives, neighbours and members of other connected organisations) who share a primary purpose: the education of young students.

In this context, there is a daily coexistence which is not foreign to the multiple social contradictions. For that reason, it is not totally possible to achieve the participation of different community sectors or any other joint initiative in disaster prevention if the schools are not visualised as areas for the development of educational communities (Campos 1999). In this respect, Pereda (2003) states that it is not a matter of the school opening up to the community but of building a network of relationships where school, family and other organisations belong as members of the community.

---

[1]Thanks are due to the inhabitants of the Municipality of Tetela del Volcán (Morelos, Mexico) for their kindness and willingness to participate in this study. Also to the educational community for opening the institution's door to learn about their psychosocial dynamics and in particular to the members or collaborators of the student brigades. To the National Council of Science and Technology of Mexico (CONACYT) for the support to the Ph.D. and Masters grantees for the development of their respective research projects: 'Intergenerational construction of the perception of volcanic risk' and 'Construction of a community strategy for the prevention of volcanic risk', with the mentorship of Dr. Esperanza López-Vázquez, taking into account that we take up some parts that were jointly developed. Likewise, the advisers and reviewers are acknowledged for their contributions to the qualification of this work.

According to this conception of educational community, volcanic risk prevention as a collective construction demands the recognition and appreciation of the relationships that communities have culturally and historically built within their territory. Therefore, processes must be generated from the communities' daily life, their language and their world view, which is where the need arises for an education that values the right to participate by placing people in new roles which, for instance, offer the opportunity to gain experience of handling real-life problems, such as those relating to the environment (Freire 2007). Knowledge is a dynamic and permanent construction that is produced on the basis of other knowledge, so a dialectic relationship is generated between the person, society and the environment (Núñez-Hurtado 2005).

For Calixto (2012) environmental education strengthens awareness of the human beings' responsibilities regarding the continuity of different forms of life on earth and helps train social, critical and participating subjects to address socio-environmental problems. From this perspective, formal and non-formal education groups are called upon to construct scenarios of dialogue between popular knowledge and scientific knowledge, leading to a critical knowledge, guiding healthy relationships between living beings, which contributes to the solution of existing environmental problems and to the development of sustainable strategies and possibilities for coexistence. Going beyond the transmission and memorisation of concepts and generating contextualised processes in the communities all proves to be necessary in the processes of volcanic risk prevention.

13.2 Research Context

Mexico is a country with active volcanoes with different levels of activity. Some of them are considered high risk from the level of danger and the amount of materials that they exhale from their craters. There are sixteen active volcanoes, among which is Popocatépetl, which is categorised as threatening a high level of danger, although its periods of activity are not continuous since it manifests long periods of rest. It is located between the states of Morelos, Puebla and Mexico. According to Macías/Siebe (2005) Popocatépetl is the most dangerous volcano in the country in the sense that more than twenty million people are vulnerable to the effects of a major eruption, which makes the risk level of the different zones high due to its proximity and geographical position. Based on Popocatépetl's previous volcanic incidents, some zones are at greater risk of falling ash or of lahars due to the winds. Its current period of activity began in December 1994. Since then, eruptions have not stopped, although it shows periods of greater activity than others (López-Vázquez et al. 2008).

The Municipality of Tetela del Volcán, in Morelos, where this research is carried out, is located on the southern slope of Popocatépetl. The word Tetela comes from the Nahuatl root: Tetella or Tetetla, which means place where there are many stones or rocky ground (Encyclopaedia of the Municipalities and Delegations of Mexico 2005). It has a population of 18,179 inhabitants (Constitutional Government of

Tetela del Volcán, Morelos, Administration 2009–2012), distributed in the municipal head that bears the same name and three other locations. One of the highlights of the Municipality is the importance of community organisation in decision-making and collective actions (Reyes 2011). In Tetela there are several educational institutions of different levels. This work was developed in an educational institution that has approximately 650 students (Constitutional Government of Tetela del Volcán, Morelos 2009–2012) and 30 teachers, mostly from the neighbouring municipality of Cuautla.

13.3 The Student Brigade Within the School and Community Context

According to the Secretary of Public Education, the student brigades are groups of students that are integrated according to the recognition of the risks and vulnerability present in the school. There are five types of brigades: (a) evacuation, (b) search and rescue of lost and injured people, (c) first aid, (d) communication, and (e) fire prevention and combat. The function of brigade members is to support the representative or group leader in the relevant functions and activities in the internal civil protection and emergency unit, to make reports and carry out the specific activities and functions according to the type of brigade to which he/she belongs.

As for the school in which this research was carried out, as well as in other schools of the Municipality, at the end of 2013 a training course was conducted through Civil Protection and Personal Security courses carried out by the National College of Professional and Technical Education (CONALEP) through students of the degree in Hygiene and Civil Protection and teachers who support the process. This process was carried out in coordination with the Institute of Basic Education of the State of Morelos (IEBEM), the State Civil Protection Institute, the Committee on Security and Civil Protection of the local Congress and the Committees for Social Participation in Education (Municipal Government of Tetela del Volcán 2014). Five brigades were formed from this training, each with five students belonging to different school grades, mostly second grade. Training included topics on safety, civil protection, first aid, use of fire extinguishers, fire management, drills, search and rescue, and evacuation, among others.

The brigades are active, and the students who integrate them, with the support of leading teachers, have carried out activities such as evacuation drills in the institution, as well as identifying risks in the school. The formation of these brigades is recognised as an important advance in risk prevention, however, the need to strengthen this action on a continuous basis is evident. Students linked to the brigades are motivated to generate risk prevention actions, not only because of the need for this work in this school, but also because it gives them a chance to participate and hone leadership skills, as well as an opportunity to contribute to the welfare of their peers and the educational community in general.

13.4 Method

The main techniques developed in the study were: participant observation, focus groups and individual in-depth interviews.

Participant observation began in March 2013 for a period of approximately one and a half years, and was conceived from an ethnographic perspective as the process of insertion and permanence in the field (natural setting of the community). This immersion made it possible to approach the subjects' psycho-social world, in which their behaviour is based on symbols and meanings provided by the group and internalised in the process of giving meaning to the world. Notes and field diaries were used to record participant observation information, as well as videotapes of both the immediate surroundings of the educational institution and its surroundings.

Another technique developed was the focus groups, understood as a process of interaction in which opinions, attitudes, codes, values, representations, behaviours and symbolic systems, among other elements, intertwine to reveal socially con-structed senses and forms of everyday knowledge (Cervantes 2002). Two focus groups were formed, one with brigade students in the same educational institution, and another with teachers and parents of this institution. Each group consisted of eight people, with whom a conversation was generated about volcanic risk starting from the associations they made with the words 'volcano' and 'risk' presented separately as inductive words, followed by the initial questions: What does volcano mean? What memories do you have about the volcano? What emotions do they have? What does risk mean? These were linked to new concerns and feedback that emerged in the dynamic of the groups. The duration of each group session was 2 h. Their contribution was fundamental to learning the educational community's per-ceptions and behaviours surrounding volcanic risk, making it possible to read this information in the social context.

With the purpose of deepening and contrasting information from people from different sectors of the community, three in-depth individual interviews were conducted with two people who have lived with the Popocatépetl activity and the application of preventive measures, and one representative of public institutions for risk management who works in this community, adult men of different ages. Both the individual interviews and the focus groups included questions related to the perception of volcanic risk and subsequent behaviour when faced with Popocatépetl's activity.

The information was analysed based on its classification in different categories, which allowed us to interpret the vast information we obtained. In this way, we began by presenting a list of emerging categories based on the focus group of brigade students, which constitutes the axis of information analysis, from which information was integrated from other sources in the educational community. From the analysis of this focus group, the initial categories emerged: parents' perceptions, habituation, existing prevention strategies, media influence, school learning about risk, need for risk prevention, role of teachers, perceptions about brigades, expec-tations towards teachers, needs of the brigades, potentialities of the brigades and their future possibilities.

These categories were contrasted and supplemented with information from different sources (students, teachers, parents, other community representatives), emerging specific categories that reflect the reality of the brigades and their context as part of the educational community. There were also two meetings with the teachers and parents to talk with them about our findings. The reflexivity was constituted in a process of permanent analysis of the phenomenon within the research team, including the researchers' own perceptions; from there came about the elaboration with respect to the theoretical conceptual frameworks, the structure and content of this work, the same method and the potential contributions that it proposes to make to and from social psychology and community environmental education.

It should be mentioned that the members of the educational community who participated in the study gave their informed consent for participating with us and for the publication of results. Below are the results obtained, highlighting in italics the textual phrases that were said at the time of the interviews and focus groups, indicating the author of the sentences in an impersonal way. We will also briefly discuss each issue.

13.5   What We Found

For some members of the educational community the volcano activity has become part of their daily life, an aspect that generates familiarity with the phenomenon. People recognise the risk to which they are exposed because they perceive volcanic activity but it has been incorporated into their daily life: '*We all know that it is dangerous, but since we have become accustomed to living with it we are no longer afraid of it*' (Community member).

This familiarisation with the Popocatépetl activity and previous experiences have had a negative influence on evacuation behaviour: '*The last time we were on alert going to red, we heard very loud noises, but people no longer pay attention to it, because if that had happened years ago, I think we would had run away, right now it seems that we no longer… About six months ago, at that time, we heard a lot of noise, but we got used to hearing those noises, they do not go away, or better to say, we did not go*' (Mother).

13.5.1   Parents Before a Volcanic Phenomenon

In order to be able to better understand the volcanic risk perceptions of the community living around Popocatépetl, it is important to include different family members in the investigations on volcanic risk (López-Vázquez 2009).

Accordingly, value is given to the family's and parents' primary role in educating their children as members of society and their role within an educational community, including their perceptions.

Some parents recognise the risks and benefits of living in a volcanic area, the need for information on the necessary measures to prevent possible risks, and also the need to care for nature in the volcanic environment: '*The volcano for me is a story, a legend ... I respect it because you have to be very careful, because its activity is exceptional, and we must also take care of it, we must reforest, so it keeps on giving us what it has given us, lots of water, lots of it, uhm, fertility*' (Father).

Researchers such as De la Cruz-Reyna (2009) and Vela (2009) state that living in a volcanic area involves not only risks but also benefits such as soil fertility, water sources and other raw materials, among others. Other aspects also contribute to the understanding of people's decision to populate and stay in volcanic areas. According to Glockner (2012) the relationship of humans with nature goes beyond its simple use as a production resource, because the rituals that some peasants perform in volcanoes as sacred sites are signs of ethics and respect for the world in which we live.

Notwithstanding previous perceptions, there is also evidence of the uncertainty and fear generated by contemplating the possibility of a major eruption of Popocatépetl, recognising that there is a latent risk. This uncertainty can be a source of motivation for parents to contribute to the construction of alternatives for living with the volcano and minimising its risks.

13.5.2 Existing Prevention Strategies Can Still Be Improved

The volcanic activity status is a mechanism of the National Civil Protection System to keep the population informed about the different levels of danger of the volcano. This activity status has three colours: green, indicating normal activity; yellow, involving alert; and red which is the alarm signal. This mechanism is recognised by some members of the community as expressed by a student: '*There is an alert. The alert is used to inform the population of the phase in which the volcano is, in order to see if they have to evacuate.*' Likewise, the dissemination of information through leaflets and conferences provided by Civil Protection in situations of volcanic alert is recognised as a mechanism of volcanic risk prevention.

Evacuation drills and road improvements are also recognised by some members of the community as measures to prevent hazards. One student reported: '*I have seen that they do evacuation drills and I believe they even removed the speed bumps so that the evacuation routes are faster.*' Evacuations as preventive measures in case of volcanic alert can be essential strategies to safeguard the well-being of the population. However, the lack of consultation and planning with the communities affects the people's attitude regarding this measure and the decision to remain in their territories.

For some of the community members, evacuation processes carried out in the past have had a negative impact due to the theft of the belongings of those who were evacuated. '*Then they had them evacuated, there in the shelters and when they returned many people were discouraged to have left, because a lot of things were stolen… then people said that they were never going to leave again*' (Community member).

In addition, there are beliefs in the community that, through social pressure, can influence behaviour and decisions: '*Here, there is a word that is used a lot [when] people are cowardly. … I think that because many people left, those who stayed here began to call the people who left "coyones" (cowards)*' (Mother). These attributions show the need to deepen the different meanings that the community has built around the volcano and risk, as well as to improve existing prevention measures.

13.5.3   Credibility Has Been Lost in the Media

Participants highlight the influence of media such as television and radio, which have impacted the decisions of some members of the community in past events. A mother expresses: '*Most people left, I think, because of the same thing as on TV, in the media. Then it was more what they heard than what they saw—'leave now'. Then many people got ready and left.*'

There is some disagreement with the management of information given in certain cases about the volcano. A member of the community says: '*… and mainly journalists who have been here make such a scandal and the volcano does not kill, but journalism does*' (Member of the community).

13.5.4   People also Learn About the Volcano in School

School is a space that brings young people to build knowledge about the volcano. This is what students say in their testimonies: A student considers that the volcano is to be studied '*In classes that have to do with the subject, for example, in Spanish class there are legends and let's say that if it has the volcanoes, it can be talked about.*' Volcanic risk has also been addressed in biology: '*In seventh grade we did a project in biology sciences and … we conducted interviews about the volcano, about what people felt.*'

Another strategy used is the construction of teaching material: '*The geography teacher told us to make brochures so that we could distribute them.*' These initiatives are a resource for the prevention of volcanic risk, since not only information is sought, but it also has a practical application which may make possible new understandings of the relations of the community with the volcano.

13.5.5 Prevention Is a Necessity

Some members of the community recognise the need for information and preparation to reduce the impact of the volcano's activity: '*If you have prepared how to act at that moment, I believe that if you did not have a project before, in a work plan things would not work*' (Student). This need for preparation was stated by different members of the educational community: '*We must have information, to know what we can do in this situation, so we can take precautions so we can do something… because, to be honest, I am not prepared for an evacuation, for an emergency exit*' (Teacher).

Establishing volcanic risk prevention plans or strategies that enable the population to be informed about what to do in case of volcanic alert is a factor that can influence the community's perception of security.

13.5.6 Teachers in Risk Prevention

Teachers are key players in the prevention process due to their training and the influence they have as reference and authority figures for students. A student expresses: '*If there is a student who trusts a teacher a lot in… he also has a lot of influence in his/her ideas.*' Beyond the academic role they play in school, teachers are participants in a particular reality, in this case, the daily living with the volcano, and are called to participate in the construction of a safe school.

A teacher committed to his or her context, in which there is a risk or a need for prevention, can become an environmental educator, for which he/she requires training on the subject, in addition to integrating pedagogy and scientific knowledge, developing the capacity to contextualise them in culture; this implies various psychosocial aspects related to values, beliefs, attitudes and practices (Dos Santos 2012).

It should be noted that most of the teachers linked to the institution are not from Tetela, so their relationship with risk is different from the relationship built by students and teachers from this municipality. In this sense, Leal (2005) emphasises the importance of generating a common and positive identity around an educational institution based on an institutional educational project.

13.5.7 Credibility of the Brigade's Actions: An Aspect to Be Strengthened

Some of the students linked to the brigades perceive that sometimes their actions are not valued by their peers and other members of the community. According to one student, '*Older people do not want us to explain anything to them and don't like that we tell them anything. They feel they know everything, we do not have to tell them a thing.' 'Sometimes when we do the evacuation drills there are children, our peers, who just take it as game. Teachers do this too.'* These perceptions can affect their motivation, as well as their self-concept and self-esteem. *'Let's say that you get your self-esteem down because you have that initiative.'*

In certain cases, the role of young subjects may be denied or disregarded in social institutions, and the role of adult subjects is emphasised, limiting young people's possibilities of participation, or restricting them to rigid preset frames, without allowing their participation in the joint construction. This finally discourages the participation of young people, therefore justifying the conception that there is youth apathy and emphasising the fear from this population sector's demands (Leal 2005).

13.5.8 Brigade Students Recognise Their Potential Contribution

For some brigade students, these brigades are a resource to promote risk prevention in the educational community: *'It is a way to tell teachers and students more about the risks of a volcano or an earthquake, so that they can know the risks and how to evacuate, that is, to have more knowledge about more serious disasters'* (Student).

The possibility of coordinating prevention actions from the brigade can be an opportunity for young people to strengthen their capacities of leadership, management, organisation, social skills and empowerment, among others, which may contribute to their personal development, as well as contribute to prevention.

As school is an important community reference in which different actors and social sectors converge, the actions undertaken by the school brigades not only impact the immediate school context, but work led by the students can also contribute to raising awareness and generating risk prevention actions at the family and social level.

School brigades, as scenarios for the participation of the educational community, can guide new ways of understanding human relations and volcanic risk prevention, from solidarity, responsibility for protecting life and social commitment to collective well-being.

13.5.9 Brigade Students Have Expectations of Their Teachers

Some students linked to the brigades expect their teachers' support, so that from their example, they motivate students and other members of the community to participate actively in generating risk prevention actions: *'teachers begin with the example because otherwise students will not pay attention to us'* (Student). Some teachers show interest in the problem of volcanic risk and, as reference figures in the educational community, they are allies in the construction of a prevention culture. *'There are some teachers that do care and try to help us or orientate us, well, not all of them'* (Student).

Teacher training in the field of environmental education implies ethical, educational and pedagogical principles, that is to say, it implies the educational act in its totality. That is why there are no dichotomies between exercising a role of environmental trainers and the same exercise of the role of teachers (Dos Santos 2012). It is important to remember that there is not a single conception of environmental education but several references (Erice 2004; Dos Santos 2012), so it may be important to review and carry out an institutional reflection to construct an intentional project internally congruent and socially relevant.

13.5.10   What Is the Expected Future for the Student Brigade?

For students, the brigades' processes can be strengthened by generating training processes that allow them to acquire new knowledge to enable them to act effectively in an emergency, as well as establish strategies that allow them to contribute their knowledge to other students: *'What we need is to learn more things and to the first grade children, to teach them so that it becomes a follow-up so that when we leave the school, others will have this knowledge and so on'* (Student).

Students recognise that forming and maintaining brigades, rather than an isolated activity, can be an institutional process, in which young people themselves contribute to training new brigade members.

13.6   Towards Volcanic Risk Management from the Educational Community

The dynamisation of an educational community is a work of constant dialogue and reflection in the process of signifying and re-signifying the beliefs that are at the basis of their cultural frameworks. It is for this reason that one of the relevant tasks, with regard to establishing a culture of volcanic risk management from the educational community, it is to identify values from which it is possible to achieve this purpose. In this sense, a value that is considered important to work with this particular educational community is the *ethics with life*, understood as the value of the human right to dignified living conditions involving active citizen participation in the decisions to enable individual and collective well-being. This value may be possible with the deep and broad participation of social actors in the task of transforming their environmental consciousness (Cardenal 2005), which implies consideration of the human beings' relationship with their territories. Participation, as a process of collective construction of knowledge, enables the people's co-responsibility towards their life and well-being, overcoming reliance on external assistance, and from the risk situation, promoting community growth based on the agreement of principles and strategies (Ojeda 2008).

A coexisting principle with the value of ethics with life is the solidarity that means taking on commitments for collective well-being, overcoming individualism and recognising the need for the well-being of others as a condition for the general welfare. Hence the need to rethink the forms of forced or passive participation, aiming to revitalise the relationship with the world and perspectives on social reality during the process of risk management.

Another value that is considered to be a must at the basis of the risk management process in the educational community is the *collective construction of knowledge*, which involves giving value to everyday knowledge, establishing horizontal relationships between the actors of the educational process and considering them capable of acting on their environment.

It is necessary to promote the recognition of the culture and the systems of relationship between the members of the community and their surroundings, in particular their relationship with the territory in order to have an efficient and legitimate volcanic risk management which takes into account that it is there where they develop relationships with nature. At the same time it allows students to dialogue with other types of knowledge with a view to achieving a wide view of the world, but in connection with their real living conditions (Freire 2007).

Reality complexity demands the integration of different perspectives that allow an approach to their understanding. For this reason, volcanic risk prevention processes are nourished by the dialogue of popular and scientific knowledge (Ojeda 2011). According to Torres (2011) researchers and the community are carriers of scientific knowledge and popular knowledge, finding consensus and dissension, but above all intersections.

These two values can form a basis for initiating a process of community environmental education around volcanic risk, taking advantage of the relationships of the communities with 'Don Goyo', another recognised name for Popocatépetl, derived from history and culture that define a link with nature on which their world view is based (King 2010).

13.7 Concluding Remarks

It is possible to understand volcanic risk from a psychosocial perspective, based on identifying and interpreting perceptions and behaviours, resulting from an ancestral and historical construction of relations between settlers and territories, with cultural, educational and political implications, among others, beyond the sensory reception of the physical manifestations of the volcano activity. Based on the above, the volcanic phenomenon has multiple interpretation possibilities, where the risk is only one of them.

However, in order to contribute to volcanic risk management, social psychology and community environmental education can be more relevant, if other possible readings are involved, such as familiarity with the volcano, the psychosocial stress of living in a risk area (López-Vázquez et al. 2008; López-Vázquez 2009), some factors related to the capacity to control the situation (López-Vázquez/Marván 2012), the relationship between members of a community and their relationship with environmental elements such as volcanoes, among others, from which it is possible to understand this community's meanings and practices as a basis for preventive actions.

The educational community is an area of special interest in relation to volcanic risk, considering both the different actors involved in it and the relationships established around the educational process, which give it a vital role in the processes of risk interpretation, as well as in collective decisions and actions related to environmental education, to build a risk management culture integrated with the strengthening of citizen participation and the enhancement of community resources.

As an added value in the process of knowledge of perceptions and behaviours and the generation of collective strategies for risk prevention, we pinpoint the possibility of evaluating the educational institution's work in the particular community environment exposed to volcanic risk, including mobilisation of their projects and their educational processes, curriculum plans, and relationships among social actors, in the search to increase environmental awareness and social responsibility, respecting and giving value to the psychosocial dynamics of a locality.

Student motivation can be verified with regard to their wanting to make changes in their community from their role as brigade members, especially related to participation, including the opportunity to propose and activate strategies for risk prevention. In the same way, their integral training is strengthened through the development of skills and attitudes. In this sense, student brigades can be considered community resources for volcanic risk management, which can be enhanced with the support of all actors in the educational community.

School brigades, beyond being groups formed by students trained to respond to a possible emergency, can be scenarios of collective values, building new ways of understanding human relations and risk prevention to increase the students' self-confidence and motivate their active participation in different contexts: family, school and community.

It is necessary to continue developing research and action research in educational communities exposed to volcanic risk, in which socio-demographic aspects such as gender, origin, age and others are also involved in the analysis, which may contribute to a more detailed analysis of the situation. In the same way, it is important to carry out other studies from social, community and environmental psychology in relation to the possibilities of enriching the educational community's participation in institutional support in emergency and disaster risk prevention.

References

Calixto, R. (2012). Investigación en educación ambiental. *Revista Mexicana de Investigación Educativa*, 17(55), 1019–1033. http://www.redalyc.org/articulo.oa?id=14024273002.

Campos, A. (1999). Educación y prevención de desastres. *Red de Estudios Sociales en Prevención de Desastres en América Latina*. http://www.desenredando.org/public/libros/2000/eypd/EducacionYPrevencionDeDesastres-1.0.1.pdf.

Cardenal, F. (2005). Educación popular y ética (pp. 15–24) Managua: Decisio. http://www.file:///C:/Users/MiPc/Downloads/decisio10_saber2.pdf.

Cervantes, C. (2002). El grupo de discusión en el estudio de la cultura y la comunicación. Revisión de premisas y perspectivas. *Revista Mexicana de Sociología*, 64(2), 5–36. https://doi.org/10.2307/3541495.

Constitutional Government of Tetela del Volcán, Morelos, Administration 2009–2012. Plan Municipal de Desarrollo Tetela del Volcán, Morelos 2009–2012.

De la Cruz-Reyna (2009). El entorno volcánico en México. *Revista de Arqueología Mexicana*, 16 (95), 34–39.

Dos Santos, M. (2012). La formación de profesores (as) y de educadores (as) ambientales: acercamientos y distanciamientos. In Flores, R. (Coord.). *Experiencias Latinoamericanas en Educación Ambiental*. México, Monterrey, N. L.: Colección Altos Estudios.

Erice, M. (2004). Las competencias ambientales en la formación profesional. En: Nuevas tendencias en investigaciones en Educación Ambiental. Doctorado Interuniversitario en Educación Ambiental. Madrid, España. Organismo Autónomo Parques Nacionales, Ministerio de Medio Ambiente. www.oei.es/historico/decada/portadas/nuevas_tendencias.pdf.

Enciclopedia de Los Municipios y Delegaciones de México (2005).

Fischhoff, B., Slovic, P., Lichtenstein, S., Read, S., Combs, B. (2000). How Safe Is Safe Enough? A Psychometric Study of Attitudes Toward Technological Risks and Benefits. In P. Slovic (1 Eds.), *The Perception of Risk*, (80–103). London, Earthscan Publications Ltd.

Flores, R. (2012). Investigación en educación ambiental. *Revista Mexicana de Investigación Educativa*, 17 (55), 1019–1033. http://www.redalyc.org/articulo.oa?id=14024273002.

Freire, P. (2007). *La Educación como práctica de la Libertad*. México: Siglo XXI.

Glockner, J. (2012). *Los volcanes sagrados*. México D.F.: Santillana Ediciones Generales.

King, P. (2010). El nombre del Popocatépetl. Religión popular y paisaje ritual en la Sierra Nevada, Veracruz: Gobierno del Estado de Veracruz,. www.revistas.unam.mx/index.php/antropologia/article/download/33161/pdf.

Krichesky, M. (2006). Escuela y comunidad: desafíos para la inclusión educativa. http://www.bnm.me.gov.ar/giga1/documentos/EL005385.pdf.

Leal, F. (2005). Lo psicosocial en contextos educativos: Consideraciones conceptuales y empíricas a partir de una experiencia en liceos de alta vulnerabilidad. Límite. *Revista de Filosofía y Psicología*, 12, 51–104. http://www.redalyc.org/pdf/836/83601203.pdf.

López-Vázquez, E. & Marván, M.L. (2012). Volcanic risks, locus of control, stress and coping strategies. *Journal of Risk Analysis and Crises Response*, 2(1), 3–12. https://doi.org/10.2991/jracr.2012.2.1.1.

López-Vázquez, E., Marván, M., Flores-Espino, F., Peyrefitte, A. (2008). Volcanic Risk Exposure, Feelings of Insecurity, Stress, and Coping Strategies in México. *Journal of Applied Social Psychology*, 38(12), 2885–2902. https://doi.org/10.1111/j.1559-1816.2008.00417.x.

López-Vázquez, E. (2009). Risk perception and coping strategies for risk from Popocatépetl Volcano, México. *Geofísica internacional*, 48(1), 301–315.

Ojeda, E. (2011). Representaciones sociales de comunidad en la Parcialidad Indígena de Jenoy. Tesis de maestría en Etnoliteratura, Departamento de Humanidades y Filosofía, Pasto: University of Nariño.

Macías, J., & Siebe, C. (2005). Popocatépetl's crater filled to the brim: significance for hazard evaluation. *Journal of Volcanology and Geothermal Research*, 141, 327–330.

Municipal Government of Tetela del Volcán (2014). Capacitan a estudiantes en temas de seguridad y PC. *Boletín de Comunicación Social*, No. 131, 16 de enero de 2014.

Núñez-Hurtado, C. (2005). Educación Popular. Educación popular: una mirada de conjunto (pp. 3–14). Guadalajara: Decisio. http://www.infodf.org.mx/escuela/curso_capacitadores/educacion_popular/decisio10_saber1.pdf. Accessed july 15, 2014.

Ojeda, E. (2008). Problemática psicosocial en comunidades residentes en la zona de amenaza volcánica alta del volcán Galeras: Una perspectiva para el acompañamiento psicosocial desde la Psicología comunitaria. *Universidad y Salud*, 8(1), 45–64.

Pereda, C. (2003). Escuela y comunidad. Observaciones desde la teoría de sistemas sociales complejos. Revista Electrónica Iberoamericana sobre Calidad, Eficacia y Cambio en Educación, 1(1), 1–24. http://www.ice.deusto.es/RINACE/reice/vol1n1/Pereda.pdf. Accessed July 16, 20014.

Reyes, M. (2011). La organización local y los Recursos de Uso Común en Tetela del Volcán. Actores, espacios de decisión y sistemas de gobierno. Tesis para obtener el grado de Maestra en Instituciones y Organizaciones. Directora: Dra. Luz Marina Ibarra. Universidad Autónoma del Estado de Morelos.

Sjöberg, L. (2000). Factors in risk perception. *Risk analysis*. 20(1), 1–11. https://doi.org/10.1111/0272-4332.00001.

Slovic, P., 2000. Perception of risk. En. P. Slovic (Ed.), *The perception of risk*, 220–231. London: Earthscan Publications Ltd.

Torres, A. (2011). Educación popular, trayectoria y actualidad. Caracas: Universidad Bolivariana de Venezuela. https://dalbandhassan.files.wordpress.com/2011/04/educacion-popular-a-torres.pdf.

UNICEF & Estrategia Internacional para la Reducción de Desastres (EIRD) (2009). Escuela segura en territorio seguro. http://www.eird.org/publicaciones/escuela-segura.pdf. Accessed july 15, 2014.

Vela, E. (2009). Editorial del Dosier Los Volcanes de México. *Revista de Arqueología Mexicana*, 16, 95, 30–31.

Chapter 14
Behaviours Related to Increased Risk of Bullying Among Schoolmates

Hannia Cabezas

Abstract Abusive behaviour among school peers is a global issue that is present irrespective of culture, country, socio-economic level or kind of educational centre, and Costa Rica is no exception to this reality. This chapter seeks to identify risky behaviours that lead to bullying, and describes the main characteristics of the victims chosen by the aggressors, who perceive their victims as shameful and fragile persons. Also discussed is how the victims are affected in the short and long term.

Keywords Abuse among peers · Abusers · Bullying · Intimidation

14.1 Introduction

As described in earlier studies (Olweus 1983; Serrano 2006; Cerezo 2006; Menesini 2009), abusive behaviour between peers is a worldwide phenomenon that is present in many nations, irrespective of culture, country, social class or educational centre, and Costa Rica is no exception to this reality.

In 2007, the percentage of abuse among peers in Costa Rican groups was 19% (Cabezas 2007). Participants in this study were 371 students aged between 13 and 16 years. The rate found was comparable to the European average that to date varies between 15 and 20% in countries such as Spain and Norway.

Subsequently, a new study conducted with 916 primary school boys and girls aged 9–14 years found that the number of abusers reached 14% (Cabezas 2011). Within this group, 21% were boys and 9% were girls. Male abusers used greater physical force against their peers whereas female abusers used more covert methods such as threats, insults or blackmail in order to get something from their victims.

Hannia Cabezas, M.Sc., Teacher-Researcher, University of Costa Rica. Email: hanniac@gmail.com.

© Springer International Publishing AG, part of Springer Nature 2018
Ma. L. Marván and E. López-Vázquez (eds.), *Preventing Health and Environmental Risks in Latin America*, The Anthropocene: Politik—Economics—Society—Science 23, https://doi.org/10.1007/978-3-319-73799-7_14

By 2012, the number of abusers had risen to 20% and recently it reached 26% (Cabezas/Monge 2014). In this most recent investigation, a sample of 1,115 male and female students was selected with ages between 8 and 15 years; at the same time, a relationship was established between family and social environments surrounding the educational centres, and a greater or lesser reported incidence of peer bullying.

It is evident that the ways in which students relate to each other is changing. Threatening behaviours become stronger in the school setting; and, in classrooms, hitting, kicking, pinching, pushing, threatening, sneering, and insulting behaviours prevail as routine forms of interacting together and also of surviving. Because of this, attention given to this phenomenon may contribute to its early detection and help to ameliorate the devastating consequences for those who are victimized by their peers.

The various forms of aggression used by students are influenced by particular traits observed in potential victims who show 'greater anxiety or insecurity than the rest. Additionally they may be hesitant, sensitive and quiet. When they feel attacked, they normally respond by crying (at least in the lower grades) and running away. Similarly, they suffer from low self esteem and have a negative opinion of themselves and their situation' (Olweus 1998, p. 50).

It becomes increasingly import to call attention to other acts that may be affecting human behaviour such as acts committed against property and the environment or the lack of respect shown on highways. Such acts lead to violence and indirectly affect the nuclear family, the community, the school, and boys and girls in general.

Even though such activities would seem to be foreign to educational centres, they are very interrelated by forming part of the familial systems which address the roles played by individuals within their family, social or work environments, and by contributing to how these settings influence the individual's feelings and, later, the development of their life goals. This leads us to believe that neighbourhood violence should be considered a highly predictive factor for the development of abuse between peers, thus making the space in which the children grow up a causative factor of abusive behaviour in the classroom.

14.2 School Bullying

During the 1970s, Dan Olweus studied the phenomenon of bullying; research that was taken up again decades later by scientists in countries such as Japan, Great Britain, Canada, the United States and Spain. Thus, the researchers developed new hypotheses to demonstrate other criteria and allow greater understanding of this theme.

In general, bullying refers to intimidation by means of insistence, demands, threats or force.

Internationally, bullying is recognized as aggressive behaviour demonstrated between school-aged youth. Such behaviour is deliberate and may cause physical and emotional harm. The phenomenon of ill-treatment by abuse of power is also known as victimization or abuse among peers.

Olweus (1998, p. 25) defined it as 'the situation of abuse and intimidation in which one student is attacked or becomes a victim when repeatedly exposed over time to negative actions taken by one or more other students.'

Three important concepts are derived from this characterization, which help to identify the phenomenon with greater clarity:

a. Aggressive behaviour is carried out with the intention of causing physical and psychological harm. This form of activity is deliberate and affects boys, girls and youth who demonstrate particular physical or psychological characteristics that make them vulnerable to the aggressors.
b. The abusive behaviours against the victims are performed repeatedly and even extend beyond the hours of the school day. The frequency of the victimization varies from several times a day to several times a week or may be spread out over a longer period of time.
c. There is evidence of a disequilibrium of power that makes the victim feel impotent and in need of help since he lacks the skills to get out of the situation without the intervention of others.

Some research (Olweus 1998; Cerezo 2006) has shown that a person is attacked when inappropriate things are said about them, when they are threatened, intimidated, or blackmailed, when their possessions are grabbed from them, when they are sent anonymous messages, or when they are hit or pushed. On occasions, such aggressions are prolonged in time and space, affecting even more severely the victims who cannot find security in school or even at home with their family.

Bullying among peers affects the capacity for learning, one's self perception, and the need to belong to a group which are essential conditions for daily life, acceptance and survival of human beings living in community.

Aggression within the classroom can occur via visible actions or via covert acts that are difficult to demonstrate. Because of their particular characteristics, social scientists call such acts 'relational aggression' or 'social cruelty' and they can set off 'physical symptoms, such as migraines or stomach pains, in both the aggressor and the victim' (Anthony/Lindert 2012, p. 24).

There are many factors that should be taken into consideration regarding boys and girls who are victimized by their peers. Some of these indicators include: physical aspect, skin colour, vocal pitch, height, or the presence of some disability. Whether such characteristics are accompanied by acceptance or rejection by the victim himself or by his peers will impact his self esteem and may drive him to show signs of weakness, fragility or timidity among other conditions.

Showing some obviously different characteristic carries with it the risk of being excluded or isolated. Being different from the norm or lacking social skills is a reason for being rejected by a peer group which denies persons with different characteristics the sense of belonging that is a fundamental human need and which allows for the sharing of emotional states, values, beliefs and situations with others such that each human being forms part of the diversity of men and women in search of their own identity.

Negative experiences between peers directly impact the victims and, in adolescence, can cause anxiety attacks, low self-esteem, and depression (Deater-Deckard 2001). Therefore, early intervention in childhood is required to prevent such adverse consequences.

The construction of one's own self-image depends largely on the actions of one's social group. This 'identity discovered through relations with others' has been called 'identity by inscription' (Horton/Hunt 1992, p. 100). This is why the judgement or perception of others with whom one interacts as well as the physical appearance of each student take on special importance within the school context and furthermore may become determining factors for peer relations, predicting the future conduct of both victims and aggressors.

Personal appearance on occasion will create false impressions in aggressors concerning their schoolmates, which may cause their rejection or being identified as stupid, easily attacked or manipulated. I conducted a study in which students were asked to answer the question 'Why did you choose this person to bully?', and they gave the following answers about their victims: Because he is stupid, because I don't like her, because he is younger, because she doesn't react, because I saw him as weaker, because she was timid, because I could manipulate him, because she is smaller than me, because it's so easy, because it's her... (Cabezas 2007).

These victim characteristics mentioned by aggressors, led them to see their victims as timid and fragile persons. These conditions are interpreted by the aggressors as weaknesses, feeding the false belief that when the victimized boys and girls show these behaviours, they will also be cognitively handicapped, which will impede their reacting with mental agility and will limit their social skills and independence. Such distorted perceptions lead aggressors to act without restraint and think they will not be denounced and therefore can freely escape from any compromising situation.

When aggressors detect signs of timidity, fragility or weakness among their peers, it leads them to make them into easy targets, objects of scorn, humiliation and isolation. These actions by peers generate in the victimized students feelings of anxiety, anguish and fear as well as feelings of rejection that lead the victims to be more conscious of their own limitations.

Short stature, skin colour, or being the youngest in a group are other risk factors that can lead to ill-treatment in school.

In Costa Rica, we have found that the different ages of students present another factor leading to classroom bullying. Ranges of ages between 9 and 13 years within a single grade level were associated with rates of violence of 70% in one single classroom (Cabezas/Monge 2014). This excessive age range was found in a primary school located in a rural zone where 43% of the boys and girls had failed grade level promotion and lagged behind their peers and where also only 15% were not working to help support their families (Moreno et al. 2002). These findings coincide with previous studies that also concluded 'school failure and grade repetition are other effects of childhood abuse' (Pincever 2008, p. 40).

Unequal social relationships increment the diverse forms of persecution towards others. Thus, older students who share the same classrooms with younger students have been called 'tiny tyrants' who 'are more aggressive and have leadership capacity whereas their victims, especially if very young, are more passive, retiring, isolated and frequently without friends' (Menesini 2009, p. 11). In this manner, the student's age becomes another risk factor that, for those who must repeat a grade level, may signify in turn that school doesn't motivate them, that they lose interest in the school, and that these feelings of frustration and anger are displaced against their weakest peers who lack the tools needed to defend themselves.

Repeating one or two grade levels also impacts the aggressors because it affects their self-esteem and leads them to the false belief that they are incapable of learning. Also, the aggressors develop feelings of loss at being displaced from the peer group in which they began their studies. This in turn generates feelings of frustration that escape through strong reactions they don't know how to control.

Studies have shown that 'bullies often state they are very proud of themselves' (Besag cited by Voors 2006, p. 59); but 'they are [generally] unsatisfied with themselves except to the extent that they can demonstrate their superiority over others.' It is as if they said 'I only feel good when I can control you' (Voors 2006, p. 59).

In general, bullies also present low self-esteem, have a blurry image of their own personal insufficiencies and do not have the capacity to face up to them—they avoid thinking about a subject they find painful. Therefore, they take recourse in using force and intimidation against others (Voors 2006).

A second investigation on the principle characteristics of victims (Cabezas 2011) found that aggressors repeated the perceived conditions of 'weakness' and 'fragility' and stated that they 'didn't like' their schoolmates whom they described as 'weaker or more fragile' or whom they bullied simply because they 'liked doing it.'

As studies have shown, social cruelty is present in children from a very early age. It is important to point this out precisely because it is in early childhood experiences and the impacts they cause 'wherein the stage is set for the formation of the child's identity and for her sense of power upon entering the turbulent waters of secondary school' (Anthony/Lindert 2012, p. 30).

The common characteristics found in the Costa Rican studies, including weakness, fragility and insecurity, lead the aggressors to ridicule their peers, to play dirty tricks on them, and to mock them. The fragility detected in the victims coincides with the contributions of Davis/Davis (2008), who found that victimization occurs with greater frequency in students who are overweight, have lower intelligence, live in poverty, or whose physical appearance has some characteristic that sets them apart from the norm.

Physical stature is a high-risk condition for students because it can make them 'easy targets for one who can't discharge his hostility against the person that provoked it: the child is a sure-fire scapegoat and it's guaranteed there will be no counter-attack' (Cerezo 2006, p. 64).

The simple presence of the victims is bothersome to the bullies, as is seen in the expressions given by Costa Rican students who said they 'disliked' their schoolmates or that they bullied them because 'they liked doing it.' For these bullies, the physical appearance of their potential victims caused feelings of discomfort or dislike that led them to engage in disruptive conduct, which was very often disproportionate.

'Intimidation is a behaviour based on opportunity. Some kids become victims rather than others because the aggressor sees that no one else defends them. Therefore, the bullies invent their own reasons to explain why the victim *merits* their aggression' (Davis/Davis 2008, p. 21). To the former must be added the toleration of bullying by the authority figures that places the victims in a position of greater vulnerability from which they cannot escape, thereby facilitating the continuation of the ill-treatment—often with greater intensity.

Boys and girls spend long periods of time in school and their relations with their peers will mark them for the rest of their lives. That is why it is important to reflect on the data mentioned and on the diverse forms of aggression used by the responsible actors.

Other important aspects to take into consideration are the insecurity, the fear, and the aggression suffered by inhabitants of the high-risk zones close to educational centres, which, although they may be categorized as conditions that are external to the educational centres, also affect the relations between peers. These high-risk zones near schools thus become another disadvantage for both bullies and their victims, since violence in their surroundings as well as that in their homes directly affects children. On repeated occasions, this theme has been the object of study via the social learning theories of Bandura (1973), whose paradigm emphasizes that parents are behavioural models and reinforce, one way or another, the behaviour of their children.

The social environment exercises a direct influence on the development of every human being; many reactions are given in response to the conditions of life or to the educational or workplace environments.

If the social milieu that students grow up in is characterized by poverty, inequality, unemployment, the lack of educational opportunities and personal dissatisfactions, these indicators can also lead to family and social violence, thus increasing the probability that altered patterns of behaviour learned in daily life are reproduced in the school. From this it can be stated that 'if a country promotes inequality and violence as a means of conflict resolution, it's likely that bullying is hiding behind every corner' (Gairín-Sallán et al. 2013, p. 25).

If it is true that the noted conditions can be learned and reproduced as forms of human behaviour, they may also release feelings of fear and isolation and cause children to believe that this is 'what they deserve,' opening the way for this role to be replayed in other life settings such as educational centres.

Each human being needs to establish his life goals. To do so requires healthy circumstances in which he can attain maximum development of his capacity to enjoy a dignified life. Therefore, approaching the theme of ill-treatment among peers is a more complicated task than it first appears, with many hidden angles to discover and straighten out in order to procure the common welfare of every student.

14.3 Reflections Concerning Some Relevant Aspects to Improve in Victims

First of all, educational programmes directed towards victims should be implemented within the school community. Such programmes should be aimed at developing improved social skills, recovering self-esteem, security, and assertiveness and at establishing effective communication among peers.

Secondly, an individualized value system applicable to each particular institution should be established such that students are informed from the start of the school year about the philosophy and beliefs that govern the educational centre and exactly what kinds of behaviour are and are not permitted within the school campus they are entering.

Once the student body has been informed of the above premises, they should be given specific guidelines concerning awards and punishments so that students may form a clear understanding of the institutional rules and become conscious of the consequences of their actions. Achieving this will require development of group or individual programmes as needed to produce changes in the students' behaviours, and this means identifying:

a. What are the stimuli that trigger aggressive conduct by the bullies?
b. What are the specific actions that occur in response to such stimuli?
c. What are the consequences that result from engaging in threatening behaviour?

It is necessary to detect the specific strengths of the victims and, based on those strengths, develop programmes to improve their interpersonal relationships and help them achieve integration in their peer group.

National studies have identified as a weakness in this group the lack of social skills, with the understanding that intervention strategies for victims must take into consideration the context within which each child is developing. This is the basis for the statement that 'social skills should be considered within a particular cultural framework as the patterns of communication vary widely between cultures and within a given culture depending on such factors as age, sex, social class, and education' (Caballo 1993, p. 4).

Developing the capacity of victims to react to specific situations or achieving their integration into school groups requires behavioural interventions that should be learned in accordance with the experience brought by the teacher.

Other areas in which these children need improvement are posture, timidity and language.

Regarding posture, victims of peer bullying demonstrate non-verbal behaviours that convey messages of fear and insecurity which, when interpreted by the bullies, triggers them to go after their victims. 'Non-verbal messages are also received unconsciously. People form impressions of others based on their non-verbal conduct without knowing exactly what it is that is agreeable or irritating about the person unless their behaviour is easily identifiable' (Caballo 1993, p. 24).

Some of the signals given by victims are evidenced by their body posture, which in turn has the effect of provoking negative responses or rejection from the aggressors. Such body language is evidenced by keeping the head down, the shoulders fallen, looking away, or curving the feet inward among other observable signs that show weakness and fragility in these children.

Help given in this respect to victims of bullying contributes to changing their perception by the group. Training in how to walk, how to stand up, how to look someone in the face and not avoid a questioner can contribute to transmitting a sensation of security and aplomb and to acquiring the ability to move easily from one place to another securely as well as having particular significance in the context within which the students develop.

Regarding timidity, it is necessary to identify and review the irrational thoughts and beliefs that victims maintain about themselves, their self-perception, and the image they believe they project to others. From this review, systematic efforts applied to eliminating negative thoughts and steering victims towards self-confidence and security will contribute to bettering their self-perception and self-esteem.

It is worth considering that each one of these situations of lack of confidence and insecurity signifies a loss for the victims such that intervention during the resulting grief process can help in understanding the disadvantage it has caused. This should not be considered as accepting any given act in particular but rather as a form of internal resolution whereby the victim acquires the necessary tools to confront the situation and tools that will permit understanding that the reason why some kids suffer from school bullying is not because there is something wrong with them.

Promoting contact between youth of both sexes who share common interests is the task of the teacher. Likewise, it is the teacher's duty to identify causes that may be generating timidity and to facilitate the process of integration of this group of students in the classroom.

Regarding language, it is important to understand that the first communication with others is established in the moment of eye contact. Victims of school bullying generally avoid making direct eye contact and thus elude a behaviour that is basic to the development of social skills and which propitiates communication and spoken language.

Some of the components of oral language are the voice, the volume, the tone, the clarity, and the velocity—conditions that seem to be altered in this group of students. On occasion, they speak hesitantly, with low volume and poor articulation—characteristics that, when hidden behind inadequate body posture, seem to sending a message to the aggressors that 'I feel inferior. Here I am. Do whatever you want to do.'

Mehrabian (2009) makes reference to the importance of non-verbal language in all human relations. He states that in the communication of emotions, verbal language is only responsible for 7%; 38% is transmitted through what he terms par-averbal communication and includes intonation, volume, pauses, and rhythm; and the other 55% is communicated via body language, including gestures, looks and movements, among others. Therefore, an effective education in this area will also facilitate good communication between peers.

14.4 Reflections Concerning Relevant Aspects to Improve in the Aggressors

To open this section, I define aggression between peers beginning with the specific acts committed by the aggressors. A review of Costa Rican studies shows that the forms of aggression most frequently used against schoolmates include: kicking, pushing, pulling hair, hitting, spitting, taking away money, pinching, stealing belongings, and in a few but not less important cases, attacking the victim using weapons or fire arms.

These acts are called primary because they leave visible marks on the victims that can be observed and measured. In this manner, evidence can be collected regarding the acts and the frequency with which they occur; measurements of observed acts show repetitions that go from various times a day to various times per week, to less frequent acts occurring one or more times per month.

Studies undertaken in Costa Rica clearly show that such aggressions are triggered by stimuli such as weakness, fragility, physical appearance or the height of the victims.

To define abuse between peers, we must also consider other actions less visible but still devastating, such as: insulting, humiliating, or disclosing a schoolmate's secret. Apart from suffering ridicule, these acts may lead the victim to become isolated, depressed, or to drop out of school. They may engender feelings of anxiety and impotence that in some cases lead to death.

It is thought that bullies use their aggressive acts to try to gain something in exchange (money, objects, lunches, etc.). Alternatively, they may commit such acts simply to annoy, to make the other kid feel bad, to meet the expectations of the rest of the group, to mock the victim, to attract attention, to feel more *macho* (manly), or for personal defence (Cabezas/Monge 2014). These behaviours may cause the aggressor to feel guilty, ashamed, happy, powerful, or worried, but such feelings do not radically alter their behaviour.

Schoolyard bullies are often taller and more muscular than their peers. Behaviourally, they are aggressive, impulsive, challenging, insensitive, and have a great need of recognition.

Skinner (1953) speaks of behaviour modification based on reinforcement, such that subjects can be taught to increase or decrease specific behaviours depending on the consequences that follow. Thus, we can say that aggression may be a learned behaviour if it brings the aggressor personal satisfaction or social recognition—as was declared by those bullies who expressed feeling happy, powerful, satisfied, relieved, or recognized for the acts they committed (Cabezas/Monge 2014).

The behaviour of school bullies is reinforced by the company of those who observe but do not directly participate in their actions; those who covertly applaud the bullies while mimicking the behaviour of the victims, mocking and segregating them from the rest without intervening to help them.

In a sample of 1,115 male and female students, data given by victims indicated that in 18% of cases, the aggressors were accompanied by others (Cabezas/Monge 2014), which led this group to feel powerful and important and thus to continue ill-treating their peers.

Learning to be a schoolyard bully cannot solely be attributed to the immediate, circumscribed reality of the classroom but is also related to more global dynamics that involve other spheres of human activity such as family and social environments that directly influence this problem.

From a collective view that encompasses the family, the social environment, and conflict zones surrounding educational centres, evidence shows a direct relationship between social and familial circumstances and those children who suffer a greater or lesser percentage of abuse in the classroom.

In one study (Cabezas/Monge 2014) it was found that poverty, overcrowded housing and generally unhealthy family living conditions increased the frequency of inadequate reactions by parents towards their children who were punished both physically and emotionally. In some of the places studied, physical blows, insults, being shoved around or threatened were the punishments most often received by children when their father or mother sought to correct them.

If school children live in hostile home environments, those aggressive models may teach them behaviours that young bullies later reproduce in the classroom. In this respect, Serrano (2006), based on the work of McCord et al. (1961), pointed out the following:

[They] observed that the existence of a family environment characterized by punitive behaviour, threats and profound parental rejection was one of the principal factors among family correlates of aggression in children. Therefore, various researchers have indicated that, to a certain extent, aggressive punishment which frustrates the child is related to correspondingly greater aggression by the children (p. 33).

In the same line, Dutton/Golant (2006) placed emphasis on the fact that if students repeat violent conduct from the home environment, this is not due solely to the influence of their immediate environment, but rather their actions extend far beyond simple imitation because this process contributes to the formation of a violent personality. When students enter the educational centre, the process of acculturation offered by the centre influences changes in their behaviour; but prior to this, the student's personality has already been shaped. This implies that teachers must make greater efforts to approach, from within the classroom, the problem of abusive peer behaviour since they will have to struggle against conducts previously acquired in order to transform them into new skills that will facilitate better student interaction.

The professor has an essential function in the formation of the students, including taking on the subject of abusive peer behaviour that causes so much damage. Her task is to implement structured programmes from the moment boys and girls enrol in the school. The teacher must also consider whether the students have had prior traumatic experiences and whether or not they can count on the support of their mothers and fathers so that these factors do not distort their sense of identity or, for this reason, cause the children to seek to reaffirm their personality via unhealthy aspects of their socio-cultural environment and thereby justify their own violence that erupts on the school campus.

The bullies also need help. Boys and girls who abuse their peers demonstrate discrete physical, academic, family and personality characteristics that require intervention. Other important aspects they also need to improve are posture, language, and social skills.

If the victims assume a posture (body language) that exemplifies their weakness, the aggressors will seize the opportunity to intimidate them by planting themselves in front of their peers with chest thrown out, arms akimbo, legs apart and a frown on the forehead or by looking at them over their shoulder and not measuring their words. Such behaviours stimulate fear in the ill-treated peers and corrective work on this aspect would help to improve the students' interaction.

Changes also need to be made in the realm of language, because in verbal communication bullies use cutting and insulting words, taking advantage of some physical characteristic or condition of their victims. They may use words like 'four eyes,' 'nerd,' or 'chicken' to provoke fear or intimidation and impose their will by sending the message that 'I'm better than you' or 'I deserve more than you.'

This group of boys and girls also present problems in the development of social skills. It may be that their behaviour is the only way they know of interacting; therefore, helping them to identify problematic conduct and working to change it can benefit the process. Other relevant aspects that should be taken into consideration are inability to control anger and impulsivity, which sometimes form parts of their belief systems as learned components.

Other deficiencies in this group of students include a lack of empathy and the ability to put oneself 'in another's shoes'—to be in their place and share their feelings. Having the ability to see the other side and to be and to feel with another person is fundamental to maintaining adequate relations with others. This can be seen as an interior journey that allows one to refresh one's memory and think carefully about things one wouldn't like to have done to oneself.

Getting in touch with one's own feelings can bring us closer to understanding the feelings of others and to develop consideration for them; therefore, this theme should be introduced and explored with all boys and girls in the school.

14.5 Conclusion

There are many kinds of behaviours that are related to increased risk of bullying among schoolmates. Different reactions to school conflicts are influenced by daily living experiences, the immediate environment, or the realities facing each student; because, since all come from different situations, each student's reaction will be influenced by their own personal experiences.

Both victims and aggressors present behaviours that facilitate bullying between peers. Among victims, fragility, the sensation of weakness, and limited social skills are converted into trigger stimuli that provoke bullies into attacking them. Such weakness can be shown unconsciously through body language involving posture or manner of looking away, or by virtually any verbal or non-verbal language victims use to interact with peers that generates annoyance or disgust in their aggressors.

On the other hand, behaviour that is aggressive or intimidating, or speaking cutting and abrasive words as used by boys and girls that ill-treat their peers, all provoke fear, anxiety and isolation in those who suffer from classroom bullying. Both groups of children are afraid but they express it in different ways. Consequently, both victims and aggressors need attention and help.

Preventive and/or remedial measures should be taken in educational institutions in order to avoid classroom bullying during an important formative period in the lives of all children when they should be enjoying healthful interaction with their peers.

References

Anthony, M., & Lindert, R. (2012). *Matoneo entre niñas. Un libro indispensable para padres y educadores*. (Bullying among girls. An indispensable book for parents and teachers.) Bogotá, Colombia: Panamericana Editorial. English edition: Anthony, M., & Lindert, R. (2010). *Little girls can be mean* (trans: Montoya, M. J.). New York: St. Martin's Griffin.

Bandura, A. (1973). *Aggression: A social learning analysis*. Englewood Cliffs: Prentice Hall.

Caballo, V. E. (1993). *Manual de evaluación y entrenamiento de las habilidades sociales*. (Social skills training and evaluation manual.) Madrid: Siglo Veintiuno de España Editores, S.A.

Cabezas, H. (2007). Detección de conductas agresivas 'Bullying' en escolares de sexto a octavo año, en una muestra costarricense. (Identification of aggressive behaviour 'bullying' in a sample of Costa Rican students from 6th through 8th grades.) *Revista Educación*, 31, 123–133 (2007). https://doi.org/10.15517/revedu.v31i1.1257.

Cabezas, H. (2011). Los niños rompen el silencio. Estudio exploratorio de conductas agresivas en la escuela costarricense. (The children break their silence. An exploratory study of aggressive behaviour in Costa Rican schools.) *Revista Educación*, 35, 139–151 (2011). Retrieved from https://revistas.ucr.ac.cr/index.php/educacion/article/view/25161.

Cabezas, H., & Monge, M. (2014). Influencia del entorno donde se ubica el centro educativo en la presencia del acoso en el aula. (Influence of the neighbourhood surrounding the school on the presence of classroom bullying.) Revista Electrónica *Actualidades Investigativas en Educación* 14(3), 1–22 (2014). https://doi.org/10.15517/aie.v14i3.16154.

Cerezo, F. (2006). Violencia y victimización entre iguales. El bullying: estrategias de identificación y elementos para la intervención a través del Test Bull-S. (Violence and victimization among peers. Bullying: strategies for identification and elements for intervention via the Test Bull-S.) Revista Electrónica de *Investigación Psicoeducativa*, 4(2), 333–352.

Deater-Deckard, K. (2001). Annotation: Recent research examining the role of peer relationships in the development of psychopathology. *Journal of Child Psychology and Psychiatry and Allied Disciplines*, 42(5), 565–579.

Davis, S. & Davis, J. (2008). Crecer sin miedo: Estrategias positivas para controlar el acoso escolar (Growing up without fear: Positive strategies for controlling school bullying.). Bogotá: Editorial Grupo Norma. English edition: Davis, S. & Davis, J. (2007). *Schools were everyone belongs: Practical strategies for reducing bullying* (2nd. ed.) (trans: Montaña, E. C.). Champaign: Research Press.

Dutton, D.G. & Golant, S.K. (2006). *El golpeador. Un perfil psicológico*. Buenos Aires: Paidós. English edition: Dutton, D. G. & Golant, S. K. (1995). *The batterer: A psychological profile* (trans: Negretto, A.). New York: Basic Books.

Gairín-Sallán, J., Armengo-Asparó, C., Silva García, B.P. (2013). El 'bullying' escolar. Consideraciones organizativas y estrategias para la intervención. (School bullying. Organizational considerations and strategies for intervention.) *Educación XX1*, 16, 19–38 (2013). https://doi.org/10.5944/educxx1.16.1.715.

Horton, P.B. & Hunt, C.L. (1992). *Sociología*. (Sociology.) México: McGraw-Hill.

McCord, W., McCord, J. Howard, A. (1961). Familial correlates of aggression in nondelinquent male children. *Journal of Abnormal and Social Psychology*, 62, 79–93 (1961). https://doi.org/10.1037/h0045211.

Mehrabian, A. (2009). *Nonverbal communication* (3rd printing). Piscataway, NJ: Transaction Publishers.

Menesini, E. (2009). El acoso en la escuela: desarrollos recientes. (Bullying in school: recent developments.) In F. Mazzone & Q. Mazzonis (Eds.), *Educación en contextos de violencia y violencia en contextos educativos. Reflexiones y experiencias desde las acciones de intercambio en Italia* (pp. 3–26). (Education in violent contexts and violence in educational contexts. Reflections and experiences from exchange programmes in Italy.) Rome: Beniamini Group.

Moreno, C., Rodríguez, M. L., Chamorro, M. J. (2002). *CARTAGO: Dimensión, Naturaleza y entorno socioeconómico del trabajo infantil y adolescente.* (CARTHAGE: Extent, nature and socioeconomic characteristics of childhood and adolescent labour.) International Programme for the Eradication of Child Labour; International Labour Organization. (63pp). Heredia, Costa Rica: INEINA, CIDENAF-UNA.

Olweus, D. (1983). Low school achievement and aggressive behaviour in adolescent boys. In D. Magnusson, & V. Allen (Eds.), *Human development. An interactional perspective* (pp. 353–365). New York: Academic Press.

Olweus, D. (1998). *Conductas de acoso y amenaza entre escolares (Bullying and threatening behaviour among students.* Madrid: Ediciones Morata. English edition: Olweus, D. (1993). *Bullying at school. What we know and what we can do* (trans: Filella, R.). Oxford: Blackwell.

Pincever, K. (Coord.) (2008). *Maltrato Infantil. El Abordaje Innovador del programa Ieladeinu. Aprendizajes de una experiencia integral comunitaria.* (Child Abuse. An innovative approach through the Ieladeinu program. Lessons from an integrated community experience.) Buenos Aires: Lumen.

Serrano, I. (2006). *Agresividad Infantil.* (Childhood Aggression.) Madrid: Ediciones Pirámide, Grupo Anaya, S.A.

Skinner, B. F. (1953). *Science and human behavior.* New York: Macmillan.

Voors, W. (2006). *Bullying: El acoso escolar, el libro que todos los padres deben conocer* (Bullying: School bullying, the book that all parents should know.) Buenos Aires: Oniro. English edition: Voors, W. (2000). *Parent's book about bullying: Changing the course of your child's life* (trans: Bravo, J. A.). Center City Minnesota: Hazelden.

Chapter 15
A Study on the Environmental Behaviour of Chilean Citizens

Nicolás C. Bronfman, Pamela C. Cisternas
and Esperanza López-Vázquez

Abstract The way in which societies have evolved has resulted in serious deterioration of the current state of the environment. If we want to stop it, it is essential to understand the variables that influence people's pro-environmental behaviours. In this chapter we describe the results of a study that characterized the pro-environmental behaviour of the citizens of Santiago, the capital of Chile, through the implementation of the Value-Belief-Norm model developed by Stern et al. (Hum Ecol Rev 6:81–98, 1999). Six dimensions of pro-environmental behaviour were evaluated: energy conservation, water conservation, consumption, biodiversity, waste and transport. The results suggest that the population has an ecological vision for the environment, being aware of the current environmental problems and recognising the responsibility that falls on the government, enterprises and households regarding the protection of the environment. Implications for enhancing pro-environmental behaviour are discussed.

Keywords Environmental behaviour · Environmental attitudes
Environmental management

15.1 Introduction

Individual behaviour and rapid economic growth have resulted in an indiscriminate use of natural resources. This has had a negative impact on the quality of the environment, leading to loss of biodiversity and a significant increase in air, soil and

Dr. Nicolás C. Bronfman, Researcher, Department of Engineering Sciences, Universidad Andres Bello, Santiago, Chile and National Research Centre for Integrated Natural Disaster Management, Santiago, Chile. Email: nbronfman@unab.cl.

Dr. Pamela C. Cisternas, Assistant Researcher, Department of Engineering Sciences, Universidad Andres Bello, Santiago, Chile and National Research Centre for Integrated Natural Disaster Management, Santiago, Chile. Email: pamecisternaso@gmail.com.

Dr. Esperanza López-Vázquez, Research Professor, Centre of Transdisciplinary Research in Psychology, Universidad Autónoma del Estado de Morelos, México. Email: esperanzal@uaem.mx.

© Springer International Publishing AG, part of Springer Nature 2018   223
Ma. L. Marván and E. López-Vázquez (eds.), *Preventing Health and Environmental Risks in Latin America*, The Anthropocene: Politik—Economics—Society—Science 23, https://doi.org/10.1007/978-3-319-73799-7_15

water pollution levels, among other serious environmental problems. One can therefore surmise that societal behaviours have directly affected the health and quality of life of individuals.[1]

The increase in environmental problems, along with the necessity to protect and restore biodiversity, have become the focus of governmental, social and scientific institutions. Governments have absorbed a large part of this task by establishing public policies that regulate the impacts of human activity. However, these efforts prove to be insufficient if they are not accompanied by social transformation towards greater environmental responsibility, helping people understand the effects and impacts that their behaviour, and that of society as a whole, have on the environment. To achieve this, it is crucial that we identify and understand the factors that inhibit or promote pro-environmental behaviour in people.

With the aim of contributing to the development of more effective and efficient environmental protection policies, this study describes the environmental behaviour of the inhabitants of Santiago, the capital of Chile. This is achieved by applying the Value-Belief-Norm model that was developed by Stern et al. (1999), and by evaluating six dimensions of pro-environmental behaviour: conservation of energy and water, consumption, biodiversity, transportation and waste.

15.2   National Context

Chile has maintained continued economic and social growth in recent decades, becoming the first Latin American member of the OECD in 2010. Most of the country's population and productivity is concentrated in Central Chile, particularly in the three main cities of the country: Santiago, Valparaiso and Concepcion.

Much of Central Chile is considered to be one of 35 biodiversity hotspots on the planet (Critical Ecosystem Partnership Fund 2014). However, many species of flora and fauna in this area are severely threatened by air, soil and water pollution, which are the result of human activity.

The high level of air pollution is one of the country's main challenges: it is estimated that over 60% of the national population is exposed to worrisome levels of fine-particle matter (Ministry of Environment of Chile 2011), with serious effects on the population's health and quality of life.

In response to the current situation, Chilean society has begun to publicly express its concern and disapproval of the environmental problems it faces. A clear example can be seen in the strong public opposition to flagship energy generation and distribution projects, which are perceived as having a significant environmental impact.

---

[1]This research was partially funded by Chile's National Science and Technology Commission (CONICYT) through the National Fund for Scientific and Technological Research (FONDECYT, Grant 1130864) and by the National Research Centre for Integrated Natural Disaster Management CONICYT/FONDAP/15110017. We also thank all the participants who collaborated in one way or another to fulfil this project.

15.3   The Value-Belief-Norm Model

The effectiveness of interventions on people's behaviour increases if factors that inhibit or promote pro-environmental behaviour are tackled. One of the most common approaches in the study of this type of behaviour is based on the moral and normative interests of individuals. The Value-Belief-Norm (VBN) model, developed by Stern et al. (1999), is an attitudinal model which suggests that individuals act pro-environmentally if they sense a moral obligation to protect other members of society, other species or the ecosystem in general (Fig. 15.1). The VBN model associates attitudes that result from personal norms and environmental behaviour (Bamberg/Möser 2007) with the New Ecological Paradigm (Dunlap et al. 2002). This generates a causal chain with five variables that lead to pro-environmental behaviour: (i) personal values, (ii) the New Ecological Paradigm (NEP), (iii) Consequence Awareness (CA), (iv) Attribution of Responsibility (AR) and (v) Personal Norms (PN).

The causal chain of the VBN model postulates that personal values influence the formation of individual environmental beliefs (NEP). These beliefs enable individuals to be aware of the consequences that their behaviours could have on the environment and also to attribute some degree of responsibility. This cognitive process is believed to elicit a sense of moral obligation towards environmental protection. This last variable then brings about pro-environmental behaviour (del Carmen Aguilar-Luzón et al. 2006).

According to the VBN model, for personal norms to activate, the individual must perceive adverse consequences that threaten his or her personal values. According to Stern (2000), when studying pro-environmental behaviour, values have a tripartite structure: biospheric, social/altruistic and egoistic. There have been doubts regarding the difference between altruistic and biospheric values, but the existence of this tripartite value structure has been empirically confirmed by Schultz (2002). Thus, people with biospheric orientation show special consideration for the costs and benefits that a given behaviour would have for the ecosystem (other species). When the perceived benefits exceed the costs, the individual is supposedly more willing to carry out pro-environmental behaviour. With people who have a social/altruistic orientation, environmental conduct depends on the cost-benefit analysis of consequences that an action might have for other people. Finally, people with an egoistic orientation consider the cost-benefit analysis of environmental conduct with regard to themselves (Calvo Salguero et al. 2008).

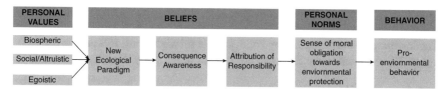

Fig. 15.1 Value-Belief-Norm model for environmental behaviour *Source* Stern (1999)

The VBN model has been successfully applied to explain environmental behaviour in different areas, such as recycling Aguilar-Luzón et al. (2006, 2005), support for environmental movements (Stern et al. 1999), protection of local biodiversity by landowners (Johansson et al. 2013), pro-environment consumer behaviour (Stern 1999) and conservation behaviour (Kaiser et al. 2005), and to study the variables that foster commitment to protect biodiversity among young people (Menzel/Bögeholz 2010).

15.4 Study Objectives

The increase in environmental problems, along with the need to protect and restore biodiversity have become the focus of governmental, social and scientific institutions. All effort, however, proves insufficient if unaccompanied by social transformation towards greater environmental responsibility. To achieve this, it is crucial that we identify and understand the factors that inhibit or promote pro-environmental behaviour in people. With the aim of contributing to the development of more effective and efficient environmental protection policies, the primary objective of this study is to describe the environmental behaviour of the inhabitants of Santiago, capital of Chile, and to study potential differences throughout socio-demographic variables. To this effect, we use the Value-Belief-Norm model, which was developed by Stern et al. (1999).

15.5 Methodology

15.5.1 Field of Study

The study was conducted among 32 urban communes in the Province of Santiago, capital of Chile. The population is 4,914,641 (Instituto Nacional de Estadísticas 2015), making it the country's most densely populated city. It contains approximately 35% of the national urban population. The average age is 34 and 52% of the population is female.

15.5.2 Participants

The instrument was pre-tested by conducting a survey among 150 people above the age of 18, residents of the 32 urban communes in the study. Results obtained in the pre-test validated the instrument, which was finally deployed between September and November 2013. The statistically representative sample from the city of Santiago consisted of a total of 1,537 cases that were segmented by socio-economic level (High, Medium-High, Medium, and Medium-Low) and age. The non-response rate was 19%. The average age of the sample was 44 (standard deviation = 17 years), within a range of 18–80 years, and 62% were female. Participants did not receive any compensation for their participation in the study.

15.5.3 Working Materials

The survey was structured into three parts. The first part of the survey is connected to the VBN model (see Table 15.1). Personal values were measured using seven items that have been adapted from Schwartz's (2003) work. The phrasing was as follows: 'The following is a list of brief characteristics describing a person's values. To what extent do you feel represented by each one of them?' As a response, the participants were asked to rate each item using the following scale: '1. This does not represent me at all, 2. This represents me slightly, 3. It's all the same to me, 4. This represents me somewhat, and 5. This represents me a lot'. NEP was measured using ten items that were adapted from Dunlap et al. (2002). Consequence Awareness was measured using six items that were adapted from Stern et al. (1995). Finally, Attribution of Responsibility was measured using five items from Gärling et al. (2003), as were Personal Norms. Participants were asked to rate their degree of agreement for each item, using a 5-point scale: *1. 'I strongly disagree'—5. 'I totally agree'.*

The second part of the survey contained items associated with Pro-Environmental Behaviour (see Table 15.1) which are based on the work of Kaiser et al. (1998, 2003). Six dimensions of behaviour were defined: energy conservation (7 items), water conservation (5 items), consumption (5 items), bio-diversity (4 items), transport (5 items) and waste (8 items). Participants were asked to rate the frequency with which they carry out the activities on a 5-point scale: *1. Never, 2. Hardly ever, 3. Sometimes, 4. Almost always, and 5. Always.* The third and final part of the survey contained socio-demographic questions.

15.5.4 Data Analysis

The data obtained was analysed by age group and socio-economic level, using IBM SPSS 17.0 software. The Tukey test was applied to analyse differences by age group and socio-economic level.

Table 15.1 Percentage distribution of participant responses for each item in the VBN model

	1 (%)	2 (%)	3 (%)	4 (%)	5 (%)
BIOSPHERIC VALUES ($\alpha = 0.79$)					
PV1. A person who believes everyone should take care of the environment	2	4	4	28	62
PV2. A person who respects the environment and believes that we should live in harmony with other living beings	1	2	4	29	64
ALTRUISTIC VALUES ($\alpha = 0.70$)					
PV3. A person who considers it important to help those around him	1	2	5	30	62
PV4. A person who believes in equal treatment for all people, including strangers	3	2	7	30	58
EGOISTIC VALUES ($\alpha = 0.70$)					
PV5. A person who makes decisions and likes to be the leader	16	12	22	28	22
PV6. A person who considers it important to have a lot of money	32	15	30	14	9
PV7. A person who considers it important to influence people and their actions	26	14	26	19	15
ECOLOGICAL VISION ($\alpha = 0.64$)					
NEP1. In recent times, the population has grown faster than the planet can sustain	4	5	12	21	58
NEP2. Earth has very limited space and resources (e.g. like a spaceship)	7	6	9	23	55
NEP3. Human beings have the right to alter the environment to suit their needs	24	12	17	20	27
NEP4. Plants and animals have the same right to live as human beings	3	5	12	17	63
NEP5. Nature is strong enough to withstand the impact generated by the modern lifestyle	32	17	20	15	16
NEP6. The balance of nature is very delicate and easily disrupted	2	5	16	18	59
NEP7. Most environmental problems can be solved by applying more and better technology	13	8	21	21	37
NEP8. Humans will learn enough about nature to be able to control it	20	14	24	19	23
NEP9. The deterioration of the environment is not as bad as is often said	55	17	13	7	8
NEP10. If things continue in the same direction, we shall soon experience a major environmental disaster	6	6	14	17	57
CONSEQUENCE AWARENESS ($\alpha = 0.69$)					
CC1. Environmental protection benefits us all	0	0	2	9	89
CC2. Environmental protection will improve everyone's quality of life	0	0	2	12	86
CC3. Environmental protection will lead to a better world for me and my family	0	0	3	12	85

(continued)

Table 15.1 (continued)

	1 (%)	2 (%)	3 (%)	4 (%)	5 (%)
CC4. Environmental deterioration directly affects my health (e.g. air pollution)	1	1	3	14	81
CC5. Environmental deterioration coming from my neighbourhood affects people worldwide	6	6	12	17	59
CC6. Over the next ten years, thousands of plant and animal species will become extinct	4	3	13	18	62
ATTRIBUTION OF RESPONSIBILITY ($\alpha = 0.60$)					
AR1. Everyone is responsible for protecting the environment	2	0	3	11	84
AR2. The government is primarily responsible for protecting the environment	8	5	7	14	66
AR3. Enterprises are primarily responsible for reducing environmental deterioration	2	2	3	12	81
AR4. My home is responsible for reducing environmental deterioration	3	2	6	21	68
AR5. All households are responsible for reducing environmental deterioration	1	1	4	17	77
AR6. I am not willing to cooperate to reduce environmental deterioration if others don't do the same	64	12	9	4	11
PERSONAL NORMS ($\alpha = 0.70$)					
NP1. I feel morally obliged to protect the environment	2	1	7	18	72
NP2. Environmental problems must not be ignored	0	1	2	11	86
NP3. I believe it's important that people protect the environment	0	0	2	10	88
NP4. The government should demand greater environmental protection	1	1	2	9	87
NP5. Companies must reduce their impact on environmental deterioration	1	0	2	7	90

Source The authors

15.6 Results

15.6.1 VBN Model Variables

Table 15.1 summarises the participants''marks for each item in the VBN model. In the *Personal Values* scale, results show that over 90% of participants felt somewhat represented by Biospheric Values (concern for the suffering of all living beings) and Altruistic Values (concern for the suffering of other people due to environmental problems). Meanwhile, about 20% of participants identified with the Egoistic Values (concern for one's own suffering due to environmental problems). In this context, we can see that people above the age of 45, belonging mostly to higher socio-economic levels, feel most represented by Biospheric and Altruistic Values.

In contrast, those under the age of 45 scored higher on the Egoistic Values scale, meaning they consider important having a lot of money, being leaders, and having influence on people and their actions.

Generally, the population scored a high percentage of answers in favour of biospheric and altruistic values, showed strong consequence awareness, high attribution of responsibility and personal norms that are connected with environmental care.

Regarding the *Ecological Vision* scale, our results suggest that the majority of participants has a pro-environmental view and is aware of current environmental problems. In this context, it is people over the age of 45, from lower socio-economic levels, who agree the most with the idea that the population has grown faster than the planet can sustain, and that human beings have no right to modify the environment.

As before, those above the age of 45, from all socio-economic levels, maintained the highest scores in the dimension, suggesting that the balance of nature is delicate and easily upset. Our results also show that people belonging to lower socio-economic levels tended to disagree most with the idea that humans can control nature through knowledge and the implementation of technology.

In the *Consequence Awareness* scale, over 90% of participants considered themselves highly aware of the consequences associated with failing to protect the environment. However, awareness of the consequences associated with failing to protect the environment decreased with socio-economic status and age.

On the *Attribution of Responsibility* scale, almost all participants acknowledged the responsibility of government, businesses, society and households with regard to environmental protection. However, it was those belonging to higher socio-economic levels who were most aware of the responsibility of all involved actors in environmental protection. Most participants were also willing to cooperate to reduce environmental deterioration, regardless of other people's behaviour. Regarding the *Personal Norms* scale, our results indicate that a substantial majority of participants assigned high marks to these items, which suggests a high sense of moral obligation to protect the environment.

Regardless of socio-economic status or the age of participants, there was complete agreement on the fact that the government should require and provide greater environmental protection, and that companies should reduce their environmental impact. That said, it is people above the age of 45, belonging mostly to Middle and Upper-Middle socio-economic levels, who perceive a higher moral obligation to protect the environment.

15.6.2 Dimensions of Environmental Behaviour

Table 15.2 shows the results for the six dimensions of behaviour studied. On the *Energy Conservation* scale, a great majority of participants behave in a manner which is supportive of energy conservation, but people above the age of 45, from

Table 15.2 Percentage distribution of participant responses for each item, in each dimension of behaviour

	Never (%)	Hardly ever (%)	Sometimes (%)	Almost always (%)	Always (%)
ENERGY CONSERVATION (α = 0.60)					
PC1. In winter, I set the heating on so that I can wear light clothes inside my house	9	9	19	17	46
PC2. In winter, I open the windows for long periods of time to ventilate my house	13	13	23	18	33
PC3. In winter, I switch off the heating in my house at night	3	2	6	14	75
PC4. In winter, when I leave my home for more than 30 min, I turn off the heating	3	2	6	16	73
PC5. I take full advantage of natural light	0	2	4	20	74
PC6. I turn off lights that are not in use	1	2	9	22	66
PC7. I unplug appliances that are not in use	14	11	21	16	38
CONSUMPTION (α = 0.68)					
EAC1. I buy biodegradable detergents for my laundry	22	10	19	12	37
EAC2. I buy organic products	24	13	25	19	19
EAC3. I buy rechargeable batteries	23	12	21	14	30
EAC4. I buy energy-saving light bulbs	7	4	16	15	58
EAC5. I buy products in returnable or reusable packaging	7	5	20	24	44
BIODIVERSITY (α = 0.56)					
B1. When I spend the day outdoors I leave places as clean as they were when I arrived	1	0	5	16	78
B2. I visit national parks and/or nature reserves	21	21	26	12	20
B3. I take my pet to the vet for check-ups	20	8	12	13	47
B4. I collect plants, seeds and natural objects when I visit natural areas	53	13	11	7	16
WATER CONSERVATION (α = 0.63)					
WC1. I make sure leaky faucets are repaired quickly	2	2	12	19	65
WC2. I let the shower run until the water reaches a pleasant temperature	18	6	18	24	34
WC3. I make sure I turn off the tap while I brush my teeth	5	5	17	15	58
WC4. I wait until I have a full load before washing clothes in the washing machine	3	3	13	18	63
WC5. I try to take short showers (less than 5 min)	12	10	23	16	39

(continued)

Table 15.2 (continued)

	Never (%)	Hardly ever (%)	Sometimes (%)	Almost always (%)	Always (%)
MOBILITY AND TRANSPORT ($\alpha = 0.46$)					
EAU1. On short distances (less than 10 blocks), I prefer walking or cycling	11	5	16	12	56
EAU2. I share my car	20	12	19	10	39
EAU3. I drive in such a way that fuel consumption is minimised	5	6	17	20	52
EAU4. I refrain from using the car on much polluted days	23	10	21	12	34
EAU5. I honk when I'm driving	33	22	25	11	9
WASTE ($\alpha = 0.88$)					
EGR1. When I go shopping, I use fabric bags instead of plastic bags	40	13	19	9	19
EGR2. I reuse plastic bags (from the supermarket)	7	3	10	19	61
EGR3. I separate food waste to make compost (composting)	82	2	3	2	11
EGR4. I separate paper and cardboard for recycling	65	2	7	4	22
EGR5. I separate beverage cans for recycling	71	2	6	2	19
EGR6. I separate glass containers for recycling	63	2	6	4	25
EGR7. I separate batteries for recycling	74	2	6	3	15

Source The authors' own elaboration

every socio-economic level, engage the most in energy conservation during the winter: *they turn off the heating when they leave the house for more than 30 min*, and *they leave the windows open for long periods of time to ventilate the house*. Meanwhile, data shows that the habit of turning off lights and unplugging appliances that are not in use increases with age. Similar results were obtained on the *Water Conservation* scale, where a majority of participants claim to behave in a way that is supportive of water conservation.

On the Environmental Values scale (biospheric, altruistic, and egoistic), egoistic values tended to get the lowest scores and there was not much difference across different socio-economic levels or age groups. As for the *Consumption* scale, people above the age of 30 who belong to higher socio-economic levels are more likely to use environmentally-friendly products. There was an exception for returnable or reusable packaging, which is mostly purchased by people from lower socio-economic levels.

In the *Biodiversity and Natural Resources* dimension, virtually all participants claim they always or almost always leave outdoor locations as clean as they were when they found them, regardless of their socio-economic level or age group. Generally, people over the age of 30 from higher socio-economic levels visit parks and/or nature reserves more frequently. A low percentage of participants (over 20%), generally above the age of 45, claim to collect plants and seeds when visiting natural areas.

Results in the *Mobility and Transportation* scale showed that most people who drive cars claim to behave in an environmentally friendly manner. In this regard, young people under 30 share their car more often and prefer walking or biking for short distances. Furthermore, people above the age of 45 make the greatest effort to drive in such a way that fuel consumption is minimised.

Finally, the dimension of behaviour for *Waste* showed that a small proportion of participants separate waste for recycling. About 10% of participants, mostly people over 60, declared that they *separate food scraps or organic waste to produce compost*. Between 15 and 30% of participants claimed that they always or almost always separate paper, cardboard, beverage cans, glass containers, batteries, and electrical and electronic equipment for recycling. It is generally people above the age of 60 belonging to the highest socio-economic levels who claim they perform this task the most.

15.7 Discussion

Generally speaking, we conclude that most of the study's population possesses highly favourable individual predispositions to engage in more responsible environmental behaviour. This is reflected in the fact that most of the participants displayed highly ecological attitudes in support of the environment. They are well aware of the problems that affect the environment and the consequences associated with not protecting it. The surveyed population also acknowledged the responsibility of government, businesses, society, and their own homes, showing a high sense of moral obligation towards the protection of the environment.

Among the variables of the VBN model, personal norms have the greatest influence on the different dimensions of environmental behaviour. The same can be observed in studies that focus on biodiversity protection (Menzel/Bögeholz 2010; Johansson et al. 2013) and recycling Aguilar-Luzón et al. (2005, 2006). The most frequently reported pro-environmental behaviours in which participants engaged are related to energy and water conservation. This could be explained by the fact that these behaviours are associated with financial savings and do not necessarily involve extra effort, such as taking advantage of natural light, turning off the heating and washing clothes in full loads. Similar behaviour can be seen in the habit of consuming more environmentally friendly products, particularly energy-efficient light bulbs and returnable or reusable containers, which also generates savings in the household budget.

Regarding the mobility and transportation dimension, we conclude that most people who drive cars try to behave in an environmentally-friendly manner. Furthermore, people under 30 reported that they prefer walking or cycling for short distances, which is to be expected since they are mostly students and/or have fewer resources than their elders.

Finally, regarding waste, we can see that a very small proportion of participants claim to separate their rubbish for recycling. People over the age of 60 who belong to higher socio-economic levels claim they perform this task the most. This could be because: (A) Recycling activities require significant changes in people's habits and routines, which involves a conscious effort to break old habits and create new ones. This can be discouraging for the adoption of environmental behaviours when motivation to break those habits is insufficient. (B) There are contextual factors (community expectations, government regulations, financial incentives, available technology, available infrastructure, etc.) that can significantly influence environmental conduct (Stern 2000). The lack of adequate infrastructure for the collection and/or disposal of previously separated waste, or a municipal collection system that doesn't differentiate collected waste (by mixing previously separated waste), could explain such a low environmental conduct. The lack of adequate infrastructure for the collection and/or disposal of solid waste is more pronounced in lower socio-economic sectors. (C) A third explanation may reside in the personal abilities of the population (Stern 2000). It is possible that citizens have insufficient knowledge about what recycling is, how to separate waste at home or where to discard it, and that difficulties to do so hinder their motivation.

In light of the obtained results, we conclude that a majority of participants possess individual characteristics that are highly conducive to maintaining responsible environmental behaviour in the dimensions of energy and water conservation, consumption, biodiversity and transportation. Nevertheless, the same does not apply to dimensions associated with waste management, where the majority of the study's population claims it does not engage in pro-recycling behaviour. We therefore conclude that, in this latter dimension of behaviour, there would be greater scope for the development of future policies, plans and programmes that lead to more responsible and environmentally-friendly social behaviour. It is necessary to develop and implement plans and environmental education programmes that promote and strengthen pro-environmental behaviour in waste management. The implementation of these plans and programmes must be accompanied by the necessary infrastructure for the appropriate collection and storage of waste.

References

Aguilar-Luzón, M.C., García-Martínez, J.M.Á., Monteoliva-Sánchez, A., de Lecea, J.M.S.M. (2006). El modelo del valor, las normas y las creencias hacia el medio ambiente en la predicción de la conducta ecológica. *Medio Ambiente y Comportamiento Humano*, 7(2), 21–44.

Aguilar-Luzón, M.C., Monteoliva, A., García, J.M.A. (2005). Influencia de las normas, los valores, las creencias ambientales responsables y la conducta pasada sobre la intención de reciclar. *Medio Ambiente y Comportamiento Humano: Revista Internacional de Psicología Ambiental*, 6(1), 23–36.

Bamberg, S., & Möser, G. (2007). Twenty years after Hines, Hungerford, and Tomera: A new meta-analysis of psycho-social determinants of pro-environmental behaviour. *Journal of Environmental Psychology*, 27(1), 14–25.

Calvo Salguero, A., Aguilar Luzón, M.C., Berrios Martos, P. (2008). El comportamiento ecológico responsable: un análisis desde los valores biosféricos, sociales-altruistas y egoístas. *Revista Electrónica de Investigación y Docencia* (REID)(1), 11–25.

Critical Ecosystem Partnership Fund (2014). Chilean Winter Rainfall-Valdivian Forests. http://www.cepf.net/resources/hotspots/South-America/Pages/Chilean-Winter-Rainfall-Valdivian-Forests.aspx. Accessed 4 February 2013.

del Carmen Aguilar-Luzón, M., García-Martínez, J.M.Á., Monteoliva-Sánchez, A., de Lecea, J.M. S.M. (2006). El modelo del valor, las normas y las creencias hacia el medio ambiente en la predicción de la conducta ecológica. *Medio Ambiente y Comportamiento Humano*, 7(2), 21–44.

Dunlap, R.E., Van Liere, K.D., Mertig, A.G., Jones, R.E. (2002). New trends in measuring environmental attitudes: measuring endorsement of the new ecological paradigm: a revised NEP scale. *Journal of Social Issues*, 56(3), 425–442.

Instituto Nacional de Estadísticas. (2015). Actualización de población 2002–2012 y proyecciones 2013–2020. Retrieved from http://www.ine.cl/estadisticas/demograficas-y-vitales.

Johansson, M., Rahm, J., & Gyllin, M. (2013). Landowners' participation in biodiversity conservation examined through the Value-Belief-Norm theory. *Landscape Research*, 38(3), 295–311.

Kaiser, F.G. (1998). A General Measure of Ecological Behavior. *Journal of Applied Social Psychology*, 28(5), 395–422.

Kaiser, F.G., Doka, G., Hofstetter, P., Ranney, M.A. (2003). Ecological behavior and its environmental consequences: a life cycle assessment of a self-report measure. *Journal of Environmental Psychology*, 23(1), 11–20.

Kaiser, F.G., Hübner, G., Bogner, F.X. (2005). Contrasting the Theory of Planned Behavior With the Value-Belief-Norm Model in Explaining Conservation Behavior. *Journal of Applied Social Psychology*, 35(10), 2150–2170.

Menzel, S., & Bogeholz, S. (2010). Values, Beliefs and Norms that Foster Chilean and German Pupils' Commitment to Protect Biodiversity. *International Journal of Environmental and Science Education*, 5(1), 31–49.

Ministry of Environment of Chile. (2011). *Informe del Estado del Medio Ambiente*.

Schultz, P. (2002). New Environmental Theories: Empathizing With Nature: The Effects of Perspective Taking on Concern for Environmental Issues. *Journal of Social Issues*, 56(3), 391–406.

Stern, P.C. (1999). Information, incentives, and pro-environmental consumer behavior. *Journal of Consumer Policy*, 22(4), 461–478.

Stern, P.C. (2000). Toward a coherent theory of environmentally significant behavior. *Journal of Social Issues*, 56(3), 407–424.

Stern, P. C., Dietz, T., Abel, T., Guagnano, G. A., Kalof, L. (1999). A value-belief-norm theory of support for social movements: The case of environmentalism. *Human Ecology Review*, 6(2), 81–98.

Chapter 16
Risk Perception and Antecedents of Safe Behaviour in Workers at a Garment Factory in Mexico

Lorena R. Pérez Floriano and Julieta Amada Leyva Pacheco

Abstract This chapter is based on field research. The job hazards in a garment factory were identified, and then the risk perception of members of the production area was assessed. To do that, we developed and validated the Textile Industry Risk Perception Scale, which is composed of two factors: (a) Perceived risk of accident and (b) Perceived risk of ergonomic hazards. The differences and similarities between these factors and their relation to productivity, occupational grouping and gender were analysed. We found some significant differences: Operators perceived greater risk than supervisors and indirect personnel. In addition, the greater their productivity, the less likely it is that operators will perceive that they are at risk of accidents at work. We discuss the importance of identifying job hazards for personnel training systems, and the development of a culture of safety in workplaces.

Keywords Occupational hazards · Working conditions · Psychosocial risks

16.1 Introduction

Workplace illnesses and injuries continue to plague workers around the world.[1] According to the International Labour Organization (2015), every year there are close to 337 million workplace accidents and 2.3 million deaths around the world, i.e., there are close to 6,300 deaths per day. The majority of these deaths are the result of illnesses resulting from repeated exposure to hazards at work (ILO 2015, http://www.ilo.org/global/topics/safety-and-health-at-work/lang–en/index.htm).

Dr. Lorena R. Pérez Floriano, Researcher-Professor, El Colegio de la Frontera Norte. Email: lperezfloriano@gmail.com.

Julieta Amada Leyva Pacheco, M.Sc., Teacher, Universidad Estatal de Sonora-Hermosillo. Email: jleyvap@yahoo.com.mx.

[1]The data for this study stem from the Master's thesis of the second author.

© Springer International Publishing AG, part of Springer Nature 2018
Ma. L. Marván and E. López-Vázquez (eds.), *Preventing Health and Environmental Risks in Latin America*, The Anthropocene: Politik—Economics—Society—Science 23, https://doi.org/10.1007/978-3-319-73799-7_16

The garment manufacturing, dressmaking and tailoring sector is not deemed to be a high-risk industry, which is why the number of accidents and illnesses reported by government bodies is underestimated. This means that the private sector continues to believe that the working conditions within this sector are adequate, focusing only on fulfilling the most basic standards and regulations (Mata 2004).

González (2006) argues that, in Mexico, some business owners within the textile industry are neither concerned about safety, nor do they deem it to be important. Furthermore, workers do not have enough information regarding the risks they face, not to mention the fact that their participation in safety measures is not sufficient. However, a number of studies show that there is a relationship between the incidence of musculoskeletal disorders, such as Carpal Tunnel Syndrome or arthritis (Melo 2012), and the incidence of respiratory disorders among workers in the textile industry (Díaz/Schlaen 1994). The majority of the employment opportunities available in this industry are almost always filled by women, who usually work in precarious situations where there is no workplace safety, little or no participation in workplace organization (including health and safety) and, above all, economic conditions that perpetuate poverty (Guadarrama et al. 2012).

Hazard identification is the first step in creating a healthy and safe working environment. The opinion of risk managers within a given industry is of vital importance as it is they who decide what measures need to be taken to manage health and safety in the workplace. However, if risk managers have no information about workers' perceptions and attitudes towards the risks they face, their opinions become somewhat limited in scope. Therefore, the evaluation of risks is a fundamental factor within the safety management process when choosing prevention and protection measures (risk management) that guarantee the safety and well-being of workers (Reinhold/Tint 2009). In this study, we will be referring to potential danger as a latent environmental factor that threatens the status of an individual or society in general. We understand risk as the likelihood of a potential risk having a negative impact on humans and/or their environment.

This study aims to relate risk perceptions to the physical and health hazards of an industry. The chapter is organized in the following fashion: Firstly, a theoretical framework encompassing industrial hazards and risks in the industry is presented, as well as contributions from a risk perspective from the emerging area of occupational and health psychology. Secondly, it continues with the presentation of a case study of a company from the garment manufacturing industry (n = 271) and presents the development and validation of the Risk Perception Scale for the Textile Industry, which encompasses two factors: (a) Perception of Risk of Accident and (b) Perception of Ergonomic Risks. Thirdly, the results of the analysis of these risk perception factors and their relationship to gender and occupational level are relayed. Lastly, conclusions and recommendations aimed at professionals within the spheres of academia, applied psychology and other areas interested in promoting safe and healthy workplaces are presented.

16.2 Occupational Risks and Hazards

Risks are conceived in different ways. Risk perception studies examine the opinions that people express when questioned, in a variety of different ways, about how they evaluate dangerous substances and activities (Slovic 1987; Slovic et al. 1982). Health and occupational psychology focuses on linking these opinions to psychosocial factors (e.g., organizational stressors) industrial psychology (job design and individual differences), and, finally, organizational psychology (group activities and a culture of industrial safety within organizations).

In the United States and Europe, psychosocial risks are currently a major focus as the economies in these regions have moved from manufacturing to service providers. In Mexico and other developing countries, economies have shifted from agriculture to manufacturing, which is why the traditional analysis of hazardous waste is of the utmost importance. In the aforementioned economic regions, these industries generate large quantities of toxic residue and waste that directly and indirectly affect people's health.

Industrial activity has led to damage to both the environment and people's health. The pollution stemming from productive processes can be seen inside and outside facilities, contaminating the air, water and soil, which are indispensable natural resources for the livelihood of every species, including humans. The health risks posed by industrial pollution range from minor to mortal.

Over the past number of decades, numerous research projects have identified and studied a wide range of workplace risks (physical, chemical, biological, psychosocial and physiological) that can lead to accidents (e.g., Hollmann et al. 2001; Salminen et al. 1993). These accidents frequently result in workplace injuries and illnesses, which have the potential to decrease productivity and generate high costs, in addition to damaging the reputation of the company (Sheu et al. 2000). Employees who have been injured can suffer not only from pain and discomfort, but also much more serious physiological and psychological problems. If workers suspect that they may be being exposed to physical harm, this can cause stress, which, by itself, is associated with coronary disease, depression and even cancer (Reinhold/Tint 2009). In Mexico, the Federal Labour Law stipulates that employers are responsible for health, safety and the prevention of workplace risks (Federal Labour Law 2012, p. 98). Quintero/Romo (2001) state that risk management is exceedingly important in the regulation of labour relations by companies in Mexico, and this can be seen in the advent of trade unions, collective contracts and labour disputes.

Human error is the one of the leading causes of workplace accidents. Human error can be attributed to people underestimating or overestimating risks, in such a way as to make workplace risk perception an important regulatory element of occupational safety (Carbonell/Torres 2010).

Most of the studies undertaken by health and safety researchers have focused on reducing 'objective' risk, through an essentially quantitative approach to risk analysis. However, people do not undertake quantitative risk analysis when evaluating workplace hazards, but rather they do the opposite, assuming 'subjective' risk evaluations (Arezes/Miguel 2008).

16.3 Risk and Productivity

Within organizations, productivity demands can become psychosocial risks. Within industries, the major determining factor for worker safety is the behaviour of the latter, but this element, by itself, 'cannot turn a dangerous workplace into a safe one'. Furthermore, if the company's management is not committed to eliminating workplace hazards and concentrating on the health and safety of its workers, the focus will remain on productivity and speed. This leads to a breakdown in preventive behaviours and attitudes (Asfahl 2000: 1).

The differences in risk perception can be explained by the organizational culture present in each group or department, given that different occupational groups have different visions of what organization means, and, as such, they have different meanings for production processes (Pérez-Floriano/Gonzalez 2007; Schein 1996). It is a fact that safety elements are an integral part of any industrial process, which is why safety procedures and policies must take precedence over aspects of the process, such as working speed (an important element for the group charged with supervising production) and profitability (a priority for an organization's management).

Other theoretical approaches to workplace hazards focus on psychosocial risks, which are aspects of the workplace design, organization and management, in addition to social frameworks that have the potential to cause psychological, social or physical harm (Cox/Rial-Gonzalez 2000: 14). Psychosocial risks are deemed to be: job stress, burnout, mobbing, sexual harassment and physical violence, among others.

In the last few years, psychosocial risks and risks to health have become more of a focal point for employers, politicians, legislators and workplace health and safety professionals. This is due to the relationship between these risks and globalization, not to mention the burden this has on jobs today (Kompier 2006). Exposure to psychosocial risks has the potential to cause serious harm. The most significant risks in this new millennium can be grouped into five categories: (a) new forms of employment and a lack of job security; (b) an ageing workforce; (c) the intensification of workloads; (d) high emotional demands in the workplace; and (e) the poor balance between family and work. There are numerous studies that focus on the effect of workplace risks, such as psychosocial factors and their association with physical and mental health effects in almost every occupation and working environment (e.g., Cox/Rial-Gonzalez 2000). Psychosocial risks manifest themselves as organizational stressors; the pressure to complete tasks quickly, with no breaks, will have a negative impact on people's health and well-being.

The purpose of risk analysis is twofold: Ideally, it establishes the relationship between workplace hazards and people's health. It then evaluates the risk to health as a result of exposure to hazards. Physical hazards are, in occupational health psychology terms, stressors that have the potential to cause damage, such as stress (considered to be a harmful psychosocial risk). Furthermore, the notion of fear has major implications on managing risks in the workplace. Burke et al. (2011) undertook a meta-analysis in which they found that the 'dread' factor has an impact on the effectiveness of industrial safety programmes. These studies focused on situations where there was a high level of risk and that required different training compared to jobs where there was a low level of risk. The finding of Burke and colleagues was post hoc as the researchers who undertook the original studies had not contemplated the importance of the 'dread' factor.

16.4 Risk Perception and Its Implications for the Industry

The relevant academic literature indicates that risk perceptions are more informative than objective risk indicators in predicting behaviour. This is because the majority of workers have difficulty understanding and evaluating probabilistic calculations (Brewer et al. 2007; Morrow/Crum 1998).

Risk perception and safety procedures can vary considerably depending on the culture found within each company, and even the subculture of different occupations within an organization will affect perceptions of risk. Members of some industrial sectors can underestimate certain corporations that are generally exposed to high levels of risk as they tend to minimize them, while in other situations, they overestimate them. There are communities and occupations where accepting risks is part of the job. Furthermore, a worker's perception of risks and their relationship with the said risk can vary significantly depending on the culture of safety within the organization (Kouabenan 2009).

For example, Arezes/Miguel (2008) found that in industrial environments with high noise levels, people who perceived a risk of losing their hearing, given the risk posed by the noise levels, used ear protectors more frequently than people who did not perceive this threat. The same authors state that they deem it reasonable to presume that the perception of risks in the workplace can influence people's behaviour and, as such, their exposure to some occupational risks. Research into risk perception has focused mainly on preventing risks as a result of accidents. Therefore, there are gaps in the research with regard to understanding how people perceive and react to the long-term risks which potentially affect their health because those risks are not clearly visible or evident. As such, it is fundamentally important to study and understand how workers perceive the risks to which they are potentially being exposed.

Pérez-Floriano/Gonzalez (2007) among others (see Burke et al. 2011), have discussed the fact that there are numerous differences in risk perception within subgroups (including documented differences within the same culture, workplace environment and occupation), demonstrating that the social construction of risk beliefs are endogenous to the environment. These differences can lead to inaccurate interpretations of the sources of potentially hazardous risks. When risks are not properly evaluated, they can lead to: risky behaviour and improper activities with regard to the source of the risk; inadequate decisions regarding safety measures; and, both ordinary and catastrophic workplace accidents (Rundmo 1996). Moreover, tackling risks in the workplace depends not only on being able to perceive and deal with them, but also on the level of knowledge and the correct ways of avoiding them (Hoyos 1995).

Experts state that, through communication and social interaction processes, people are capable of eventually understanding the properties of hazards, including the risks associated with different types of events and hazardous exposure. When hazardous events and exposure can result in potentially fatal consequences, such as fires or exposure to toxic chemical products, it is expected that this will lead to fear and other negative emotions and feelings. The 'dread' factor and awareness regarding vulnerability to injuries and illnesses, as well as the feelings experienced, must be the main motivating factors for training in how to handle the said hazards (Burke et al. 2011). In summary, the judgements of different stakeholders regarding the level of risk of a number of hazards helps us to identify where and how to focus policies, procedures and training, which are fundamental to improving the systems and safety culture of the occupational groups under scrutiny.

16.5 Ergonomic Risk, Repetitive Motion Injuries, and Risk of Accident

The pollution found within factories, stemming from production processes, emissions, industrial waste and hazardous waste management, is considered to be a minor danger. Sometimes, managers and technical personnel do not know or do not want to recognize the long-term impact that processes have on workers' health and safety (Lezama 2004). Repeated exposure to these types of hazardous objects and situations is considered to lead to ergonomic risks and can result in injury and illness.

Society's perceptions of the environmental threats that industries represent has been modified as a result of industrial accidents. For example, in Mexico, the industrial accident at the *Pasta de Conchos* coal mine in Coahuila in 2006 led to the deaths of 65 miners. Mining involves a latent risk of accidents or illnesses stemming from the different gases that miners inhale; however, there are workers who minimize this risk, while other recognize that there is an implicit risk and are afraid of suffering an accident (Tejeda/Pérez-Floriano 2011).

The 2013 collapse of the Rana Plaza garment factory in Dhaka, Bangladesh, where more than 1,100 textile workers died and about 2,500 were injured, exposed the precarious conditions to which garment workers are exposed. Industrial accidents can affect the lives of hundreds of thousands of people; however, some risks are considered to be 'secondary effects or necessary evils, and, in many cases, these are risks that are not perceived by the same people who are directly exposed to them as a result of the introduction of technologies that involve unknown dangers' (Lezama 2004: 17).

Despite the catastrophic industrial accident in the Rana Plaza garment factory in Bangladesh, the garment manufacturing industry is, debatably, considered to be one of the cleanest industries in environmental terms; however, it is possible to identify a number of effects, including emissions that lead to: sound pollution; air pollution stemming from dust generated during the cutting process within the plants; the presence of solid waste from fabric, oil and fat residues; and the excessive use of resources, such as electricity and water, among others (Parada/Pelupessy 2006).

16.6 Methods

In the field research, the following activities were undertaken: archival data analysis, group interviews, direct observation, open interviews, and surveys. The goal of these activities was to identify the major health risks facing people working in the production process. As a measurement tool, a specific questionnaire was used, covering the following variables: age, gender, education level, organizational tenure, daily productivity, and position. The data compiled helped to validate the Risk Perception Scale for the Textile Industry.

16.6.1 Participants

The target population comprised those workers within the production area of XY textile factory. 405 questionnaires were distributed to workers from different production areas, with a response rate of 66.91%. The convenience sample (n = 271) was composed of 67.89% (=184) women and 32.11% (n = 87) men. The average age was 33.58 years old. Production operators were deemed to be all those people involved directly in the production process, in sewing operations, inspection, auditing, garment spotting, distribution of materials to production modules (materials). A group of workers (indirect personnel), involved in several jobs relating to production, was distinguished from operators, including: production supervisors, line managers, quality auditors and mechanics. Administrative personnel are people who coordinate a production department, which is why they are considered to be supervisors.

16.6.2 Risk Analysis and Measurement of Risk Perception Within the Textile Industry

A risk analysis was undertaken at a textile factory, which included data from the Mexican Social Security Institute (*Instituto Mexicano del Seguro Social* in

Spanish), academic literature, interviews, and a focus group with managers from the factory. The goal of this process was to identify the hazards facing production personnel at the factory.

The purpose of identifying workplace hazards was to use them to: measure workers' perception of risk of exposure to each of those working conditions; characterize the hazards; and analyse the effects that hazards have on workers' subsequent behaviour. The participants identified potential hazards to which people from the sewing and garment spotting areas were exposed, and the risk of each of these hazards was evaluated on a scale from 1 '*the level of risk is low*' to 7 '*the level of risk is very high*'.

The risk perception scale was subjected to a factor analysis using the Kaiser Meyer-Olkin method (KMO) and Bartlett's test of sphericity. The factor analysis included the extraction of factors or major axes and oblique rotation (Oblimin); to determine the number of factors, the Kaiser Guttman rule (Nunnally/Bernstein 1994) was employed to obtain a reduction in the variables relating to one or more factors that explain their lineal combinations (Nunnally/Bernstein 1994). From this analysis, two factors emerged, explaining 50.46% of the variance of the workplace hazards or inherent safety conditions for the working environment. However, the risks of accident and from environmental conditions, and ergonomic risks or risk from repeated exposure, are different in the factorial solution found. The resulting risk factors were the following:

Factor 1 Perception of Risk of Accident: risks stemming from dangers that are conducive to accidents and hazards from environmental elements. This factor comprises the evaluation of some risks from physical agents, such as the ignition of chemical agents and the exposure to toxic substances within the factory. This factor also encompasses unexpected situations resulting from safety conditions that lead to workplace accidents. The dangers identified within this factor were: exposure and/or handling of flammable substances; exposure and/or handling of toxic substances; electric shock from spilling water or any other liquid on the sewing machines, garment spotting guns, other machinery or equipment; the clothing or hair of the operator becoming caught in the sewing machine; fires resulting from a dirty or poorly-maintained sewing machine; damage to eyes from broken needles ejected from the machine; trips and falls; insufficient illumination; fires resulting from the overheating of equipment, tools and machine motors; insufficient ventilation; not using safety equipment for handling heavy boxes.

Factor 2 Perception of Ergonomic Risks: this factor characterizes ergonomic risks, i.e., those situations stemming from the interaction between the person and a system, e.g., injuries resulting from repetitive movements and the implicit health risks of the environmental conditions, such as noise and dust inhalation. The hazards identified in this factor were: deteriorated eyesight; heat from physical effort; being exposed to a lot of noise for long periods; inhaling lint; damage to eyes resulting from a blast of compressed air; muscular injuries resulting from repetitive motion; and heat stroke (exposure to intense heat).

Although the list of hazards identified is not exhaustive, it is representative of the main safety elements and hazardous situations that affect the health of the workers. The risk of 'piercing fingers with needles' was eliminated as it fell into both factors.

16.7 Results of Risk Perception Analysis of Work at the Textile Factory

The differences between risk perception by occupational level and by gender were analysed. Visits were made to the factory to observe the production processes and carry out informal interviews with members of the organization. It was found that the operators are exposed to dust particles (what they commonly refer to as 'lint'), which detach themselves from the material and stick to their face and clothes. Some operators were using masks for protection, while others were not. Furthermore, they were exposed to the noise and heat stemming from the motor of the sewing machines used in the production process, and they were not using earplugs.

16.7.1 Risk Perception and Gender

Multiple studies have found that men tend to judge risks to a lesser extent than women (for example, Fischer et al. 1991; Finucane et al. 2000; Sjöberg 2003; Slovic 1999; Weber et al. 2002). Various hypotheses have been presented to explain these differences, one of which focuses on biological and social factors. For example, they state that women are more concerned about human health and safety because they are mothers and have been educated to promote and maintain life.

With the variables from the factor analysis, we proceeded to undertake a statistical analysis to find out whether there were significant differences between men and women, even though the average risk perception from both genders was high; on a scale of 1 to 7, the averages for both factors were above 5.

However, given the fact that most of the operators are women, an analysis was undertaken to find out whether there was a significant difference between men and women with regard to each of the risks found in the production process. We found that women are significantly more concerned than men about the following hazards: muscle injuries because of repetitive movements; fires as result of overheating of tools, equipment or machine motors; exposure to extreme heat; trips and falls; and the inhalation of lint.

In addition to gender analysis, risk perception by occupational strategy was assessed for both factors (accident-environmental risks and ergonomic risks). Significant differences were found among the three occupational groups: production operators, administrators and supervisors. The operators perceived a higher level of

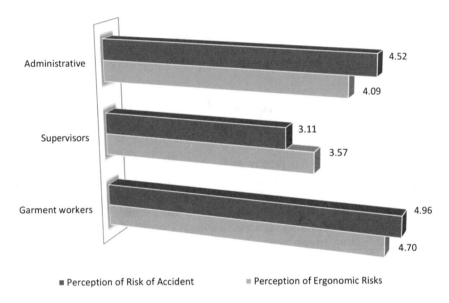

Fig. 16.1 Risk factors by occupational level (own elaboration). *Source* The authors

risk than the other two groups, and it is noteworthy that ergonomic risks, according to the participants, pose the greatest risk to the health of the workers (see Fig. 16.1).

16.7.2 Descriptions of Perception of Risk Factors by Occupational Level

The average of each of the risks from the perception scale by occupational level, including the experts in this analysis, is shown in Fig. 16.2. Production operators, administrative and indirect personnel and experts gave the highest ratings to the risk of muscle injuries stemming from repetitive movements, while for supervisors the greatest risk came from lint inhalation. However, the operators ranked the risk from inhaling lint higher than the supervisors. It is important to underscore the fact that the supervisors gave the lowest score to each of the risks. It is the supervisors who set the pace in the workplace and, as such, the production demands. Finally, it should be noted that the experts were those people who took part in the focus group, and who, once the decision had been reached as to which risks would be included in the final survey, received a list of these risks and ranked them. The group of 'experts' were those who took part in the group interview and who filled out a different form from the other groups. They evaluated and put the risks in order based on their potential harm to workers' well-being.

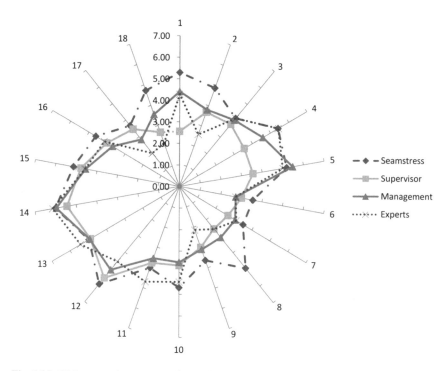

Fig. 16.2 Risk perception average for each hazard by occupational level (own elaboration). *Source* The authors

List of Hazards:

1. Heat stroke from physical exertion
2. Trapping clothing, hair and/or clothing accessories in the sewing machine
3. Falls and slips
4. Electric shock from dropping water or other liquid on sewing machines, garment dismantling guns, or other machines or equipment
5. Wear and tear
6. Exposure to lint for long periods of time
7. Exposure to flammable substances
8. Exposure to toxic substances
9. Exposure to excessive noise and for long periods of time
10. Heat stroke (exposure to intense heat)
11. Inadequate lighting
12. Fire from overheated engines, equipment or tools
13. Fire caused by lack of cleaning and maintenance of the sewing machine
14. Eyes injuries caused by projectiles, such as broken needles
15. Eye injuries caused by compressed air blown in the face

16. Muscle injuries due to repetitive movements
17. Failure to use safety equipment to handle heavy boxes
18. Inadequate ventilation

16.7.3 Production, Authority and Risk Perception

In keeping with the numerous studies that analyse the relationship between productivity and a culture of industrial safety, we found that productivity demands lead to a lesser focus on safety measures. The people who reported making more tops on a daily basis also perceived fewer workplace risks. This would indicate that the more tops produced, the less attention workers pay to workplace hazards. The relationship between productivity and ergonomic risks, although pointing towards a negative trend, was not significant.

16.8 Conclusions

This study indicates that occupational characteristics are more important in predicting risk than demographic factors, such as gender. The degree of exposure to lint is rated considerably higher by the operators than by the supervisors and experts (management). This is in keeping with that suggested by academics in the area of cultures of safety. They have shown that placing an emphasis on productivity leads to individuals ignoring and/or minimizing the real and potential risks found in industrial environments. We can see this in our results. The operators perceive a greater level of risk compared to the other occupational groups, given that the former face the implicit possibility of bodily harm, which, frequently, can have an irreversible impact. Identifying industrial hazards will lead to the creation of workplace practices and policies that are safer for workers and their environment.

Although the gender differences in both factors included in the scale were not significant, it is worth reviewing whether there are any differences in perception of those hazards considered to be 'important' given their potential to negatively impact the health of the people who work in this industry and, in particular, in this factory. The risk of inhaling lint, heatstroke (in a city where the temperature can reach more than 40°C in summer) and muscle injuries resulting from repetitive movements were statistically significant.

In conclusion, understanding the risk perceptions and safety attitudes of workers is necessary in developing a culture of safety, where each person accepts responsibility for working safely (Fleming et al. 1998). Identifying workplace risks lies at the core of industrial safety, and given that safety, just like quality and productivity, is inherent in each and every manufacturing process, when safety aspects are ignored, workers will infer that safety is not a priority, simply by omission, leading in the eyes of the workers to a weak industrial safety climate (Zohar/Luria 2004).

Hazard identification and characterization is a powerful tool to better understand the risks posed by the industrial sector. At the same time, understanding risk perception can lead to more accurate recommendations regarding training for workers in safe practices and the improvement of health and safety conditions to eliminate and control the risks found in the workplace. Furthermore, analysing risk perception can contribute to the understanding of risk management and safe working conditions (Rundmo 1996). These perceptions can be used for training, personnel development, wages and salaries, and even personnel selection.

References

Arezes, P.M., & Miguel, A.S. (2008). Risk perception and safety behavior: a study in an occupational environment. *Safety Science, 46*, 900–907.

Asfahl, C.R. (2000). Seguridad industrial y salud, 4a ed., México: Prentice Hall.

Brewer, N.T., Chapman, G.B., Gibbons, F.X., Gerrard, M., McCaul K.D., Weinstein, N.D. (2007). Meta-analysis of the relationship between risk perception and health behavior: The example of vaccination. *Health Psychology, 26*, 136–145 (2007). https://doi.org/10.1037/0278-6133.26.2.136.

Burke, M.J., Salvador, R.O., Smith-Crowe, K., Chan-Serafin, S., Smith, A., & Sonesh, S. (2011). The dread factor: how hazards and safety training influence learning and performance. *Journal of Applied Psychology, 96*(1), 46–70.

Carbonell, A., & Torres, A. (2010). Evaluación de percepción de riesgo ocupacional. *Ingeniería Mecánica, 13*(3), 18–25.

Cox, T., & Rial-Gonzalez, E. (2000). Risk management, psychosocial hazards and work stress. J. Rantanen & S. Lehtinen (Eds.), *Psychological Stress at Work*, Helsinki: Finnish Institute of Occupational Health.

Díaz, X., & Schlaen, N. (1994). *La salud ignorada: trabajadoras de la confección.* Santiago: Ediciones Centro de Estudios de la Mujer.

Finucane, M.L., Slovic, P., Mertz, C.K., Flynn, J., & Satterfield, T.A. (2000). Gender, race, and perceived risk: The 'white male' effect. *Health, Risk & Society, 2*(2), 159–172.

Fischer, G.W., Morgan, M.G., Fischhoff, B., Nair, I., Lave, L.B. (1991). What risks are people concerned about? *Risk Analysis, 11*(2), 303–314.

González, O.X. (2006). 'La seguridad industrial para la mediana empresa textil en el área de la confección', (tesis de maestría) México, IPN, Unidad Profesional Interdisciplinaria de Ingeniería y Ciencias Sociales y Administrativas, sin pie de imprenta.

Guadarrama, O.R., Hualde, A.A., López, E.S. (2012). Precariedad laboral y heterogeneidad ocupacional: una propuesta teórico-metodológica. *Revista Mexicana de Sociología, 74*(2), 213–243.

Hollmann, S., Heuer, H., Schmidt, K.H. (2001). Control at work: a generalized resource factor for the prevention of musculoskeletal symptoms? *Work & Stress 15*, 29–39.

International Labour Organization (ILO) (2015). Safety and health at work. http://www.ilo.org/global/topics/safety-and-health-at-work/lang–en/index.htm.

Ley Federal del Trabajo (2012). *Diario Oficial de la Federación.* México y H. Congreso de la Unión.

Hoyos, C.G. (1995). Occupational safety: Progress in understanding the basic aspects of safe and unsafe behaviour. *Applied Psychology, 44*(3), 233–250.

Kouabenan, D.R. (2009). Role of beliefs in accident and risk analysis and prevention. *Safety Science, 47*, 767–776.

Kompier, M. (2006). Work-related stress and health-risks, mechanism and countermeasures. *Scandinavan Journal of Work Environment and Health*, 32(6) 413–419.

Lezama, C. (2004). *Percepción del riesgo y comportamiento ambiental en la industria*. México: El Colegio de Jalisco/Consejo Estatal de Ciencia y Tecnología de Jalisco/ Centro de Investigaciones y Estudios Superiores en Antropología Social.

Mata, N.M. (2004). Desarrollo de un plan de seguridad ocupacional para la industria de la confección. (Tesis de maestría), México: Instituto Politécnico Nacional, Sección de Estudios de Posgrado e Investigación, sin pie de imprenta.

Melo Junior, A.S. (2012). The risk of developing repetitive stress injury in seamstresses, in the clothing industry, under the perspective of ergonomic work analysis: a case study. *Work: A Journal of Prevention, Assessment and Rehabilitation*, 41, 1670–1676.

Morrow, P.C., & Crum, M.R. (1998). The effects of perceived and objective safety risk on employee outcomes. *Journal of Vocational Behavior*, 53, 300–313 (1998). https://doi.org/10.1006/jvbe.1997.1620.

Nunnally, J.C., & Bernstein, I.H., (1994). *Psychometric theory* (3rd ed.), New York: McGraw-Hill.

Parada, A., & Pelupessy, W. (2006). Los efectos ambientales de la cadena global de prendas de vestir en Costa Rica. *Revista Iberoamericana de Economía Ecológica*, 3, 63–79.

Pérez-Floriano, L., & Gonzalez, J.A. (2007). Risk, safety and culture in Brazil and Argentina: The case of Transinc Corporation. *International Journal of Manpower* 28, 403–417.

Quintero, R.C., & Romo, A.M.L. (2001). Riesgos laborales en la maquiladora: La industria Tamaulipeca. *Frontera Norte*, 13 (2), 11–46.

Reinhold, K., & Tint, P. (2009). Hazard profile in manufacturing: determination of risk levels towards enhancing the workplace safety. *Journal of Environmental Engineering and Landscape Management*, 17(2), 69–80.

Rundmo, T. (1996). Associations between risk perception and safety. *Safety Science*, 24, 197–209.

Salminen, S., Saari, J., Saarela, K.L., Rasanen, T. (1993). Organizational factors influencing serious occupational accidents. *Scandinavian Journal of Work, Environment and Health*, 19: 257–352.

Schein, E.H. (1996). Three cultures of management: The key to organizational learning. *Sloan Management Review*, 38(1), 9–20.

Sheu, J.J., Hwang, J.S., Wang, J.D. (2000). Diagnosis and momentary quantification of occupational injuries by indices related to human capital loss: analysis of a steel company as an illustration. *Accident Analysis and Prevention*, 32, 435–443.

Sjöberg, L. (2003). The different dynamics of personal and general risk. *Risk Management*, 5(3), 19–34.

Slovic, P. (1987). Perception of Risk. *Science*, 236, 280–285.

Slovic, P. (1999). Trust, emotion, sex, politics, and science: survey the risk-assessment battlefield. *Risk Analysis*, 19(4), 689–701.

Slovic, P., Fischhoff, B., Lichtenstein, S. (1982). Why study risk perception? *Risk Analysis*, 2(2), 83–93.

Tejeda, N., & Pérez-Floriano, L. (2011). La amplificación social del riesgo: evidencias del accidente en la mina Pasta de Conchos. *Comunicación y Sociedad*, 15, 71–99.

Weber, E., Blais, A.R., & Betz, N. (2002). A domain-specific risk-attitude scale: measuring risk perceptions and risk behaviors. *Journal of Behavioral Decision Making*, 25, 263–290.

Zohar, D., & Luria, G. (2004). Climate as a social-cognitive construction of supervisory safety practices: scripts as proxy of behavior patterns. *Journal of Applied Psychology*, 89(2), 322.

About the Editors

Ma. Luisa Marván is a psychologist with an M.Sc. in Psychobiology and a Ph.D. in Biomedical Sciences from the National Autonomous University of Mexico. She is a full-time researcher at the Institute of Psychological Research at Universidad Veracruzana in Mexico, where she carries out research in the field of Health Psychology, and teaches courses to postgraduate students. She belongs to the Academic Team 'Psychology, Health and Society', and her main interests are psychosocial variables related to women's reproductive health, including risk factors and preventive behaviours. She has published more than 70 scientific articles, more than 10 informative articles (divulgation) in national and international journals, 7 book chapters, and 1 book. Several of her publications have been in collaboration with colleagues from the United States of America or Spain. She has been recognized by the Mexican National Researchers System, reaching the maximum level that this institution grants in 2015.

Address: Instituto de Investigaciones Psicológicas, Universidad Veracruzana. Av. Dr. Luis Castelazo Ayala s/n Col. Industrial Ánimas C.P. 91190 Xalapa, Ver.

Email: mlmarvan@gmail.com
Website: www.uv.mx

© Springer International Publishing AG, part of Springer Nature 2018
Ma. L. Marván and E. López-Vázquez (eds.), *Preventing Health and Environmental Risks in Latin America*, The Anthropocene: Politik—Economics—Society—Science 23, https://doi.org/10.1007/978-3-319-73799-7

Esperanza López-Vázquez completed a doctorate and a Master's degree in Social Psychology at the Université Toulouse Le-Mirail in France, and undergraduate studies in Social Psychology at the Universidad Metropolitana-Xochimilco in Mexico. She is the Head of Development and Research of the Centre of Transdisciplinary Research in Psychology at the Universidad Autónoma del Estado de Morelos in Mexico, where she carries out research in the field of Social Psychology and Environmental Psychology. She teaches undergraduate and postgraduate courses.

Dr. López-Vázquez was the founder president of the Society for Risk Analysis Latin-America in 2008 and the founder president of the Mexican Association for Disaster Prevention and Attention in 2011. She has published more than 15 scientific articles, 3 book chapters and was the editor of a scientific journal's special issue about *Risk Perception and Social Trust*. She has been recognized by the Mexican National Researchers System since 2001. Among her research interests are the study of risk perception and psycho-social factors involved in the coping responses of people exposed to natural hazards and environmental risks, as well as the risk perception and environmental behaviour of vulnerable populations.

Address: Centre of Transdisciplinary Research in Psychology, Pico de Orizaba No. 1 Col. Volcanes. 62350 Cuernavaca, Morelos, México.

Email: esperanzal@uaem.mx
Website: www.uaem.mx

About the Authors

Tania Romo-González is a Pharmaceutical Chemist Biologist at the Universidad Veracruzana (UV) and has a Doctorate degree in Biomedical Sciences from the National Autonomous University of Mexico. She currently works as a leading researcher in the area of Biology and Integral Health at the Institute of Biological Research of the UV, where she develops several research projects in the lines of (1) Psychoneuroimmunology, health and wellness and (2) lifestyles and health. Since 1999 she has coordinated workshops for undergraduate students and university professors to promote healthy lifestyles. Tania has published more than 32 scientific articles in international journals on the mind-body-health-disease relationship and the promotion of healthy lifestyles.

Address: Av. Dr. Luis Castelazo Ayala, Industrial de las Animas, 91190 Xalapa Enríquez, Ver.
Website: https://www.uv.mx/iib/dra-tania-romo-gonzalez-de-la-parra/

Raquel González-Ochoa is a psychologist at the Universidad Veracruzana (UV), with a Masters in Eriksonian Psychotherapy from the Eriksonian Center, A.C., and is currently pursuing a doctorate in Psychology at the UV. During the last five years she has been participating in different projects aimed at prevention and the promotion of health and well-being. She has been facilitator of the workshops: 'Learning to relate healthily to myself and others', 'Learning to maintain my attitude of service', 'Self-knowledge and care of the soul', and of the seminars: 'Human Development' and 'Emotional pedagogy and psycho-educational models for integral formation'.

Address: Av. Dr. Luis Castelazo Ayala, Industrial de las Animas, 91190 Xalapa Enríquez, Ver.
Website: www.uv.mx

Gabriel Gutiérrez-Ospina is a Medical Surgeon at the National Autonomous University of Mexico (UNAM), with a Master's and Ph.D. in Physiological Sciences from the same university. He is the head of the Laboratory of Systems Biology, and Chief Operating Officer of the Institute of Biomedical Research of the UNAM. He has been a Qualifying Jury in the evaluation of the projects registered

© Springer International Publishing AG, part of Springer Nature 2018
Ma. L. Marván and E. López-Vázquez (eds.), *Preventing Health and Environmental Risks in Latin America*, The Anthropocene: Politik—Economics—Society— Science 23, https://doi.org/10.1007/978-3-319-73799-7

in the Support Programme for Research and Technological Innovation Projects of the UNAM and the National Council of Science and Technology. He has published more than 60 research articles, which cover the analysis of biological systems, mainly the nervous system and the vertebrate reproductive system.

Address: Tercer Circuito Exterior, Apartado Postal 70228, Ciudad Universitaria, 04510 México CDMX.
Website: http://www.biomedicas.unam.mx/personal-academico/gabriel-gutierrez-ospina/

Rosalba León-Díaz holds an undergraduate Biology Degree from the Universidad Veracruzana (UV), and a Master's and Ph.D. degree in Biological Sciences from the National Autonomous University of Mexico. Currently she is a postdoctoral researcher in the area of Biology and Integral Health at the Institute of Biological Research of the UV. Her main research interest is to develop tools for the diagnosis and treatment of communicable and chronic-degenerative diseases, in addition to promoting healthy lifestyles.

Address: Av. Dr. Luis Castelazo Ayala, Industrial de las Animas, 91190 Xalapa Enríquez, Ver.
Website: www.uv.mx

Carlos Larralde (1938–2015) was Medical Surgeon at the National Autonomous University of Mexico (UNAM), and graduated in Pathology before being awarded a Ph.D. degree by the University of Washington. He was an Emeritus Professor of the Institute of Biomedical Research at UNAM. His research covered various immunological aspects of human and experimental disease, especially in cysticercosis, AIDS and breast cancer, mainly aimed at the diagnosis, prevention and pathogenesis.

Yamilet Ehrenzweig is a psychologist with a Master's in Health Psychology and in Psychology and Community Development from the Universidad Veracruzana, and a Ph.D. in Education from the University of La Salle. She worked as a researcher at the Institute for Psychological Research at Universidad Veracruzana until 2015, when she retired. Her research focused on cognitions and beliefs related to cancer prevention.

Address: Allende No. 126, Centro, Xalapa, Veracruz, México.

Socorro Herrera-Meza holds a Ph.D. in Food Sciences from the Technological Institute of Veracruz. She is currently working as a researcher in the Institute of Psychological Research at Universidad Veracruzana. Her work deals with dietary lipids and their implications for health, Omega-3 fatty acids supplementation, metabolic syndrome in murine models and their association with food.

Address: Av. Dr. Luis Castelazo Ayala, Industrial de las Animas, 91190 Xalapa Enríquez, Ver.
Website: www.uv.mx

Grecia Herrera-Meza holds a Ph.D. in Neuroethology from the Universidad Veracruzana, and undertook postdoctoral studies at the Technological Institute of Veracruz (UNIDA). Her work deals with fatty acids effect on neurodegenerative diseases in murine models. In parallel, she studies behavioural biology in different species, as well as neuropsychology and emotional recognition in individuals from different school levels.

Address: Carretera Xalapa Veracruz Km 2 No. 341, Colonia Acueducto Animas, 91190 Xalapa Enríquez, Ver.
Website: http://ux.edu.mx/?page_id=4169#1459066646891-1cded748-f472

Arturo G. Rillo is a physician from the Faculty of Medicine of the Autonomous University of the State of Mexico, with an M.Sc. in Biomedical Science from the UNAM and a Ph.D. in Humanities from the Autonomous University of the State of Mexico. His line of research is focused on the medical humanities.

Address: Paseo Tollocan esq Jesíus Carranza s/n, Colonia Moderna de la Cruz, Toluca, Mex. C.P. 50180.
Website: www.uaemex.mx/fmedicina/

Ninfa Ramírez Durán graduated as Chemist Pharmacist, Master in Pharmaceutical Sciences and Ph.D. in Biological Sciences at the Metropolitan Autonomous University. Currently a Research Professor at the Faculty of Medicine of the Autonomous University of the State of Mexico, her research line is Medical and Environmental Microbiology. Her publications refer to the identification of pathogenic actinomycetes and the production of metabolites from environmental actinomycetes.

Address: Paseo Tollocan esq Jesíus Carranza s/n, Colonia Moderna de la Cruz, Toluca Mex, C.P. 50180.
Website: www.uaemex.mx/fmedicina/

Horacio Sandoval Trujillo graduated as a Chemist Bacteriologist Parasitologist from the National School of Biological Sciences of the National Polytechnic Institute. His doctorate is from the University Claude Bernard Lion-1, France. He is currently Professor 'C' at the Metropolitan Autonomous University, Visiting Professor at the Universities of Colima and Durango (Mexico) and Researcher at the Universidad de los Andes (Merida, Venezuela). He was awarded the Bicentennial Distinction Medal granted by the University of the Andes, Merida, Venezuela. He is former President and Founder of the International Research Group on Pathogenic Actinomycetes (GIIAP).

Address: Calzada del Hueso 1100, Col. Villa Quietud, Delegación Coyoacán, C.P. 04960, Ciudad de México.
Website: http://www.uam.mx/u_xoc

Roseane de Fátima Guimarães Czelusniak holds an undergraduate Kinesiology and Physical Education degree from the Pontifical University Catholic of Parana

(PUCPR); a Master's in Physical Education specializing in Physical Activity and Health from Federal University of Parana (UFPR); and a Doctoral degree in Paediatrics/Child and Adolescent Health from the University of Campinas (UNICAMP). She is currently working at the Salesian University Centre of Sao Paulo (UNISAL) as Head of the Department of Physical Activity and Teacher of the undergraduate course. Her research focuses on physical activity, sedentary behaviour and health in children and adolescents.

Address: Rua Baronesa Geraldo de Rezende, 330. Jardim Auxiliadora. CEP: 13075-270. Campinas. Sao Paulo, Brazil.
Website: www.unisal.br

Jorge Luis Arellanez Hernández holds a Psychology Ph.D. degree from the National Autonomous University of Mexico (UNAM). He is currently working at the Institute of Psychological Research at Universidad Veracruzana. His research interest is related to the identification of risk/protection factors of drug consumption in vulnerable populations (children, adolescents, women, international migrants). His other research specialismis the analysis of the mental health within international migrant populations.

Address: Av. Dr. Luis Castelazo Ayala, Industrial de las Animas, 91190 Xalapa Enríquez, Ver.
Website: www.uv.mx

Iris W. Cátala-Torres (Psy.D, M.S., Clinical Psychology, Carlos Albizu University; MRC, Rehabilitation Counselling, University of Puerto Rico; CRC, Certified Rehabilitation Certification, Commission on Rehabilitation Counsellor Certification) is a Counsellor at the Nursing School of the Medical Sciences Campus University of Puerto Rico (MSC-UPR). Dr. Cátala has supervised Rehabilitation Counselling students from the Graduate School of Rehabilitation Counselling, University of Puerto Rico and has trained clinical psychology students in psychosocial intervention at disasters. She is part of the Institutional Review Board for human subjects research of MSC-UPR. Her focus areas are disability issues, college students, disaster interventions, and protection of human subjects in research.

Address: Universidad de Puerto Rico, Recinto de Ciencias Médicas, Asuntos Estudiantiles. Escuela de Enfermería, Apartado 365067, San Juan, PR 00936-5067.
Website: www.rcm.edu

Jesús Manuel Macías holds a Ph.D. in Geography from the National Autonomous University of Mexico (UNAM). He is a Researcher at the Centre for Research and Higher Studies on Social Anthropology (CIESAS-Mexico). He has participated in scientific advisory committees of the federal and State Governments on Civil Protection and Emergency Management. He was founder of the network of social studies for disaster prevention in Latin America (La Red) and the first President of the Scientific Advisory Committee of the National Civil Protection System on social sciences in Mexico (SINAPROC). He coordinates the Permanent Seminar of

Social Vulnerability to Disasters at CIESAS and works on inter-institutional research projects. He is a member of the National System of Researchers and of the Board of the International Research Committee of Disasters of the International Sociological Association (RC39).

Address: Calle Juárez, 87. Col. Tlalpan, Ciudad de México, C.P. 14000, México.
Website: http://www.ciesas.edu.mx/macias-medrano-jesus-manuel/

Eric C. Jones (Ph.D., Anthropology, University of Georgia) is an Assistant Professor in the School of Public Health at the UT Health Science Center at Houston. He has focused on how social relationships impact individual and group well-being and recovery after extreme events. His general research interests concern heterogeneity and collective action, often under conditions of environmental or social change but also in improving formal and informal science education. He is co-author of the edited volumes *The Political Economy of Hazards and Disasters*, and *Social Network Analysis of Disaster Response, Recovery and Adaptation*.

Address: 1851 Wiggins Rd, HSN 487, El Paso, TX 79968.
Website: https://sph.uth.edu/campuses/el-paso/

A. J. Faas has a Ph.D. in Anthropology from the University of South Florida. He is Assistant Professor of Anthropology at San José State University. His research interests centre on exchange practices, social organization, governance, and culture in environmental crises; he has principally conducted research in Mexico, Ecuador, and the United States. Faas edited the 2016 special issue of *The Annals of Anthropological Practice* on 'Continuity and Change in the Applied Anthropology of Risk, Hazards, and Disaster' and the 2017 special issue of the *Journal of Latin American and Caribbean Anthropology* on 'Twenty-First Century Dynamics of Cooperation and Reciprocity in the Andes.'

Address: San Jose State University, Clark Hall 404L, One Washington Square, San Jose, CA 95192.
Website: https://sph.uth.edu/faculty/?type=faculty&campus=El+Paso&division= EHGES

Arthur D. Murphy has a Ph.D. in Anthropology from the Temple University. He teaches Anthropology at UNC—Greensboro. He has carried out research in Mexico, Panama, Ecuador, and Bolivia. Dr. Murphy has served on the faculty at Georgia State University, Baylor University, the University of Georgia, the University of Guadalajara, the National University of Mexico, and CIAD-Hermosillo. His publications include a number of academic articles and chapters, as well as the books *Social Inequality in Oaxaca, The Mexican Urban Household* and *Complejos Bioculturales de Sonora: Pueblos y Territorios Indígenas*. He co-edited *Latino Workers in the Contemporary South* and *The Political Economy of Hazards and Disasters*.

Address: Department of Anthropology, The University of North Carolina at Greensboro, 426 Graham Building, PO Box 26170, Greensboro, NC 27412-5000.
Website: https://anthropology.uncg.edu/faculty-staff/murphy/

Christopher McCarty has a Ph.D. in Anthropology from the University of Florida. He was trained as a cultural anthropologist, and has specialized in social network research with a focus on the application of personal network analysis to a broad range of topics. McCarty created a software program called Egonet that is used by researchers across many disciplines, increasingly in health care research. He has extensive experience in survey research, having built one of the largest university-based survey research centres in the world, housed in the UF Bureau of Economic and Business Research (BEBR), which he directs. He is chair of the UF Anthropology Department, and has served as a programme officer for the NSF Cultural Anthropology Program.

Address: Bureau of Economic and Business Research, Ayers Technology Plaza, 720 SW 2nd. Ave Ste 150, PO Box 117148, Gainesville, Florida 32611.
Website: https://anthro.ufl.edu/mccarty/

Graham A. Tobin has a Ph.D. in Geography from the University of Strathclyde. He is professor of geography at the University of South Florida. His research specialities include natural hazards, water resources policy and environmental contamination. He has published several books, chapters, and many refereed papers and other articles. He has been department chair at two universities and an associate vice chancellor, and he has served on committees at national, university, college, and department levels. Tobin has also held office in professional organizations, including co-executive directorship of the annual Applied Geography Conference.

Address: School of Geosciences, University of South Florida, 4202 E. Fowler Avenue, NES 107, Tampa, FL 33620-5550, USA.
Website: http://hennarot.forest.usf.edu/main/depts/geosci/faculty/gtobin/

Linda M. Whiteford has a Ph.D. in Anthropology from the University of Wisconsin, Milwaukee; and a Master's degree in Public Health from the University of Texas. She is Professor of Anthropology at the University of South Florida, an applied medical anthropologist, and past President of the Society for Applied Anthropology. She has fieldwork experience in Ecuador, Bolivia, Costa Rica, Mexico, Guatemala, and Nicaragua focusing on waterborne and water-washed diseases, disaster planning, and mitigation strategies using partici-patory methods. She is author of *Primary Health Care in Cuba*: *The Other Revolution*, author or editor of seven other books, and author of more than 50 other peer-reviewed articles.

Address: Department of Anthropology, University of South Florida, 4202 E. Fowler Avenue, SOC 107, Tampa, FL 33620-8100, USA.
Website: http://anthropology.usf.edu/faculty/whiteford/https://www.bebr.ufl.edu/about/faculty-staff/christopher-mccarty-phd

Úrsula Oswald Spring is a researcher at the Regional Centre for Multidisciplinary Research at UNAM, National Researcher SNI III and coordinates the Gender and Equity Programme. She has studied medicine, psychology, philosophy, languages, anthropology and ecology in Madagascar, Paris, Zurich and Mexico. She was Secretary of Environmental Development in Morelos and is a member of the Intergovernmental Panel on Climate Change and has served as Secretary General of the International Peace Research Association. Alone or in collaboration she has written 55 books, 376 scientific articles and book chapters on gender, security, environment, water, climate change, disasters, food and development.

Address: Priv. Río Bravo 1, Col Vistahermosa, Cuernavaca, 62290, Morelos, Mexico.
Website: www.afres-press.de/html/download_oswald.html

Dayra Elizabeth Ojeda Rosero is a Psychologist, Specialist in University Teaching, Specialist in Social Management and Magister in Ethnoliterature from the Universidad de Nariño in Colombia; She has a Ph.D. in Psychology from the Universidad Autónoma del Estado de Morelos, México. Currently, she works as a full time teacher at the Psychology Department from the Universidad de Nariño. She is a researcher of the Convivencia y Entornos Psicológicos (CONEPSI) group and the Instituto Andino de Artes Populares (IADAP). Her research focuses on social and community psychology, in emergency and disaster risks as well as in relationships and intergenerational processes. She has worked in various social projects with regional, national and international institutions.

Address: Cra. 40 No. 12 A 34, Mariluz 1, Pasto (Nariño, Colombia).
Website: http://psicologia.udenar.edu.co/

Melissa Ricaurte Cepeda is a psychologist at the University of Nariño in Colombia and has an MA in Psychology from the Universidad Autónoma del Estado de Morelos. She is currently working at the Centre for Studies on Health (CESUN) at the University of Nariño as postgraduate professor and public health group researcher. She has worked on research and intervention projects with institutions such as the Inter-American Development Bank (IDB), Nariño Governorate and Pasto Mayor's Office.

Address: Universidad de Nariño—Ciudad Universitaria Torobajo, Calle 18 Cr 50, Bloque Tecnológico, Piso 2.
Website: http://cesun.udenar.edu.co/

Hannia Cabezas Pizarro graduated from the University of Costa Rica (UCR), Faculty of Education, with the following degrees: Bachelor in Mental Retardation, Master of Education in Educational Administration, and Master of Science in Integral Rehabilitation. She is a professor at the School of Guidance and Special Education at the UCR. Her major areas of research include autism spectrum disorder and aggressive behaviour among school aged children.

Address: Escuela de Orientación y Educación Especial, Facultad de Educación, Ciudad Universitaria Rodrigo Facio, San José, San Pedro Montes de Oca, Costa Rica.
Website: www.facultadeducacion.ucr.ac.ac.cr

Nicolás Bronfman Cáceres holds an industrial engineering degree and a Ph.D. in Engineering Sciences from the Pontificia Universidad Católica de Chile. He is currently working as Vice Chancellor at the Andres Bello University and as Associate Researcher in the National Research Center for Integrated Natural Disasters Management (RCINDIM/CIGIDEN). His main specialization areas are risk perception, trust on authorities, quantitative risk analysis and natural hazards mitigation.

Address: Universidad Andrés Bello, Antonio Varas 880, Providencia, Santiago, Chile.
Website: www.unab.cl

Pamela Cisternas Ordóñez holds an industrial engineering degree and Master's in Environmental Management and Sustainability from the Andres Bello University. She is currently studying for a Ph.D. in engineering sciences focusing on social preparedness for natural disasters and working at the Research Center for Integrated of National Research Center for Integrated Natural Disasters Management (RCINDIM/CIGIDEN).

Address: Universidad Andrés Bello, Antonio Varas 880, Providencia, Santiago, Chile.
Website: www.unab.cl

Lorena R. Pérez-Floriano received her Ph.D. in Industrial-Organizational Psychology at the California School of Professional Psychology. She is a research professor in Mexico at El Colegio de la Frontera Norte, Department of Social Studies. Lorena's work has focused on analyzing the relationship of culture with organizational behaviours and the social construction of workplace danger. Lorena also researches why things often go wrong in organizations, from organizational corruption to job burn-out. Her most recent work examines stigmatized occupations and their relationship with cultural beliefs, death awareness, stress, and performance.

Address: Km 18.5 carretera escénica Tijuana - Ensenada, San Antonio del Mar, Tijuana, Baja California, México, C.P. 22560.
Website: www.colef.mx

Julieta A. Leyva Pacheco received her B.Sc. in Industrial Engineering at the University of Sonora, and her MA in Environmental Management at El Colegio de la Frontera Norte. She is currently a full-time professor at the State University of Sonora, Mexico in the environmental engineering programme. Julieta teaches and consults in industrial safety and risk perception for domestic and multinational corporations.

Address: Avenida Diez No. 150. Prados del Sol. Hermosillo, Sonora. C.P. 83100.

Index